LONDON MATHEMATICAL SOCIETY STUDENT TEXTS

Managing Editor: Professor C. M. Series, Mathematics Institute, University of Warwick, Coventry CV4 7AL, United Kingdom

London Mathematical Society Student Texts 43

FOURIER ANALYSIS ON FINITE GROUPS AND APPLICATIONS

Audrey Terras
University of California, San Diego

CAMBRIDGE
UNIVERSITY PRESS

CAMBRIDGE UNIVERSITY PRESS
Cambridge, New York, Melbourne, Madrid, Cape Town, Singapore, São Paulo

Cambridge University Press
The Edinburgh Building, Cambridge CB2 2RU, UK

Published in the United States of America by Cambridge University Press, New York

www.cambridge.org
Information on this title: www.cambridge.org/9780521451086

First published 1999
Reprinted 2001

A catalogue record for this publication is available from the British Library

Library of Congress Cataloguing in Publication data

Terras, Audrey.
Fourier analysis on finite groups and applications / Audrey
Terras.
p.   cm. – (London Mathematical Society student texts; 43)
ISBN 0-521-45108-6 (hc.)
1. Fourier analysis.   2. Finite groups.   I. Title.   II. Series.
OA403.5.T47   1999
515′.2433 – dc21   98-36455
                                                    CIP

ISBN-13 978-0-521-45108-6 hardback
ISBN-10 0-521-45108-6 hardback

ISBN-13 978-0-521-45718-7 paperback
ISBN-10 0-521-45718-1 paperback

Transferred to digital printing 2005

To the Julias.

... Sometimes when I am alone in the dark, and the
universe reveals yet another secret, I say the names
of my long lost sisters, forgotten in the books that
record our science –
Aglaonice of Thessaly,
Hypatia,
Hildegard,
Catherine Hevelius,
Maria Agnesi
– as if the stars themselves could

remember . . . .

Caroline Herschel (1740–1848) by Siv Cedering
in *Math. Intelligencer.* Vol. 7, No. 4 (1985), p. 72

Contents

Preface

My thanks go to lots of creatures – many of them at USCD. In particular, I want to mention MSRI for supporting me during several productive
visits, my graduate and undergraduate students at USCD who helped with
various versions of this book, and the ever patient (NOT!) POSSLQ. I particularly want to thank Aubin Whitley for help with the proof reading and index-
ing. And I would like to thank Ellen Tirpak and the staff at TechBooks as
well as Alan Harvey at Cambridge University Press. Thanks also to the AMS,
DIMACS, and IMA for sponsoring some inspirational conferences. I also must
thank the authors of the various computer programs I used – beginning with

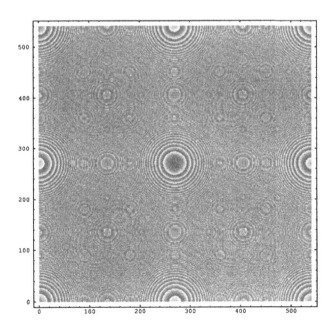

the patient POSSLQ whose friendly word processor runs on an Atari ST and ending with the authors of ChiWriter for the PC, which is still my favorite, after sampling many competitors. Finally, I want to thank those who sent me corrections for earlier versions – especially Aloys Krieg, and Jason Rosenhouse.

Part I
Finite Abelian Groups

It is unfortunate that so many scientists have been conditioned
to believe that, say, 10^{30} particles can *always* be well
approximated by an infinite number of points.

D. Greenspan [1973]

In this age of computers, it is very natural to replace the continuous with the
finite. One thinks nothing about replacing the real line $\mathbb{R}$ with a finite circle
(i.e., a finite ring $\mathbb{Z}/n\mathbb{Z}$) and similarly one replaces the real Fourier transform
with the fast Fourier transform (FFT). Then computers become happier about
our computations. Moreover, the prerequisites for our book become much less
formidable than they were for our earlier volumes on harmonic analysis on
symmetric spaces (Terras [1985, 1988]). In fact, some (such as Greenspan
[1973]) might arguethat, since the universe is finite, it is more appropriate to
use finite models than infinite ones. Greenspan decides to "deny the concept
of infinity" despite feeling "unmathematical" and "un-American" in doing so.
Physicists have also begun considering dynamicalsystems over finite fields (see
Nambu [1987]).

Here our goal is to consider finite analogues of the symmetric spaces such
as $\mathbb{R}^n$ and the Poincaré upper half plane, which were studied in Terras [1985,
1988]. We will discover finite analogues of all the basic theorems in Fourier
analysis, both commutative and noncommutative, including the Poisson summation formula and the Selberg trace formula. One motivation of this study
is to develop an understanding of the continuous theory by developing the finite model. Here in finite-land, we will have no worries about convergence
of integrals or interchange of summation and integration. Such worries often
obscured the continuous theory in a myriad of analytic details.

In fact, finite Fourier "series" are quite analogous to the classical series
that Fourier used to analyze periodic functions such as that describing the

temperature on a circular ring. Instead we must think of a finite set of points arranged on a circle and a discrete diffusion process.

It appears that mathematicians actually considered the Fourier transform (DFT) on the finite circle (defined in Chapter 1) before Fourier's work on Fourier series. See Heideman et al. [1984] who discuss the history. They have found that Clairaut considered the discrete Fourier transform in 1754, while Gauss found the fast version (FFT) in 1805 (two years before Fourier's paper and 160 years before Cooley and Tukey [1965]). Clairaut was applying the discrete Fourier transform to the determination of orbits. Gauss applied the DFT to number theory, that is, the quadratic reciprocity law via Gauss sums. Dirichlet also made use of this theory in his proof that there are infinitely many primes in any arithmetic progression.

By the early 1900s certain mathematicians knew how to do Fourier "series" on finite groups (e.g., Frobenius, Schur) and, later, on compact groups and locally compact groups (Cartan, Weyl, Weil, Pontryagin,...). The classical Fourier series are series of sines and cosines (or complex exponentials). In the analogue for a finite group G, the sines and cosines are replaced by the matrix entries of irreducible unitary representations π of G, that is, $\pi : G \rightarrow U(n)$ such that $\pi(gh) = \pi(g)\pi(h)$, where $U(n)$ is the group of unitary $n \times n$ matrices. When the group G is abelian, as in Part 1, we have $n = 1$.

Engineers and applied mathematicians (as late as the 1960s) have preferred to develop the subject independently using their own vocabulary, for example, saying "Hadamard transform" instead of Fourier transform on the additive group $(\mathbb{Z}/2\mathbb{Z})^k$ of vectors of 0s and 1s. That this transform has applications or recreations associated with it could come as a surprise to the pure mathematician working on representations of finite groups. For example, Figure I.1 is a picture of a code associated to the Fourier transform on $(\mathbb{Z}/2\mathbb{Z})^5$. This code was used in the transmission of data from the 1969 *Mariner* Mars probe (see Posner's article in Mann [1968] and our discussion of codes below). For the recreational aspects of this transform, see Ball and Coxeter [1987].

I have tried to make this book accessible to nonexperts such as advanced undergraduates, beginning graduate students, and scientists outside of mathematics. The book does mix up subjects that are most often kept in sterile separate compartments – number theory, group theory, graph theory, Fourier analysis, statistics, and coding theory. We have used it in undergraduate and graduate number theory courses at U.C.S.D. The undergraduates found it challenging. Any reader must be willing to do the exercises.

What are the prerequisites? The main one is a familiarity with linear algebra. See, for example, Gilbert [1976], Herstein [1964], or Strang [1976]. You will

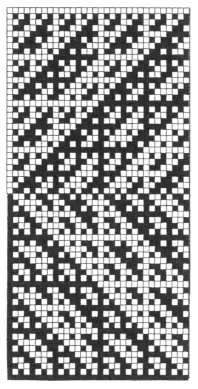

Figure I.1. The (32, 6) biorthogonal Reed–Muller code from Posner's article in Mann [1968, p. 23].

also need to know a tiny bit about finite groups, rings, and fields as in Gallian [1990], Gilbert [1976], and Herstein [1964]. We will give a brief treatment of the necessary number theory (congruences and the finite rings $\mathbb{Z}/n\mathbb{Z}$), but it might help to have a number theory book handy such as Hua [1982], Ireland and Rosen [1982], Rosen [1993], or Stark [1978]. You don't need to know anything about classical Fourier analysis. However, if you would like to compare with standard books on Fourier's world, look at Powers [1993, Chapter 1], Dym and McKean [1964], or Terras [1985, Chapter 1].

I will often make use of the rather costly computer packages Mathematica (see Wolfram [1996]) and Matlab (see MathWorks [1995]) to experiment, draw pictures, etc. If I were to write this book again, I would also use the computational group theory package GAP (see Schönert et al. [1995]) – a free program that can be found by surfing the net. GAP stands for groups, algorithms, and programming.

What sort of applications will we be considering? Here are a few examples:

1. construction of graphs that are good expanders,
2. reciprocity laws in number theory,
3. a study of graph analogues of Kac's question:
   "Can you hear the shape of a drum?" (see Kac [1966]),
4. error-correcting codes,
5. Ehrenfest model of diffusion,
6. switching functions,
7. random walks on graphs, and
8. vibrating systems and chemistry of molecules.

As I began to write this book I was inspired by the books of Diaconis [1988], Lubotzky [1994], and Sarnak [1990]. I also found the following papers inspirational, among others: Arthur [1989], showing an example of the trace formula on finite groups; Brooks [1991] and Buser [1988], giving connections between spectral theory on graphs and Riemannian manifolds; Chung and Sternberg [1992, 1993], on the spectra of buckyballs; and, finally, Lubotzky, Phillips, and Sarnak [1988], giving examples of Ramanujan graphs.

I have attempted to write a "user-friendly" book. To me, this means an abandonment of the style that starts with definition 1.1.1. and ends with corollary 10.66.5. However, I don't mean to say I will abandon proofs. Hardy [1940] notes that "all physicists, and a good many mathematicians, are contemptuous about proof." Then Hardy tells a story about some failures in analytic number theory where it is easy to make wrong guesses and where there are theorems "which have never been proved and which any fool could have guessed." Although finite Fourier analysis does not on the surface appear to be as difficult as analytic number theory, we will also stumble upon some "theorems" like those described by Hardy.

There are still not many books that attempt to do Fourier analysis both on abelian and nonabelian finite groups – with applications. The problem may be that the applied mathematician needs what Jessie MacWilliams calls "the antithesis of classical algebra" [see her coding theory survey article in Mann, 1968]. This means that one may well need explicit matrix entries rather than a coordinate-free approach. MacWilliams went on to write a limerick [see Mann, 1968]:

> Delight in your algebra dressy
> But take heed from a lady named Jessie
> Who spoke to us here
> of her primitive fear
> That good codes just might be messy.

And so it goes with many applications of group representations. Thus we will avoid the methods which Curtis and Reiner [1966] call "the second stage in the development of representation theory." This stage was begun by Emmy Noether in 1929. According to Curtis and Reiner, Noether's approach "resulted in the absorption of the theory into the study of modules over rings and algebras."

Cast of Characters

The definitions below describe most of the characters needed from a basic algebra course. See Dornhoff and Hohn [1978], Dummit and Foote [1991], Gallian [1990], Gilbert [1976], Herstein [1964], Hungerford [1974], Lang [1984, 1987], Strang [1976], or van der Waerden [1991] for more details.

The Abstract

A *group G is a* set with a binary operation ($\cdot$) which gives a unique $x \cdot y = xy \in G$ for every $x, y \in G$ such that
- the operation is associative (i.e., $(xy)z = x(yz)$);
- there is an element e (the identity) in G such that $ex = xe = x$ for all $x \in G$;
- for each $x \in G$ there is a $y \in G$ such that $xy = yx = e$.

A *homomorphism mapping group G* into group H is a function $f : G \to H$ such that for all $x, y \in G$, $f(xy) = f(x)f(y)$.

A *set of generators S* of a group G means that the smallest subgroup of G containing S is G itself. Write $G = \langle S \rangle$. In the case that S has one element we say that G is cyclic. If G is finite, the number of elements in $\langle a \rangle$, $a \in G$, is the order of a.

The quotient space G/H for a subgroup H of the group G consists of cosets $gH = \{gh | h \in H\}$. It forms a group iff (if and only if) H is a normal subgroup, that is, $Hg = gH$ for all $g \in G$.

A *ring R* is a set with two binary operations, addition ($+$) and multiplication ($\cdot$), such that
- R is a commutative group under $+$ with identity 0;
- multiplication is associative;
- distributive laws hold: $x(y + z) = xy + xz$ and $(x + y)z = xz + yz$, for all $x, y, z \in \mathbb{R}$.

A *field F* is a ring such that the set of nonzero elements forms a group under multiplication (with 1 as the identity).

A *vector space V* over a field F is an abelian group under $+$ such that there is an operation of scalar multiplication taking $a \in F$, $v \in V$ to $av \in V$

such that for all $x, y \in V, a, b \in F$:

- $a(x + y) = ax + ay$;
- $(a + b)x = ax + bx$;
- $a(bx) = (ab)x$;
- $1x = x$.

A linear map from vector space V to vector space W is a function $f : V \rightarrow W$ such that $f(x + y) = f(x) + f(y)$ and $f(ax) = af(x)$ for all $x, y \in V$ and $a \in F$.

A basis of an n-dimensional vector space V is a set of vectors $v_1, \ldots, v_n \in V$ such that every vector $v \in V$ can be expressed as a linear combination $v = \sum_{j=1}^{n} a_j v_j$, for unique scalars $a_j \in F$.

The matrix of a linear transformation or map $L : V \rightarrow V$, where V is an n-dimensional vector space with basis $v_1, \ldots, v_n$, is the $n \times n$ array of scalars $(m_{ij})_{1 \le i, j \le n}$, where

$$Lv_j = \sum_{i=1}^{n} m_{ij} v_i, \quad \text{where } m_{ij} \in F.$$

An eigenvalue of a linear transformation $L : V \rightarrow V$ is a scalar $\lambda \in F$ such that $Lx = \lambda x$ for some nonzero vector $x \in V$. And x is called an eigenvector.

A (simple, undirected) graph X is a set of vertices V and edges E connecting pairs of vertices. The edges are undirected and each pair x, y of vertices has at most one edge connecting it. Then we say x and y are adjacent. The degree of a vertex x is the number of edges coming out of that vertex.

The adjacency matrix A of a graph X with n vertices $x_1, \ldots, x_n$ is the $n \times n$ matrix A with entry $a_{ij} = 1$ if vertex i is adjacent to vertex j and $a_{ij} = 0$, otherwise.

The Concrete

$\mathbb{Z} = \{\ldots, -3, -2, -1, 0, 1, 2, 3, \ldots\}$ = the ring of integers.

$\mathbb{Q} = \{p/q \mid p, q \in \mathbb{Z}, q \neq 0\}$ = the field of rational numbers.

$\mathbb{R}$ = {all decimals} = {limits of Cauchy sequences of rational numbers} = the field of real numbers.

$\mathbb{C} = \{x + iy \mid x, y \in \mathbb{R}\}$ = the field of complex numbers; $i = \sqrt{-1}$.

$\mathbb{T} = \{z \in \mathbb{C} \mid |z| = 1\}$ = the unit circle = the 1-torus.

$\mathbb{F}_q$ = the finite field with $q = p^r$ elements; p = prime.

$K^n = \{v = {}^t(v_1, \ldots, v_n) \mid v_j \in K, \text{ all } j = 1, \ldots, n\}$ = n-dimensional vector space over a field K, vectors v being column vectors and ${}^t v$ denoting transpose of v.

$GL(n, K) = \{g \in K^{n \times n} \mid \det g \neq 0\}$ = the general linear group of all invertible $n \times n$ matrices over the field K.

$\mathbb{Z}/n\mathbb{Z}$ = the finite circle = the quotient group of integers modulo n consisting of equivalence classes of integers where $a, b \in \mathbb{Z}$ are equivalent if n divides $b - a$ and we write $a \equiv b \pmod n$.

$U(n) = \{g \in GL(n, \mathbb{C}) \mid {}^t\bar{g}g = I\}$ = unitary group, where I = identity matrix and ${}^t\bar{g}$ is the matrix obtained from g by transposing and then replacing each entry by its complex conjugate.

$O(n) = \{g \in GL(n, \mathbb{R}) \mid {}^tgg = I\}$ = orthogonal group.

$K[x]$ = the ring of polynomials with coefficients in a field K.

$K[x]/(g(x))$ = the quotient ring of polynomials modulo $g(x)$.

R^* = the multiplicative group of units in a ring R ($a \in R$ is a unit if $a^{-1} \in R$).

$L^2(X)$ = vector space or Hilbert space of all complex-valued functions on a finite set $X = \{f : X \to \mathbb{C}\}$, considered as a vector space over $\mathbb{C}$ of dimension $|X|$ and inner product

$$\langle f, g \rangle = \sum_{x \in X} f(x)\overline{g(x)}.$$

S_n = the symmetric group of permutations of the set $\{1, 2, 3, \ldots, n\}$.

A_n = the alterating group of even permutations in S_n.

Chapter 1

Congruences and the Quotient Ring of the Integers mod n

Monitor to Tegan: "Their language is the language of numbers
and they have no need to smile."
Monitor to the Doctor: "Yes, Doctor, you were right. Our
numbers were holding the fabric of the universe together."

Dr. Who in Logopolis

Congruences

In this first section, we review a little elementary number theory. We consider congruences mod n and the ring $\mathbb{Z}/n\mathbb{Z}$. We assume that the reader is familiar with some notions from elementary number theory, for example, divisibility of integers and unique factorization into primes. For more information, see Hua [1982], Ireland and Rosen [1982], Rosen [1993], or Stark [1978]. Some references for algebra are Dornhoff and Hohn [1971], Gallian [1990], Gilbert [1976], and Herstein [1964].

Definition. Suppose that n is a positive integer. Then for any integers a, b we say *a is congruent to b modulo n*, written

$$a \equiv b \pmod{n} \Leftrightarrow n \text{ divides } (a - b) \tag{1}$$
$$\Leftrightarrow a - b \in n\mathbb{Z} = \text{the ideal of integer multiplies of } n$$
$$\Leftrightarrow a \text{ and } b \text{ have the same remainder upon division by } n.$$

Gauss introduced the congruence notation [in *Disquisitiones Arithmeticae*, 1799]. Consider two integers to be the same if they are congruent modulo n. We will fix n throughout this paragraph. The elements of the quotient ring $\mathbb{Z}/n\mathbb{Z}$ are defined to be the equivalence classes you get upon making this identification. Thus $\mathbb{Z}/n\mathbb{Z}$ is in 1-1 correspondence with the set $\{0, 1, 2, \ldots, n - 1\}$. Note

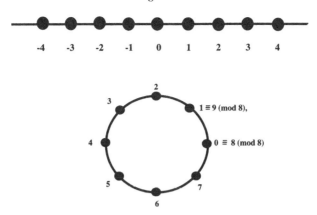

Figure I.2. Rolling up the line of integers into a finite circle.

that here we identify 0 with n. Thus we are taking the integers usually thought of as in a line and rolling that line up into a circle. See Figure I.2.

So we may view $\mathbb{Z}/n\mathbb{Z}$ as the finite circle. Note that we can use other sets of representatives for $\mathbb{Z}/n\mathbb{Z}$ (e.g., $\{1, 2, \ldots, n\}$). In fact we can replace any number j by $j + an$ for some $a \in \mathbb{Z}$.

Define addition and multiplication on $\mathbb{Z}/n\mathbb{Z}$ by using $+$ and $\times$ in $\mathbb{Z}$ and then taking the remainder of the result upon division by n. Since $\mathbb{Z}/n\mathbb{Z}$ is finite, it is easy to write tables for addition and multiplication. The entry in the ith row and jth column stands for $i + j$ (mod 7) in the addition table (Table I.1) and $i * j$ (mod 7) in the multiplication table (Table I.2).

Exercise. Complete Tables I.1 and I.2. Then the tables for $\mathbb{Z}/12\mathbb{Z}$.

Clearly the addition tables are pretty predictable. Each row is obtained from the one above it by moving everything over 1 and then moving the stuff hanging out at the end back to the beginning.

Table I.1. *Addition mod 7*

+ mod 7	0	1	2	3	4	5	6
0	0	1	2	3	4	5	6
1	1	2	3	4	5	6	0
2	2	3	4	5	6	0	1
3					0		
4			0				
5		0					
6	0						

Table I.2. *Multiplication mod 7*

* mod 7	0	1	2	3	4	5	6
0	0	0	0	0	0	0	0
1	0	1	2	3	4	5	6
2	0	2	4	6	1	3	5
3	0	3			5		
4	0	4		5			
5	0	5	3				
6	0	6					

It is not hard to see that $\mathbb{Z}/n\mathbb{Z}$ forms a commutative group under addition. It is closed under $+$ and $-$, contains 0, and $+$ is associative and commutative. In fact, it is a cyclic group generated by 1, since any element $a \pmod{n}$ is a sum of a ones.

Moreover, there is a way to visualize this additive group $G = \mathbb{Z}/n\mathbb{Z}$ as the *Cayley graph* obtained as follows. Let $S = \{1, -1 \pmod{n}\}$. This is a set of generators of G. Take the vertices of the graph to be the elements of G. Draw an edge between two vertices v and w if $w \equiv v+s \pmod{n}$, $s \equiv \pm 1 \pmod{n}$. For $n = 8$, we get the graph shown in Fig. I.3, which is just the finite circle graph.

Cayley graphs should actually be directed graphs having edges labeled with the appropriate generator of the group. Since we are taking a symmetric set S of generators of G (i.e., $s \in S$ implies $-s \in S$), we will leave off the directions and draw only one edge between each pair of vertices. We will say more about Cayley graphs in Chapter 3 and elsewhere. References for Cayley graphs are Biggs [1974], Bollobás [1979], and Gallian [1990]. There are connections with finite-state machines or automata [see Dornhoff and Hohn, 1978].

The main application of congruences that we will consider is to Fourier analysis. Replace the real line $\mathbb{R}$ or the circle $\mathbb{T} \cong \mathbb{R}/\mathbb{Z}$ with the finite circle.

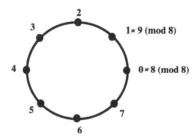

Figure I.3. Cayley graph for additive group $\mathbb{Z}/8\mathbb{Z}$ with generating set $S = \{\pm 1 \pmod 8\}$.

Then the usual Fourier transform or Fourier series can be approximated with the finite Fourier transform.

Of course congruences have numerous applications in computing, error-correcting codes, and cryptography. See Rosen [1993], Schroeder [1986], Knuth [1981], and Szabó and Tanaka [1967]. We will discuss some of these applications later.

What about the multiplication table for $\mathbb{Z}/7\mathbb{Z}$? If you leave out 0 (mod 7), you have a group under multiplication usually denoted $(\mathbb{Z}/7\mathbb{Z})^*$. To see this, you need to see that every $x \not\equiv 0$ (mod 7) has an inverse y (mod 7) so that $xy \equiv 1$ (mod 7). For example, $5x \equiv 1$ (mod 7) has the solution $x \equiv 3$ (mod 7). So 3 behaves like $1/5$ modulo 7. We say that $\mathbb{Z}/7\mathbb{Z}$ is a field because you can divide by nonzero elements. Here zero means the equivalence class of integers divisible by 7.

By contrast $\mathbb{Z}/12\mathbb{Z}$ is not a field. It is only a ring. For example, $2x \equiv 1$ (mod 12) has no solution in integers x. So you cannot divide by 2 in $\mathbb{Z}/12\mathbb{Z}$. A ring is a set in which you can add, subtract, and multiply, but not necessarily divide, with the usual laws of algebra holding. We leave it to you to consider the abstract concepts of ring and field as in Dornhoff and Hohn [1978], Gallian [1990], Gilbert [1976], and Herstein [1964].

Note that if we want to be strictly legal, we should show that our definition of equivalence classes of integers mod n makes sense. That means we should show that

$$a \equiv b \,(\mathrm{mod}\, n) \quad \text{and} \quad c \equiv d \,(\mathrm{mod}\, n)$$

implies

$$a + c \equiv b + d \,(\mathrm{mod}\, n) \quad \text{and} \quad a * c \equiv b * d \,(\mathrm{mod}\, n).$$

This is left as an exercise.

Theorem 1. $\mathbb{Z}/n\mathbb{Z}$ is a field, that is, closed under addition, subtraction, multiplication, and division by nonzero elements if and only if n is a prime.

Proof.

$$n = \text{prime} \quad \text{implies} \quad \mathbb{Z}/n\mathbb{Z} = \text{field.}$$

We have to show that if n is prime, we can divide by nonzero elements of $\mathbb{Z}/n\mathbb{Z}$. Suppose that $a \not\equiv 0$ (mod n). This implies that the greatest common divisor[†] $(a, n) = 1$. But one knows (see the exercise below) that then there are

---

[†] Here "greatest common divisor" or g.c.d.(a, n) means the largest positive integer dividing both a and n.

integers x and y so that

$$1 = xa + yn.$$

This implies that $ax \equiv 1 \pmod{n}$, that is, that $x \pmod{n}$ is the reciprocal of $a \pmod{n}$.

Exercise.
a) Obtain a constructive proof for the existence of integers x, y such that

$$\text{g.c.d.}(a, n) = xa + yn,$$

   using the euclidean algorithm. You can find this in most elementary number theory books (for example, K. Rosen [1993]).
b) Obtain an existence proof for the x, y in Part a using the facts that

$$I = \{ax + ny \mid x, y \in \mathbb{Z}\}$$

   is an ideal in the ring $\mathbb{Z}$ and that all ideals in $\mathbb{Z}$ are principal, that is, of the form $d\mathbb{Z}$, for some $d > 0$ which is the least positive element in I. In this case, $d = \text{g.c.d.}(a, n)$. This sort of proof occurs in I. Herstein [1964, p. 18].

$$\mathbb{Z}/n\mathbb{Z} = \text{field} \quad \text{implies} \quad n = \text{prime}.$$

Now we have to show that if n is not a prime then there are integers $a \not\equiv 0 \pmod{n}$ such that the equation $ax \equiv 1 \pmod{n}$ has no solution in integers x. This is easy. If n is not prime, then $n = ab$ with $1 < a, b < n$. Suppose that $ax \equiv 1 \pmod{n}$ has a solution. Then $ax - 1 = abq$, for some $q \in \mathbb{Z}$ and $1 = ax - abq = a(x - bq)$. This says a divides 1, which is impossible. ∎

So $\mathbb{Z}/p\mathbb{Z}$ is a finite field with p elements (also called $\mathbb{F}_p$) if p is a prime. Any finite field has $q = p^r$ elements for some prime p and some positive integer r. See Dornhoff and Hohn [1978], Gallian [1990], Gilbert [1976], Herstein [1964], or Small [1991]. One way to prove this is to show that a finite field must contain some $\mathbb{Z}/p\mathbb{Z}$ as a subfield and then that it is a finite-dimensional vector space over this subfield. See Chapter 3.

The example $\mathbb{Z}/n\mathbb{Z}$ of a quotient ring should be compared with the example of quotient rings formed from $\mathbb{Q}[x]$ = the ring of polynomials with rational coefficients and indeterminate x. The fields $\mathbb{Q}[x]/f(x)\mathbb{Q}[x]$, for $f(x)$ an irreducible polynomial, give all the field extensions of $\mathbb{Q}$ of finite degree. See Herstein [1964].

We could also replace the field of rationals here with a finite field and obtain any finite field as a quotient of polynomial rings over $\mathbb{F}_p$.

Our next task is to consider the structure of a ring like $\mathbb{Z}/12\mathbb{Z}$. This comes out of something called the Chinese remainder theorem, which dates back to Sun Tsu, in the first century A.D. It has numerous computer applications, for example, in writing programs to multiply large integers. First we define the *direct sum of two rings* R and S by

$$R \oplus S = \{(r, s) \mid r \in R, s \in S\}. \tag{2}$$

Then define addition and multiplication of vectors componentwise:

$$(r, s) + (t, v) = (r + t, s + v),$$
$$(r, s) * (t, v) = (r * t, s * v).$$

The result is that $R \oplus S$ is also a ring. You can similarly define the direct sum of any number of rings.

*Theorem 2 (**The Chinese Remainder Theorem**).* Suppose that the moduli m_1, $\ldots, m_r$ are pairwise relatively prime, that is, g.c.d.$(m_i, m_j) = 1$ for $i \neq j$. Let $m = m_1 m_2 \cdots m_r$. Then we have the following ring isomorphism:

$$\mathbb{Z}/m\mathbb{Z} \cong (\mathbb{Z}/m_1\mathbb{Z}) \oplus (\mathbb{Z}/m_2\mathbb{Z}) \oplus \cdots \oplus (\mathbb{Z}/m_r\mathbb{Z}).$$

Proof. The isomorphism T is defined by

$$T(x \bmod m) = (x \bmod m_1, \ldots, x \bmod m_r).$$

Exercise. Check that T is indeed a ring isomorphism, that is, T is a 1-1, onto map that preserves $+$ and $*$. Note that since both sides of the isomorphism have the same number of elements, it suffices (by the pigeon hole principle[†]) to show that the map is either 1-1 or onto. See Dornhoff and Hohn [1978, p. 12]. The easiest is 1-1. You must also check that T is well defined. ∎

Note that this is not the proof from the year 1. That proof shows that the map is onto. In fact, there is a song that explains the ancient construction. See Hua

---

[†] The pigeonhole principle says if you have a set S of n pigeons and a set T of n pigeonholes and a function $f : S \to T$, if f is 1-1, no pigeons can go to the same pigeonhole and so the map must be onto. Similarly if f is onto, it must be 1-1.

[1982, p. 30]:

> Three people walking together, 'tis rare that one be seventy,
> Five cherry blossom trees, twenty one branches bearing flowers,
> Seven disciples reunite for the half-moon,
> Take away (multiple of) one hundred and five and you shall know.

Here the problem is to solve the simultaneous congruences:

$$x \equiv 2 \ (\text{mod } 3),$$
$$x \equiv 3 \ (\text{mod } 5),$$
$$x \equiv 2 \ (\text{mod } 7).$$

The meaning of the song is: Multiply by 70 the remainder of x when divided by 3, multiply by 21 the remainder of x when divided by 5, and multiply by 15 (the number of days in half a Chinese month) the remainder of x when divided by 7. Add the three results together and then subtract a multiple of 105 and you get the smallest solution, 23:

$$2 \times 70 + 3 \times 21 + 2 \times 15 = 233 = 23 + 2 \times 105.$$

Where did 70 come from? It is a multiple of 5 and 7 which is congruent to 1 mod 3. Similarly 21 is a multiple of 3 and 7 which is congruent to 1 mod 5. And 15 is a multiple of 3 and 5 that is congruent to 1 mod 7.

It is possible to use the Chinese remainder theorem to add and multiply large integers. See Knuth [1981], Richards [1980], Rosen [1993], or Schroeder [1986], for example. Winograd showed in 1965 how the Chinese remainder theorem could help to do rapid addition (see Dornhoff and Hohn [1978, Sections 5.11 and 5.12]).

Suppose you have a really stupid computer that can only handle the numbers from 1 to 15. You can use the Chinese remainder theorem with moduli 3 and 5. Then each number from 1 to 15 has a place in the rectangle shown in Table I.3.

Table I.3. $\mathbb{Z}/15\mathbb{Z} \cong \mathbb{Z}/3\mathbb{Z} \oplus \mathbb{Z}/5\mathbb{Z}$

	1	2	3	4	5
1	1	7	13	4	10
2	11	2	8	14	5
3	6	12	3	9	15

Table I.4. *How to create Table I.3. Move numbers up a multiple of 3 and left a multiple of 5 places*

	1	2	3	4	5	1	2	3	4	5
1	1									
2		2								
3			3							
1				4						
2					5					
3						6				
1							7			
2								8		
3									9	
1										10
2										
3										

Note that you can fill in the boxes as follows. Make the big table as shown in Table I.4 and move the numbers up a multiple of 3 and left a multiple of 5 places.

Note. This encoding is useful because it behaves well with respect to $+$ and $\times$. So you can multiply numbers ≤ 15 by multiplying much smaller numbers that are ≤ 5. See Knuth [1981, Vol. II, pp. 268–301] for a discussion of how fast we can multiply.

We can use the Chinese remainder theorem to find another visualization of $\mathbb{Z}/15\mathbb{Z}$. Instead of a finite circle or cycle graph, we can consider the product of two finite cycle graphs, which is a finite torus graph. See Figure I.4 for a picture of a continuous torus obtained by rolling up a square piece of material. A finite torus is shown in Figure I.5.

Question. Why are the following rings of order 2^n all different:

$$\mathbb{Z}/2^n\mathbb{Z}, \quad (\mathbb{Z}/2\mathbb{Z})^n, \quad \mathbb{F}_{2^n}?$$

The additive group of the ring on the left is cyclic $\mathbb{Z}/2^n\mathbb{Z} = \langle 1 \pmod{2^n} \rangle$. See the Cast of Characters or formula (5) in the next section where you have to replace g^n with ng. In the additive groups of the other two rings every nonzero element has order 2, meaning that $2x = 0$ for all $x \neq 0$. The ring on the right

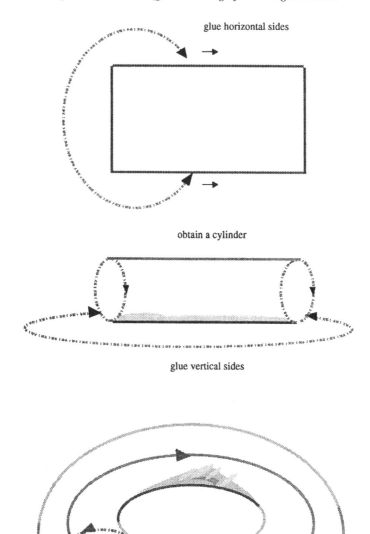

glue horizontal sides

obtain a cylinder

glue vertical sides

obtain a torus

Figure I.4. Continuous torus or doughnut formed from rolling up a rectangle.

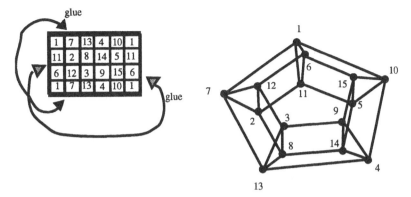

Figure I.5. Finite torus formed from a Cayley graph of $\mathbb{Z}/15\mathbb{Z} \cong \mathbb{Z}/3\mathbb{Z} \oplus \mathbb{Z}/5\mathbb{Z}$.

is a field and thus has no zero divisors $ab = 0$ with a and b not zero. But the other two rings have zero divisors.

Invertible Elements (for Multiplication) or Units of the Ring $\mathbb{Z}/n\mathbb{Z}$ – Euler's Phi Function

We study the multiplicative group of integers $a \pmod n$ with g.c.d.$(a, n) = 1$.

Definition. The group of *units* of the ring $\mathbb{Z}/n\mathbb{Z}$ is

$$(\mathbb{Z}/n\mathbb{Z})^* = \{a \pmod n \mid \text{g.c.d. } (a, n) = 1\} \tag{3}$$
$$= \{a \pmod n \mid ax \equiv 1 \pmod n \text{ has a solution } x \in \mathbb{Z}\}.$$

Exercise.
a) Prove the last equality. You need to use the property of the greatest common divisor that we used in the proof of Theorem 1.
b) Prove that $(\mathbb{Z}/n\mathbb{Z})^*$ is a group under multiplication.

Examples. We have the following multiplicative groups: $(\mathbb{Z}/p\mathbb{Z})^* = \{1, 2, 3, \ldots, p - 1\}$ for any prime p.

$$(\mathbb{Z}/8\mathbb{Z})^* = \{1, 3, 5, 7\}, (\mathbb{Z}/12\mathbb{Z})^* = \{1, 5, 7, 11\}.$$

Definition. Euler's phi function is

$$\phi(n) = \text{the order of the group } (\mathbb{Z}/n\mathbb{Z})^*. \tag{4}$$

Table I.5. *Euler's phi function*

n	1	2	3	4	5	6	7	8	9	10	11
$\phi(n)$	1	1	2	2	4	2	6	4	6	4	10

Examples. See Table I.5.

Exercise. Compute $\phi(n)$ for $12 \leq n \leq 30$. You might want to wait until we have proved a few more facts about Euler's phi function.

Theorem 3. **Facts about Euler's Phi Function.**
1. If p is a prime, $\phi(p^n) = p^n - p^{n-1}$.
2. If g.c.d. $(n, m) = 1$, then $\phi(nm) = \phi(n)\phi(m)$. This makes the Euler phi function a *multiplicative function*.
3.

$$\phi(m) = m \prod_{\substack{p|m \\ p = \text{prime}}} \left(1 - \frac{1}{p}\right).$$

4. Suppose x is an integer with g.c.d. $(x, n) = 1$. Then

$$x^{\phi(n)} \equiv 1 \pmod{n}.$$

Proof.
1. By the definitions $\phi(p^n)$ is the number of integers a between 0 and $p^n - 1$ such that g.c.d.$(a, p) = 1$. Equivalently, we can count the numbers a between 0 and $p^n - 1$ such that p divides a and subtract this from p^n. The numbers a with p dividing a such that $0 \leq a \leq p^n - 1$ are

$$0 \cdot p, 1 \cdot p, 2 \cdot p, \ldots, (p^{n-1} - 1) \cdot p.$$

There are p^{n-1} numbers on this list. Thus there are $p^n - p^{n-1}$ elements in $(\mathbb{Z}/p^n\mathbb{Z})^*$.
2. This is really the Chinese remainder theorem. Since

$$\mathbb{Z}/mn\mathbb{Z} \cong (\mathbb{Z}/n\mathbb{Z}) \oplus (\mathbb{Z}/m\mathbb{Z}),$$

we see that $x \pmod{mn}$ is invertible if and only if $x \pmod{n}$ and $x \pmod{m}$ is invertible.
3. We leave the proof of this formula as an exercise.

4. This is a special case of Lagrange's theorem in group theory. See Gallian [1990] or Herstein [1964], for example. If G is a group with r elements, then $x^r =$ the identity of the group, for any $x \in G$. There is another proof of fact 4 to be found in most elementary number theory books (a proof which works for any finite abelian group). In fact, the word "group" seems to be on the censored list for most of these number theory books. ∎

Corollary. ***Fermat's Little Theorem.*** Suppose that p is a prime. Then for all $a \in \mathbb{Z}$, we have

$$a^p \equiv a \pmod{p}.$$

If p does not divide a, then

$$a^{p-1} \equiv 1 \pmod{p}.$$

History

Fermat stated this in 1640 and Euler generalized it in 1760. A special case of Fermat's theorem is that if p is a prime then p divides $2^p - 2$. The ancient Chinese knew this and also believed that the converse was true. They were wrong since, for example, 341 divides $2^{341} - 2$, even though $341 = 11 \cdot 31$. Today a composite integer n such that n divides $2^n - 2$ is called a *pseudoprime* (base 2). The first two pseudoprimes are 341 and 561. There are infinitely many pseudoprimes base 2. See Rosen [1993, p. 193].

There are, in fact, composite numbers n called *Carmichael numbers* such that n divides $b^n - b$ for all b. They are named after the American mathematician who discovered their properties in 1904. In fact 561 is a Carmichael number. Recently (see Cipra [1993, Vol. I]) W. R. Alford, A. Granville, and C. Pomerance have shown that there are infinitely many Carmichael numbers.

Exercise. Show that $\sum_{\substack{0 < d \\ d \mid n}} \phi(d) = n$. Here $d \mid n$ means d divides n, that is, $n = dc$ for some integer c.

Hint. You can use the fact that both sides of the inequality are multiplicative.

Notes. Property 2 makes phi a multiplicative function. There are lots of other multiplicative functions, for example,

$$\sigma_k(n) = \sum_{\substack{0 < d \\ d \mid n}} d^k.$$

Our goal for the moment is to study the structure of the multiplicative groups $(\mathbb{Z}/n\mathbb{Z})^*$ more closely. The simplest groups are the cyclic groups (e.g., the *additive* group $\mathbb{Z}/n\mathbb{Z}$).

Primitive Roots

Definition. A (multiplicative) group G is said to be *cyclic* with *generator* g if every element $x \in G$ has the form $x = g^n$, for some integer n. Write

$$\langle g \rangle = G = \{g^n \mid n \in \mathbb{Z}\}. \tag{5}$$

If the group G is additive we must replace g^n with ng. If G is a finite cyclic group with d elements, then it is easy to see that G is isomorphic to the additive group of $\mathbb{Z}/d\mathbb{Z}$.

Exercise. Prove the last statement. The map $T : G \to \mathbb{Z}/d\mathbb{Z}$ is defined by $T(g^n) = n \pmod{d}$. You need to show that this map is well defined, 1-1, and onto and carries multiplication in G to addition in $\mathbb{Z}/d\mathbb{Z}$.

Definition. The *order* of an element g in a finite group G is the number of elements in (or order of) the subgroup $\langle g \rangle$ generated by g.

Question. When is $(\mathbb{Z}/n\mathbb{Z})^*$ cyclic? When this happens, a generator $g \pmod{n}$ of $(\mathbb{Z}/n\mathbb{Z})^*$ is called a *primitive root modulo n*.
Answer. $(\mathbb{Z}/n\mathbb{Z})^*$ is a cyclic group if and only if $n = 2, 4, p^m, 2p^m$, for odd primes p.

We will content ourselves with proving that $(\mathbb{Z}/n\mathbb{Z})^*$ is cyclic for prime p. You might try to do the rest for some nontrivial exercises.

Example 1. The number 2 is not a primitive root for $p = 7$ since $2^3 \equiv 1 \pmod{7}$. However, 3 is a primitive root, since

$$3^2 \equiv 2, \ 3^3 \equiv 6, \ 3^4 \equiv 4, \ 3^5 \equiv 5, \ 3^6 \equiv 1 \pmod{7}.$$

A Mathematica Remark. Mathematica has a function called PowerMod $[a, b, n]$ which gives $a^b \pmod{n}$. It is much better to use PowerMod than to write Mod $[a\,\hat{}\,b, n]$. Why?

Example 2. By the Chinese remainder theorem,

$$(\mathbb{Z}/12\mathbb{Z})^* \cong (\mathbb{Z}/3\mathbb{Z})^* \oplus (\mathbb{Z}/4\mathbb{Z})^*.$$

So every element of $(\mathbb{Z}/12\mathbb{Z})^*$ has order 2. So there can be no primitive root modulo 12. That is, $(\mathbb{Z}/12\mathbb{Z})^*$ is *not* a cyclic group.

Exercise. Write a Mathematica program to find a primitive root mod n if it exists.

Note. Although we can prove that primitive roots exist for any prime modulus, we won't be able to provide an easy way to find them. Trial and error is the standard method. That would be rather time consuming for large moduli.

Before proving our theorem, we need the following lemma.

Lemma. Let k be a field and suppose that $f(x) \in k[x]$, that is, that $f(x)$ is a polynomial with coefficients in the field k. Let n be the degree of f. Then f has at most as many roots as the degree of f.

Proof. Use the division algorithm for polynomials, which works in any field. This says that if we are given $f(x)$ and $g(x) \in k[x]$, we can find $q(x)$ and $r(x) \in k[x]$ so that $f(x) = g(x) q(x) + r(x)$, where degree $r <$ degree g or r is the 0-polynomial. The point is that if a is a root of $f(x)$, then $x - a$ divides $f(x)$ (with remainder 0). And you get factors for each distinct root of $f(x)$. Moreover, the degree of a product of polynomials is the sum of the degrees of the factors. Thus you can have no more than n distinct linear factors. ∎

Theorem 4. $(\mathbb{Z}/p\mathbb{Z})^*$ is cyclic for any prime p. This means that there is a primitive root modulo any prime p.

Proof. We seek to show that $(\mathbb{Z}/p\mathbb{Z})^*$ has an element of order $p - 1$. If d is a divisor of $(p - 1)$, let $\psi(d)$ denote the number of elements x of $(\mathbb{Z}/p\mathbb{Z})^*$ of order d (meaning that $x^r \equiv 1 \pmod{p}$ holds for $r = d$ and no smaller positive power r). Then, if $\psi(d) \neq 0$, there is an $x \pmod{p}$ such that the set

$$\langle x \rangle = \{1, x, x^2, x^3, \dots, x^{d-1} \pmod{p}\}$$

is a cyclic subgroup of $(\mathbb{Z}/p\mathbb{Z})^*$ with *exactly d* elements. Why are there exactly d elements?

By the lemma, the equation

$$T^d \equiv 1 \ (\mathrm{mod} \ p)$$

can have at most d solutions in the field $\mathbb{Z}/p\mathbb{Z}$. Therefore the set $\langle x \rangle$ includes all the solutions to $T^d \equiv 1 \ (\mathrm{mod} \ p)$ in $\mathbb{Z}/p\mathbb{Z}$. Hence the set $\langle x \rangle$ contains all elements of $(\mathbb{Z}/p\mathbb{Z})^*$ of order d.

But the cyclic group $\langle x \rangle$ is isomorphic to the additive group $\mathbb{Z}/d\mathbb{Z}$, by an exercise above. The latter has $\phi(d)$ elements of order d (exercise). Here $\phi(d)$ is Euler's phi function. Thus $\psi(d) \neq 0$ implies $\psi(d) = \phi(d)$.

It follows that since every element of $(\mathbb{Z}/p\mathbb{Z})^*$ has an order dividing $p - 1$, we have

$$p - 1 = \sum_{\substack{0 < d \\ d \,|\, (p-1)}} \psi(d) \leq \sum_{\substack{0 < d \\ d \,|\, (p-1)}} \phi(d) = p - 1.$$

The last equality comes from an earlier exercise. It follows that the inequality must, in fact, be an equality. Thus $\psi(d) = \phi(d)$ for every divisor d of $p - 1$, including $d = p - 1$. This proves the theorem. ∎

Exercise.

a) Find all the primitive roots modulo all primes ≤ 20.

b) How many primitive roots are there mod p, for a given prime p?

Remarks. It is also possible to show that the multiplicative group of any finite field is cyclic. See Herstein [1964, p. 317]. The proof is also in Dornhoff and Hohn [1978]. In fact, a little use of field theory shortens the proof considerably.

Theorem 4 was first proved by Gauss. Gauss conjectured that 10 is a primitive root for infinitely many primes p, after "laborious calculations." In 1927, Emil Artin conjectured that if $a \neq -1$ and $a \neq$ square, then there are infinitely many primes p such that a is a primitive root mod p. The conjecture is still unproved. See L. J. Goldstein [1971] for details on the state of the conjecture twenty years ago. Other references are Shanks [1985] and Silverman [1997].

Artin actually had a more specific conjecture to the effect that if $a \neq -1$ and $a \neq$ square, then approximately 3/8 of all primes will have a for a primitive root. The heuristic probabilistic argument that led to this conjecture will be discussed soon. The conjecture needs modification for certain values of a.

Robert Baillie of the Computer Center of the University of Illinois at Urbana computed lengthy tables verifying the Artin conjecture (see Table I.6). Baillie's tables say, for example, that if $A \cong .37395\ 58136\ 19$ is Artin's constant defined

Table I.6. *A computer verification of Artin's
conjecture by Baillie*

a	# Primes $\leq 33 \times 10^6$ with a as primitive root	Artin factor, A in (6)
2	759,733	759,754
3	759,658	759,754
5	800,218	$799,741 = \dfrac{20}{19} 759,754$
6	760,037	759,754
7	760,133	759,754
8	455,894	$455,854 = \dfrac{3}{5} 759,754$
10	760,192	759,754
11	760,352	759,754
12	759,988	759,754
13	764,719	$764,655 = \dfrac{156}{155} 759,754$

by (6) below, then, since the number of primes $\leq 33 \times 10^6$ is 2,031,667, we have

$$A \cdot 2,031,667 \cong 759,754.$$

Perhaps we should explain Artin's mysterious constant. A heuristic probabilistic argument shows it to be given by the infinite product

$$A = \prod_{p \text{ prime}} \left[1 - \frac{1}{p(p-1)} \right]. \tag{6}$$

The argument goes as follows:

Consider $a = 2$. Look at the primes less than or equal to N. For every prime p, choose a primitive root g (mod p) and write $g^m \equiv 2$ (mod p) and g.c.d. $(m, p-1) = G$. What is the probability that 2 divides G? Except for $p = 2$, $p - 1$ is always even, and m is even in half the cases – that is, when 2 is a square mod p. Since G must be 1 if 2 is to be a primitive root mod p, we must delete such cases. This leaves on the average

$$\left(1 - \frac{1}{2} \right) \pi(N) \text{ primes},$$

where $\pi(N)$ is the *number of primes* $\leq N$.

What is the probability that 3 divides G? Except for $p = 3$, all primes are congruent to 1 or 2 mod 3. Thus 3 divides $p - 1$ in half the cases, while 3 divides m in one third of the cases. Elimination of the primes in which 3 divides G leaves

$$\left(1 - \frac{1}{2}\right)\left(1 - \frac{1}{3 \cdot 2}\right)\pi(N).$$

Continuing with the same argument for G divisible by 5, 7, etc., we obtain the formula above for Artin's constant. One can actually compute the infinite product with various tricks. Wrench obtains 40 significant digits, for example.

Question. Why isn't this a proof of Artin's conjecture?

Answer. We are assuming things are independent events that cannot be proved independent. See Schroeder [1986, pp. 40–42] for a similar probabilistic argument for the *prime number theorem* (proved, for example, in Davenport, [1980]), which says

$$\pi(N) \sim N/\log N, \quad \text{as } N \to \infty. \tag{7}$$

Also see Pólya [1984, Vol. III, pp. 436–45].

There is one more interesting facet of Artin's conjecture. Hooley [1667] modified the conjecture and showed that it is implied by the Riemann hypothesis for zeta functions of certain algebraic number fields. These zeta functions are analogous to the Riemann zeta function. We will discuss another sort of zeta function in Chapter 2.

H. Bilharz [1937] proved the Artin conjecture for the ring $k[x]$ of polynomials over a finite field. His proof required the Riemann hypothesis for the zeta functions associated to such rings. That was proved by A. Weil several years later. See Ireland and Rosen [1993].

The group $(\mathbb{Z}/n\mathbb{Z})^*$ is a finite abelian group. By the fundamental theorem of abelian groups,[†] any such group is a direct product of finite cyclic groups. Here, by the direct product of two groups, we mean the analogue of direct sum of rings defined before the Chinese remainder theorem, except that now there is only one operation (multiplication, in our case). You might want to write $(\mathbb{Z}/n\mathbb{Z})^*$ explicitly as a direct product of cyclic groups. First note that if

$$n = p_1^{e_1} \cdots p_r^{e_r}, \quad \text{with distinct primes } p_i,$$

[†] See Chapter 10 and Herstein [1964].

the Chinese remainder theorem tells us that

$$(\mathbb{Z}/n\mathbb{Z})^* \cong (\mathbb{Z}/p_1^{e_1}\mathbb{Z})^* \times \cdots \times (\mathbb{Z}/p_r^{e_r}\mathbb{Z})^*.$$

Here we use the symbol $\times$ to denote the direct product of multiplicative groups.

It follows from the above argument that it suffices to consider $(\mathbb{Z}/p^e\mathbb{Z})^*$ for prime p. One can prove the following proposition.

Proposition 1. If p is an odd prime and e is a positive integer, then $(\mathbb{Z}/p^e\mathbb{Z})^*$ is cyclic.

Proof Sketch. We know that there exists a primitive root g (mod p). Look at $g + p$. It is also a primitive root mod p. And $(g + p)^{p-1} \not\equiv 1$ (mod p^2) if $g^{p-1} \equiv 1$ (mod p^2). Why? Thus we can assume that $g^{p-1} \not\equiv 1$ (mod p^2).

We claim that such a g is a primitive root mod p^e. To show this, it suffices to prove that if $g^n \equiv 1$ (mod p^e), then $\phi(p^e)$ divides n.

Now $g^{p-1} = 1 + ap$, with p not dividing a. And one can show (as an exercise) that if p is an odd prime and p does not divide a, then p^{e-1} is the order of $(1 + ap)$ (mod p^e).

It follows that $n = p^{e-1}n'$. By Fermat's little theorem, $g^{n'} \equiv$ (mod p). Thus $p - 1$ divides n' and so $\phi(p^e)$ divides n. ∎

For the powers of 2, the result is as follows.

Proposition 2. $(\mathbb{Z}/2\mathbb{Z})^*$ and $(\mathbb{Z}/4\mathbb{Z})^*$ are cyclic. For $e \geq 3$, $(\mathbb{Z}/2^e\mathbb{Z})^*$ is the direct product of two cyclic groups, one of order 2 and one of order 2^{e-2}. Thus $(\mathbb{Z}/2^e\mathbb{Z})^*$ is not cyclic if $e \geq 3$.

Proof Sketch. One can show (as an exercise) that for $e \geq 3$

$$\{(-1)^a 5^b \;(\text{mod } 2^e) \mid a = 0, 1; b = 0, 1, \ldots, 2^{e-2} - 1\} = (\mathbb{Z}/2^e\mathbb{Z})^*.$$

∎

Theorem 5. There are primitive roots mod n if and only if n is of the form $2, 4, p^e, 2p^e$, where p is an odd prime.

Proof Sketch. Suppose that $n = ab$, with g.c.d. $(a, b) = 1$ and $a, b > 2$. Then $\phi(a)$ and $\phi(b)$ are both even. So both $(\mathbb{Z}/a\mathbb{Z})^*$ and $(\mathbb{Z}/b\mathbb{Z})^*$ have elements of order 2. But then the direct product of these two groups cannot be cyclic, because a cyclic group can only have one element of order 2.

We know that 2, 4, and p^e yield cyclic groups. The same holds for $2p^e$ since

$$(\mathbb{Z}/2p^e\mathbb{Z})^* \cong (\mathbb{Z}/2\mathbb{Z})^* \times (\mathbb{Z}/p^e\mathbb{Z})^*.$$

∎

Exercise. Complete the proof of the last theorem by explaining why $(\mathbb{Z}/2^e\mathbb{Z})^*$ is not cyclic for $e \geq 3$.

The next exercise introduces another important multiplicative function.

Exercise. Show that the sum of all the primitive roots mod p is congruent to $\mu(p-1)$ mod p. Here $\mu(n)$ is the *Möbius function* defined by

$$\mu(n) = \begin{cases} 1, & n = 1, \\ 0, & n \text{ not square-free, } n > 1, \\ (-1)^r, & \text{if } n = p_1 \cdots p_r, \\ & \text{where the } p_i \text{ are distinct primes.} \end{cases} \qquad (8)$$

Exercise. Show that if $n > 1$ and μ denotes the Möbius function of the preceding exercise, we have

$$\sum_{\substack{d \mid n \\ d > 0}} \mu(d) = 0.$$

Exercise. **The Möbius Inversion Formula.** Suppose that $f : \mathbb{Z}^+ \to \mathbb{C}$ and let μ denote the Möbius function defined in the exercise above. Let

$$F(n) = \sum_{\substack{d \mid n \\ d > 0}} f(d).$$

Show that

$$f(n) = \sum_{\substack{d \mid n \\ d > 0}} \mu(d) F\left(\frac{n}{d}\right).$$

A Few Remarks on Multiplicative Functions

We have given three examples of multiplicative functions $f : \mathbb{Z}^+ \to \mathbb{Z}^+$ in this section: $\phi(n)$, $\sigma_k(n)$, and $\mu(n)$. It is useful to make the following definition.

Definition. Suppose that f and g are any multiplicative functions $f, g : \mathbb{Z}^+ \to \mathbb{C}$, that is, $f(mn) = f(m)f(n)$, if g.c.d.$(m, n) = 1$. Define *convolution* $f * g$ by

$$f * g(n) = \sum_{d|n} f(d) g\left(\frac{n}{d}\right). \tag{9}$$

Exercise. Show the following properties of convolution of functions f, g, h: $\mathbb{Z}^+ \to \mathbb{C}$:

a) $f * g = g * f$,
b) $f * (g * h) = (f * g) * h$.

Exercise.

a) Define

$$\delta_a(n) = \begin{cases} 1, & n = a, \\ 0, & \text{otherwise.} \end{cases} \tag{10}$$

Show that $\delta_1 = \mu * 1$, where μ is the Möbius function and 1 denotes the constant function that has value 1 for all $n \in \mathbb{Z}^+$.
b) Prove that $\delta_a * \delta_b = \delta_{ab}$ and $\delta_1 * f = f$.
c) Use the properties of $*$ to show the Möbius inversion formula in an earlier exercise, which says that $f = \mu * (f * 1)$.

Here we are considering convolution of functions with domain the infinite discrete set $\mathbb{Z}^+$ of positive integers, which is *not* a group under multiplication. Instead it is what is called a monoid. See Dornhoff and Hohn [1978, p. 165]. Since this is really a book about *finite groups*, we will say no more about this example. In the next Chapter we will consider convolution with $\mathbb{Z}^+$ replaced by the finite additive group $\mathbb{Z}/n\mathbb{Z}$. Later we will replace $\mathbb{Z}^+$ with any finite group G. It is possible to do convolution for continuous infinite groups such as the additive groups $\mathbb{R}$ or $\mathbb{R}/\mathbb{Z}$, but one has to use series or integrals rather than sums. See Dym and McKean [1972] or Terras [1985].

A Look Forward

We have already considered systems of linear congruences. The logical next question is: How do you solve quadratic congruences? That is, we will ask the question:

For a fixed prime $p \geq 3$ and given $a \in \mathbb{Z}$, can we find $x \in \mathbb{Z}$ so that $x^2 \equiv a \pmod{p}$?

That is, we will be looking for square roots in $\mathbb{Z}/p\mathbb{Z}$. Since we know that $\mathbb{Z}/p\mathbb{Z}$ is a field which we hope is a finite model for the field of real numbers, we might expect the answer to be similar to that for the field of real numbers. Half the nonzero real numbers (the positive reals) are squares of other real numbers and half aren't. Indeed, that is the case in $\mathbb{Z}/p\mathbb{Z}$ as well.

There is a surprising theorem in this area – the quadratic reciprocity law. One of our first applications of the discrete Fourier transform will be to prove it. See Chapter 8.

An Application – Public–Key Cryptography

There are many situations in which one wants to send a message which can only be deciphered by the recipient. Public-key cryptography allows one to do this fairly easily and feel fairly secure, assuming that no one has figured out something about number theory that we don't know.

Think of your message as a number m mod pq, where p and q are very large primes. The encryption of m is just m^t (mod pq), for some power t. To decrypt one must find a power s so that

$$m^{ts} \equiv m \ (\text{mod } pq).$$

From what we now know about $(\mathbb{Z}/pq\mathbb{Z})^*$, assuming that p and q don't divide m, we know that we need to solve

$$ts \equiv 1 \ (\text{mod } \phi(pq)).$$

The easiest way to solve this linear congruence for s may be to take

$$ts \equiv t^{\phi(\phi(pq))} \ (\text{mod } \phi(pq))$$

and thus

$$s \equiv t^{\phi(\phi(pq))-1} \ (\text{mod } \phi(pq)).$$

Why? This requires one to know $\phi(pq) = (p-1)(q-1)$, for prime p, q.

What happens is that everyone who wants to receive a secret message chooses a triple p, q, t and publishes t and the product pq. Then anyone who wants to send a secret message m will compute m^t (mod pq) and send this number.

Why can't anyone figure out m from this? Well, the catch is that if p and q are large enough then no one can compute $\phi(pq)$, because, to do that, one would have to factor a very large number pq. It is much easier to find two primes with 50 digits than to factor a 100-digit number. The size of the primes p and q is dependent on the state of the art of factoring and primality testing.

References for public–key cryptography are Rosen [1993] and Schroeder [1986].

Exercise. Investigate public-key cryptography and write a Mathematica program to encode and decode messages.

Exercise. Show that for any prime p if $a \in (\mathbb{Z}/p\mathbb{Z})^n, a \neq 0, b \in \mathbb{Z}/p\mathbb{Z}$, then the number of solutions $x \in (\mathbb{Z}/p\mathbb{Z})^n$ of the equation $\sum_{j=1}^{n} a_j x_j = b$ is p^{n-1}. Another way to say this is:

$$|\{x \in (\mathbb{Z}/p\mathbb{Z})^n \mid {}^t ax = b\}| = p^{n-1}.$$

Hint. The set whose order we seek is a hyperplane in n-space over a finite field, as the elements of the set are vectors satisfying one linear equation in n unknowns. Use the standard methods of linear algebra, which work as well over a finite field $\mathbb{Z}/p\mathbb{Z}$ as over the real numbers $\mathbb{R}$.

Question. Can you generalize the last exercise replacing the linear equation with a quadratic equation

$$\sum_{j=1}^{n} a_j x_j^2 = b?$$

This will be of interest in Chapter 5.

Chapter 2

The Discrete Fourier Transform on the Finite Circle $\mathbb{Z}/n\mathbb{Z}$

> We come now to reality. The truth is that the digital computer
> has totally defeated the analog computer. The input is a
> sequence of numbers and not a continuous function. The output
> is another sequence of numbers, whether it comes from a digital
> filter or a finite element stress analysis or an image processor.
> The question is whether the special ideas of Fourier analysis
> still have a part to play, and the answer is absolutely *yes*.
>
> G. Strang [1986, p. 290]

First we consider the easiest kind of Fourier analysis – that on the additive group $\mathbb{Z}/n\mathbb{Z}$, the integers modulo n. This is an abelian group of order n and it is cyclic (generated by the congruence class 1 mod n). Thus it is the simplest possible group for Fourier analysis. Yet it seems to have the most applications. As we saw in the last chapter, it may be viewed as the multiplicative group of nth roots of unity. This can be drawn as n equally spaced points on a circle of radius 1. Thus $\mathbb{Z}/n\mathbb{Z}$ is a finite analogue of the circle (or even of the real line).

The discrete Fourier transform on $\mathbb{Z}/n\mathbb{Z}$, or DFT, arises whenever anyone needs to compute the classical Fourier series and integrals of sines and cosines. In fact, the first application of the discrete Fourier transform was perhaps A.-C. Clairaut's use of it in 1754 to compute an orbit, which can be considered as a finite Fourier series of cosines. See Heideman et al. [1984]. This work actually preceded Fourier's landmark 1807 paper. Of course, the computation of orbits using sums of trigonometric functions goes back at least to the ancient Babylonians.

When speeded up, the discrete Fourier transform becomes the fast Fourier transform or FFT (discussed in Chapter 7), which has revolutionized many aspects of modern life from weather forecasting to medicine. The FFT is usually attributed to Cooley and Tukey [1965], but Heideman et al. [1984] note that C. F. Gauss discovered the fast Fourier transform in 1805 while computing the eccentricity of the orbit of the asteroid Juno.

The DFT provides an approximation to the continuous Fourier transform, but that is not its only *raison d'être*. For there are many discrete problems that can be solved using the DFT, for example, studies of eigenvalues of adjacency matrices of graphs to be found in subsequent chapters.

Thus we will find that we can use the finite or discrete Fourier transform (DFT) on $\mathbb{Z}/n\mathbb{Z}$ to solve many sorts of problems – from practical problems in physics, statistics, and error-correcting codes to more theoretical problems in number theory and graph theory. Historically perhaps the second application was to number theory (see Chapter 8) and it was, not surprisingly, also made by C. F. Gauss to whom we owe much of the basic theory of $\mathbb{Z}/n\mathbb{Z}$ developed in Chapter 1.

What is the Finite or Discrete Fourier Transform?

First, we need to consider the space of functions to be transformed. Let G be the additive group $G = \mathbb{Z}/n\mathbb{Z}$. We need a quick review of finite-dimensional inner product spaces and in particular the space $L^2(G)$, defined by:

$$L^2(G) = \{f : G \to \mathbb{C}\} = \text{the set of all complex-valued functions on } G.$$

It is a finite-dimensional vector space with an *inner product*

$$\langle f, g \rangle = \sum_{x \in G} f(x)\overline{g(x)}, \quad \text{for } f, g \in L^2(G). \tag{1}$$

This inner product makes $L^2(G)$ a finite-dimensional *inner product space* (see the exercise below). We can get a *norm* by setting

$$\|f\| = \langle f, f \rangle^{1/2}.$$

We will view $\|f - g\|$ as the distance between f and g in $L^2(G)$. This makes $L^2(G)$ a metric space. References for inner product spaces are Byron and Fuller [1992], Lang [1983], and Strang [1976]. $L^2(G)$ is also often called a (finite-dimensional) *Hilbert space*.

Exercise. With the preceding definitions, show that $L^2(G)$ is an inner product space, that is, a vector space V over $\mathbb{C}$ with inner product $\langle f, g \rangle \in \mathbb{C}$

satisfying:

$$\langle f, g \rangle = \overline{\langle g, f \rangle},$$

$\langle f, g \rangle$ is linear in f, conjugate linear in g,

$$\langle f, f \rangle \geq 0,$$

$$\langle f, f \rangle = 0 \quad \text{iff} \quad f(x) = 0 \text{ for all } x \in G.$$

Then prove the *Cauchy–Schwarz inequality* $|\langle f, g \rangle| \leq \|f\| \|g\|$. (This requires a trick.) Show that we have the triangle inequality $\|f + g\| \leq \|f\| + \|g\|$.

Definition. We say that two functions $f, g \in L^2(G)$ are *orthogonal* iff

$$\langle f, g \rangle = 0.$$

Exercise.

a) Show that orthogonal vectors in an inner product space V must be linearly independent. That is when $v = \sum_{j=1}^{n} a_j v_j = 0$, for some scalars $a_j \in \mathbb{C}$, it follows that all the a_j must be 0.

b) Suppose that $v_1, \ldots, v_n$ are nonzero, pairwise orthogonal vectors in V. Let $v = \sum_{j=1}^{n} a_j v_j$, for scalars $a_j \in \mathbb{C}$. Show that

$$a_j = \frac{\langle v, v_j \rangle}{\langle v_j, v_j \rangle}.$$

By the exercise below, we can view $L^2(G)$ as $\mathbb{C}^n$, complex n-space, with $n = |G| =$ order of G, and we can view orthogonality in the usual sense of two vectors in Euclidean space. But we also want to make the connection with orthogonal functions such as $\sin(2\pi x)$ and $\sin(4\pi x)$, for $x \in [0, 1]$.

Exercise.

a) Show that $L^2(\mathbb{Z}/n\mathbb{Z})$ is an n-dimensional vector space over $\mathbb{C}$ with basis consisting of the delta functions $\delta_1, \ldots, \delta_n$ defined by

$$\delta_i(j) = \begin{cases} 1, & \text{if } i \equiv j \pmod{n}, \\ 0, & \text{otherwise.} \end{cases} \tag{2}$$

Here we write δ_i rather than $\delta_{i(\bmod n)}$ to simplify the notation. Hopefully it won't be too confusing. We will often identify a congruence class $x \bmod n$ with a representative $x \in \mathbb{Z}$.

b) Show that the delta functions form an orthonormal basis of $L^2(\mathbb{Z}/n\mathbb{Z})$, that is, they satisfy

$$\langle \delta_i, \delta_j \rangle = \begin{cases} 1, & i \equiv j \pmod{n}, \\ 0, & \text{otherwise.} \end{cases}$$

c) Since the $\delta_a, a = 0, 1, \ldots, n-1$ form a complete orthonormal set in $L^2(\mathbb{Z}/n\mathbb{Z})$, we can write a "Fourier" expansion of any function $f : \mathbb{Z}/n\mathbb{Z} \to \mathbb{C}$:

$$f(x) = \sum_{a=0}^{n-1} f(a)\, \delta_a(x), \quad \text{where } f(a) = \langle f, \delta_a \rangle.$$

d) Show that we have a vector space isomorphism $T : L^2(\mathbb{Z}/n\mathbb{Z}) \to \mathbb{C}^n$, defined by

$$T(f) = (f(0), \ldots, f(n-1)).$$

We can define convolution for functions in $L^2(\mathbb{Z}/n\mathbb{Z})$, where we are viewing $\mathbb{Z}/n\mathbb{Z}$ as an additive group.

Definition. Suppose $f, g : \mathbb{Z}/n\mathbb{Z} \to \mathbb{C}$. Define the *convolution* (or splat) $f * g$ by

$$(f * g)(x) = \sum_{y \in \mathbb{Z}/n\mathbb{Z}} f(y)g(x - y), \quad \text{for } x \in \mathbb{Z}/n\mathbb{Z}. \tag{3}$$

According to Tolimieri, An, and Lu [1989] "linear convolution is one of the most frequent computations carried out in digital signal processing." It is also a common technique in statistical data analysis. Brigham [1974, p. 58] says that convolution and its behavior under the DFT described below is "probably the most important and powerful tool in modern analysis." If we were looking at $G = \mathbb{R}$, we would actually be convolving functions in $L^1(G)$. But for finite groups G, $L^1(G) = L^2(G) =$ the space of all functions $f : G \to \mathbb{C}$.

Exercise. Prove the following properties of convolution of functions $f, g, h : \mathbb{Z}/n\mathbb{Z} \to \mathbb{C}$.

a) $f * g = g * f$.

b) $f * (g * h) = (f * g) * h$.

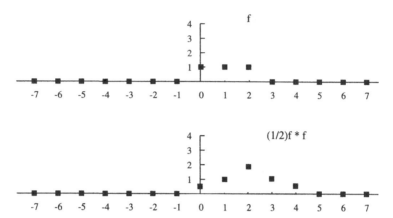

Figure I.6. Example of discrete convolution of f with $(1/2)$ f. Out of two boxes, we get a triangle with a broader base than the boxes.

c) Show that if δ_a is the delta function defined above on $\mathbb{Z}/n\mathbb{Z}$, then

$$\delta_a * \delta_b = \delta_{a+b(\text{mod } n)},$$
$$(f * \delta_a)(x) = f(x - a).$$

Example. Take $n = 15$. Look at $f = \delta_0 + \delta_1 + \delta_2$ and $g = \frac{1}{2} f$. Then it is easy to use the results of the preceding exercise to convolve f and g:

$$f * g = \frac{1}{2}(\delta_0 + 2\delta_1 + 3\delta_2 + 2\delta_3 + \delta_4).$$

We have sketched f and $f * g$ in Figure I.6. Note that it is much easier to convolve functions on the finite circle than on $\mathbb{R}$, where you have to do integrals.

There are at least two totally different ways of viewing convolution. An algebraist would call $L^2(G)$, with the operations of pointwise addition, multiplication by complex scalars, and convolution, the *group algebra*[†] of the additive group $G = \mathbb{Z}/n\mathbb{Z}$. For, by one of the exercises above, convolution of delta functions corresponds to addition in the additive group $\mathbb{Z}/n\mathbb{Z}$. The structure of such algebras was studied by J. H. M. Wedderburn and provides a development of the theory of representations of G which is parallel to ours. It was Emmy Noether [1983] who noticed in 1929 that the Wedderburn structure theory of central simple algebras could be used to study group representations. We will not follow the path of our heroine, however. It seems to make things less intuitive. Certainly many would disagree. See Curtis and Reiner [1966] or Lang [1984].

[†] An algebra is a vector space with a multiplication having the properties of $*$.

An analyst or probabilist would view convolution somewhat differently. Convolution corresponds to a sum of independent random variables. See Theorem 3 in Chapter 6 or Terras [1985, Vol. 1].

Let us now return to our discussion of the DFT – the discrete or finite Fourier transform. If you know about the classical Fourier transform (which we don't assume but see Byron and Fuller [1992] or D. L. Powers [1987], if you wish), you expect to find expansions in sines and cosines or complex exponentials, since

$$e^{ix} = \cos x + i \sin x, \quad i = \sqrt{-1}, \ x \in \mathbb{R}.$$

You won't be totally disappointed. We need the exponentials in the following definition.

Definition. Suppose $a, x \in \mathbb{Z}/n\mathbb{Z}$. Define

$$e_a(x) = \exp\left(\frac{2\pi i a x}{n}\right). \tag{4}$$

Note that $e_a(x)$ is independent of the congruence class representatives chosen for x and a.

If $\mathbb{T}$ denotes the multiplicative group of complex numbers of norm 1, then

$$e_a : \mathbb{Z}/n\mathbb{Z} \to \mathbb{T}$$

is a group homomorphism from the additive group $G = \mathbb{Z}/n\mathbb{Z}$ into the multiplicative group $\mathbb{T}$.[‡] Such a homomorphism is often called a *character* (or 1-dimensional representation) of G. Much of this book will involve representations of finite groups. But here in the abelian case, things are much simpler.

Definition. The *discrete Fourier transform or DFT* of $f \in L^2(\mathbb{Z}/n\mathbb{Z})$ is

$$\mathscr{F}_n f(x) = \mathscr{F}f(x) = \hat{f}(x) = \sum_{y \in \mathbb{Z}/n\mathbb{Z}} f(y) e_x(-y) = \langle f, e_x \rangle. \tag{5}$$

Here $e_a(x)$ is defined by (4). Clearly $\mathscr{F}_n : L^2(\mathbb{Z}/n\mathbb{Z}) \to L^2(\mathbb{Z}/n\mathbb{Z})$ is a linear map. We are interested in its properties. The following exercise shows that if we use the obvious basis of our vector space, the matrix of the DFT is rather interesting.

---

‡ Here $\mathbb{T}$ stands for torus. The doughnut is $\mathbb{T}^2$ and we also call $\mathbb{T}^n$ a torus.

Exercise. Using the basis of $L^2(\mathbb{Z}/n\mathbb{Z})$ given by the delta functions defined above, show that the matrix of the DFT is

$$F_n = \left(\omega^{-(j-1)(k-1)}\right)_{1\le j,k\le n}, \tag{6}$$

where $\omega = \exp(2\pi i/n)$ is a primitive nth root of unity. Thus, when $n = 2$, we get

$$\begin{pmatrix} 1 & 1 \\ 1 & -1 \end{pmatrix}.$$

When $n = 4$, we obtain

$$\begin{pmatrix} 1 & 1 & 1 & 1 \\ 1 & -i & -1 & i \\ 1 & -1 & 1 & -1 \\ 1 & i & -1 & -i \end{pmatrix}.$$

*Theorem 1. **Basic Properties of the DFT on the Finite Circle.***

1) $\mathscr{F}: L^2(\mathbb{Z}/n\mathbb{Z}) \to L^2(\mathbb{Z}/n\mathbb{Z})$ is a 1-1, onto, linear map.
2) **Convolution:** $\mathscr{F}(f * g)(x) = \mathscr{F}f(x) \cdot \mathscr{F}g(x)$, for all $x \in \mathbb{Z}/n\mathbb{Z}$.
3) **Inversion:**

$$f(x) = \frac{1}{n}\mathscr{F}\mathscr{F}f(-x) = \frac{1}{n} \sum_{y\in\mathbb{Z}/n\mathbb{Z}} \hat{f}(y)\exp\left(\frac{2\pi i x y}{n}\right).$$

4) **Plancherel Theorem or Parseval Equality:**

$$\langle f, f \rangle = \frac{1}{n}\langle \hat{f}, \hat{f} \rangle.$$

This says the linear map $n^{-1/2}\mathscr{F}$ gives a Hilbert space isometry of $L^2(\mathbb{Z}/n\mathbb{Z})$ onto itself.

Remarks. Fact 3 says that the exponentials $\{e_a \mid a = 0, \ldots, n-1\}$ give a complete orthogonal set in $L^2(\mathbb{Z}/n\mathbb{Z})$ and that $f(x)$ has a generalized Fourier expansion in terms of these exponentials.

Fact 4 says that if we multiply the matrix F_n of the Fourier transform from Equation (6) by $n^{-1/2}$, we obtain a *unitary matrix*

$$\Phi_n = n^{-1/2} F_n. \tag{7}$$

This means ${}^t\overline{\Phi_n}\Phi_n = I =$ the identity matrix. Note that we write the transpose on the left of the matrix and the overbar means take the complex conjugate of every entry.

The DFT takes a function $f \in L^2(\mathbb{Z}/n\mathbb{Z})$ to another function in $L^2(\mathbb{Z}/n\mathbb{Z})$. Thus the DFT is more like the Fourier integral transform of functions on $\mathbb{R}$ than the Fourier transform on $\mathbb{T}$, the unit circle. We say that the additive group $G = \mathbb{Z}/n\mathbb{Z}$ is *self-dual* because the characters e_a are parameterized by $a \in G$. We will say more about this when we consider other finite groups. For the additive group $G = \mathbb{Z}/n\mathbb{Z}$, define the *dual group*

$$\hat{G} = \{\chi : G \to \mathbb{T} \mid \chi \text{ is a group homomorphism}\}$$
$$= \left\{ e_a(x) = \exp\left(\frac{2\pi i a x}{n}\right) \,\middle|\, a \in \mathbb{Z} \right\}.$$

The group operation in $\hat{G}$ is pointwise multiplication of functions $e_a e_b = e_{a+b(\mathrm{mod}\ n)}$. Thus $\hat{G}$ is a group isomorphic to $G = \mathbb{Z}/n\mathbb{Z}$.

You might ask: What is the good of the DFT as opposed to the ordinary Fourier transform? Of course, most functions given to us by nature are just sequences of numbers, for example, the voltage from a voice over the telephone, wind direction or speed measurements, variable star brightness, rainfall data for some city, population of the world, or grey levels from an image of Mars or a human lung. We can regard the measurements as representing a function or signal $f(t)$. We will usually have equally spaced samples. It is thus natural to compute the discrete rather than the continuous transform.

You can think of $\mathbb{Z}/n\mathbb{Z}$ as a finite analogue of the circle and then discrete Fourier analysis can be used to approximate the real thing, namely, the continuous Fourier analysis of periodic functions. If a function $f(x)$ is known at $x_j = -a + \frac{j(a+b)}{n}$, $j = 1, \ldots, n$, the usual Fourier transform is approximated by

$$\int_{-a}^{+b} f(y) \exp(-2\pi i s x)\, dx \cong \delta \sum_{j=1}^{n} f(x_j) \exp(-2\pi i x_j s),$$

where

$$\delta = \frac{a+b}{n}.$$

This is really the DFT of a function on $\mathbb{Z}/n\mathbb{Z}$. To get a good approximation, of course, one must take δ sufficiently small. See Körner [1988].

Proofs of the Basic Facts about the DFT.

1) We postpone the proof that the map is 1-1, onto until after the proof of 3, which gives an explicit inverse. The linearity is clear.

2)

$$\mathscr{F}(f * g)(a)$$

$$= \sum_{b \in \mathbb{Z}/n\mathbb{Z}} (f * g)(b) \exp\left(\frac{-2\pi i a b}{n}\right)$$

$$= \sum_{b \in \mathbb{Z}/n\mathbb{Z}} \exp\left(\frac{-2\pi i a b}{n}\right) \sum_{c \in \mathbb{Z}/n\mathbb{Z}} f(c) g(b - c)$$

$$= \sum_{c \in \mathbb{Z}/n\mathbb{Z}} \sum_{d \in \mathbb{Z}/n\mathbb{Z}} \exp\left(\frac{-2\pi i a (c + d)}{n}\right) f(c) g(d), \quad \text{setting } d = b - c,$$

$$= \sum_{c \in \mathbb{Z}/n\mathbb{Z}} \exp\left(\frac{-2\pi i a c}{n}\right) f(c) \sum_{d \in \mathbb{Z}/n\mathbb{Z}} \exp\left(\frac{-2\pi i a d}{n}\right) g(d)$$

$$= \mathscr{F}f(a) \cdot \mathscr{F}g(a).$$

It was legal to make the change of variables that replaced the sum over b with the sum over d, holding c fixed, since both b and d run over all of the group in question – just in a different order. Our group is finite and there is no problem changing order of summation.

3) (From Diaconis [1988].) Both sides of the formula that we are trying to prove are linear in f. Thus it suffices to check the formula for a basis of $L^2(\mathbb{Z}/n\mathbb{Z})$. It was shown in an exercise that one such basis consists of the delta functions δ_a, $a \in \mathbb{Z}/n\mathbb{Z}$. Then the Fourier transform of δ_a is

$$\mathscr{F}\delta_a(x) = \exp\left(\frac{-2\pi i a x}{n}\right), \quad \text{for } x, a \in \mathbb{Z}/n\mathbb{Z}.$$

The formula we are trying to prove is: $\mathscr{F}\mathscr{F}\delta_a(-c) = n\delta_a(c)$, that is,

$$\sum_{b \in \mathbb{Z}/n\mathbb{Z}} \exp\left(\frac{-2\pi i a b}{n}\right) \exp\left(\frac{2\pi i b c}{n}\right)$$

$$= \sum_{b \in \mathbb{Z}/n\mathbb{Z}} \exp\left(\frac{2\pi i b (c - a)}{n}\right) = n\delta_a(c).$$

The last equality comes from the following lemma.

*Lemma. **The Orthogonality Relations of the Characters of** $\mathbb{Z}/n\mathbb{Z}$.* Let $e_a(b) = \exp(2\pi i a b / n)$, for $a, b \in \mathbb{Z}/n\mathbb{Z}$. Then

$$\sum_{b \in \mathbb{Z}/n\mathbb{Z}} \exp\left(\frac{2\pi i b a}{n}\right) = \langle e_a, e_0 \rangle = \begin{cases} n, & a \equiv 0 \;(\text{mod } n) \\ 0, & \text{otherwise} \end{cases} = n\delta_0(a).$$

Similarly we have

$$\langle e_x, e_y \rangle = \begin{cases} n, & \text{if } x \equiv y \,(\text{mod } n) \\ 0, & \text{otherwise.} \end{cases} = n\delta_x(y). \tag{8}$$

Proof. The first result is clear if $a \equiv 0 \,(\text{mod } n)$. So let us suppose that n does not divide a. Then let the sum on the left-hand side of the equality be called S. We have

$$\exp\left(\frac{2\pi i a}{n}\right) S = \exp\left(\frac{2\pi i a}{n}\right) \sum_{b \in \mathbb{Z}/n\mathbb{Z}} \exp\left(\frac{2\pi i b a}{n}\right)$$

$$= \sum_{b \in \mathbb{Z}/n\mathbb{Z}} \exp\left(\frac{2\pi i a(b+1)}{n}\right)$$

$$= \sum_{d \in \mathbb{Z}/n\mathbb{Z}} \exp\left(\frac{2\pi i a d}{n}\right), \quad \text{setting } d = b + 1,$$

$$= S.$$

Here we have used the fact that b runs through $\mathbb{Z}/n\mathbb{Z}$ as fast as $d = b + 1$ does. So we have shown that

$$\exp\left(\frac{2\pi i a}{n}\right) S = S.$$

But, if n does not divide a, $\exp(2\pi i a/n)$ is not equal to 1, and therefore $S = 0$. This completes the proof of the first equality of the lemma. The second equality follows from the first. ∎

This also completes the proof of the inversion formula for the discrete Fourier transform (part 3) of theorem 1.

Exercise. Find another proof of the lemma above using the formula for a geometric progression:

$$\sum_{j=1}^{n} w^j = \frac{1 - w^{n+1}}{1 - w}.$$

Exercise.
a) Find another proof of the inversion formula for the DFT by computing the double sum represented by $\mathscr{F}\mathscr{F}f$ directly.
b) Find yet another proof of the inversion formula for the DFT by noting that for $G = \mathbb{Z}/n\mathbb{Z}$, $L^2(G)$ is an n-dimensional inner product space (Hilbert

space) and the exponentials e_a, $a \in \mathbb{Z}/n\mathbb{Z}$, form a complete orthogonal set in $L^2(G)$ (since no other function can be orthogonal to all of them). See Byron and Fuller [1992], Lang [1983], or Strang [1976] for more information on finite- (and infinite-) dimensional Hilbert spaces. We can normalize the e_a, by setting $u_a = n^{-1/2}e_a$, $a = 0, \ldots, n - 1$. Then $\langle u_a, u_b \rangle = 0$, for $a \neq b(\mathrm{mod}\ n)$, and $\langle u_a, u_a \rangle = 1$. So we have

$$f(x) = \sum_{a=0}^{n-1} \langle f, u_a \rangle u_a(x). \tag{9}$$

This is true for any orthonormal set u_a and is proved by showing that the difference of the two sides is orthogonal to all of the u_a and thus must vanish. But (9) is just the Fourier inversion formula.

Exercise. Prove Fact 4 about the DFT – the Plancherel Theorem or Parseval Formula. One method is to prove the more general formula:

$$\langle f, g \rangle = \frac{1}{n}\langle \hat{f}\hat{g} \rangle, \quad \text{for } f, g \in L^2(G). \tag{10}$$

To see (10), note that this formula is equivalent to the inversion formula (Diaconis [1988]), for both sides of (10) are linear in f and thus it suffices to prove it for $f = \delta_a$. But then the formula is the Fourier inversion formula.

Another proof of Fact 4 comes by noticing that it is a general fact about Fourier expansions (9). You can see this by plugging (9) into the inner product $\langle f, f \rangle$. For the orthonormality of the u_a then implies

$$\sum_{a=0}^{n-1} |\langle f, u_a \rangle|^2 = \langle f, f \rangle = \sum_{a=0}^{n-1} |f(a)|^2,$$

which is the Parseval formula.

This exercise ends the proof of Theorem 1. ∎

Figures I.7–I.10 give pictures of DFTs of *even* functions of x mod n. For such functions the DFT is real valued (as it is actually a cosine transform).
Moral of Figure I.7.   The DFT erases $n - 1$ dots out of a horizontal "line" of n dots and moves the remaining dot up.
Question. What if the original line of dots is shorter? See Figures I.8 and I.9 for some examples.

Exercise. Compute the discrete Fourier transform of the constant function $f(x) = 1$, for all $x \in \mathbb{Z}/n\mathbb{Z}$.
Answer. See Figure I.7.

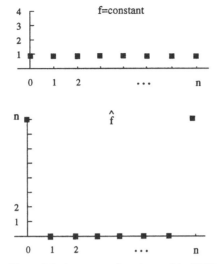

Figure I.7. A constant function and its DFT.

Remark. Figure I.7 says that if we define the *uniform probability* by $p_U(x) = 1/n$, for all $x \in \mathbb{Z}/n\mathbb{Z}$, then the DFT is $\mathscr{F}p_U(x) = \delta_0$. This implies, by Fourier inversion, that $\mathscr{F}\delta_0 = np_U$. Thus the Fourier transform of the delta function at 0 is the uniform probability multiplied by n.

Exercise. Give a proof that the DFT is 1-1 and onto by computing the determinant of its matrix.
Hint (see Lang [1987, p. 208]).
   Let V_n denote the *Vandermonde determinant*:

$$V_n = \det \begin{pmatrix} 1 & x_1 & x_1^2 \cdots x_1^{n-1} \\ 1 & x_2 & x_2^2 \cdots x_2^{n-1} \\ \vdots & \vdots & \vdots \quad \vdots \quad \vdots \\ 1 & x_n & x_n^2 \cdots x_n^{n-1} \end{pmatrix}.$$

   First prove that $V_n = \prod_{i<j}(x_j - x_i)$. To prove this, multiply each column by x_1 and subtract it from the next column on the right, starting from the right-hand side. Show that $V_n = (x_n - x_1)(x_{n-1} - x_1) \cdots (x_2 - x_1)V_{n-1}$. Use induction to complete the proof. It should be clear from this product formula that V_n is not 0 if $x_i \neq x_j$ for $i \neq j$.
   We claim that, in fact, you can use this to see that if $n = p$ is an odd prime, then if $\Phi_p = p^{-1/2}F_p$ as in Equation (7),

$$\det(\Phi_p) = \begin{cases} (-1)^m, & \text{if } p = 4m + 1, \\ -i(-1)^m, & \text{if } p = 4m + 3. \end{cases}$$

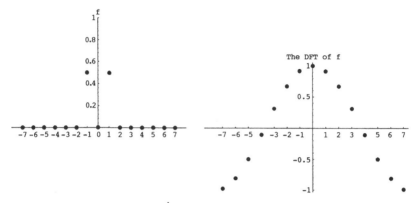

Figure I.8. A Function $f(x) = \frac{1}{2}(\delta_1(x) + \delta_{-1}(x))$ whose DFT is a sampled cosine.

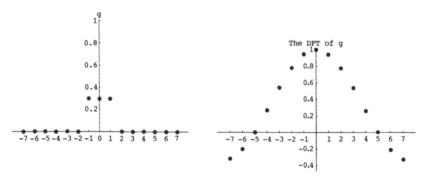

Figure I.9. The Fourier transform of $\frac{1}{3}(\delta_1(x) + \delta_0(x) + \delta_{-1}(x))$.

Table I.7. *A Short table of discrete*
Fourier transforms

$f(x)$	$\mathscr{F}f(y)$
1	$n\delta_0(y)$
$e_a(x)$	$n\delta_a(y)$
$\delta_a(x)$	$e_a(-y)$
$\frac{1}{2}(\delta_1 + \delta_{-1})(x)$	$\cos(\frac{2\pi y}{n})$
$\frac{1}{3}(\delta_{-1} + \delta_0 + \delta_1)(x)$	$\frac{1}{3}\left(1 + 2\cos\left(\frac{2\pi y}{n}\right)\right)$
$\frac{1}{2}(\delta_0 + \delta_1)(x)$	$\exp\left(\frac{-\pi i y}{n}\right)\cos\left(\frac{\pi y}{n}\right)$

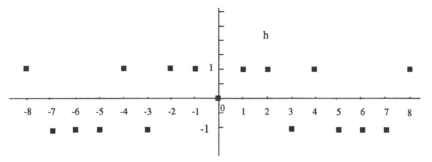

Figure I.10. A function proportional to its own DFT.

Next consider Table I.7 giving a few discrete Fourier transforms. One line of the table is probably all one really needs to remember, as the other lines are easily derived from line 2.

Another entry of such a table of DFTs comes from Chapter 14 on the uncertainty principle. Suppose $n = ab$. There we find

$$f = \frac{1}{a} \sum_{j=0}^{a-1} \delta_{jb} \quad \text{implies} \quad \hat{f} = \sum_{j=0}^{b-1} \delta_{ja}.$$

Exercise. Check that Fourier inversion holds for this function f.

Example 1. Define $f(x) = \frac{1}{2}(\delta_1(x) + \delta_{-1}(x))$. Then

$$\hat{f}(y) = \cos\left(\frac{2\pi y}{n}\right).$$

See Figure I.8.

Example 2. Define $g(x) = \frac{1}{3}(\delta_1(x) + \delta_0(x) + \delta_{-1}(x))$. Then

$$\hat{g}(y) = \frac{1}{3}\left\{1 + 2\cos\left(\frac{2\pi y}{n}\right)\right\}.$$

See Figure I.9.

We can find some more entries for Table I.7 as Gauss did while considering the equation $x^2 \equiv a \pmod{p}$, for given a mod the prime p. A central problem in number theory is the solution of Diophantine equations (named for Diophantus who lived around 300 A.D., although Bell [1990] says many would subtract 200 years). Diophantine equations are simply equations involving integers.

One important method in the study of such equations is the reduction of the equation mod p. We will find that the congruence corresponding to $x^2 = a$ is more interesting than the original Diophantine equation.

Example 3. Suppose p is an odd prime and define the *Legendre symbol* by

$$\left(\frac{x}{p}\right) = h(x) = \begin{cases} 0, & \text{if } p \text{ divides } x, \\ 1, & \text{if } p \text{ does not divide } x \text{ and } x \text{ is a square mod } p, \\ -1, & \text{if } p \text{ does not divide } x \text{ and } x \text{ is not a square mod } p. \end{cases}$$

We will see in Lemma 1 of Chapter 8 that $\hat{h}(-x) = g\, h(x)$, where $g = \hat{h}(-1) = G_p(1)$ is defined by Equation (11) below.

Definition. The *Gauss sum* $G_p(x)$ is

$$G_p(x) = \sum_{k \in \mathbb{Z}/p\mathbb{Z}} \exp\left(\frac{2\pi i x k^2}{p}\right). \tag{11}$$

Gauss's famous evaluation of the Gauss sum was obtained after more than a year's work, in a lightning flash of insight. It says (see Chapter 8) that

$$g = \begin{cases} \sqrt{p}, & p \equiv 1 (\text{mod } 4), \\ i\sqrt{p}, & p \equiv 3 (\text{mod } 4). \end{cases}$$

This means we have found a function (namely $h(x)$) that is proportional to its own DFT (see Figure I.10): Since $h(-x) = h(-1)h(x)$, we have $h(x) = \frac{\hat{h}(x)}{g h(-1)}$. That is, $h(x)$ is a finite analogue of the Gaussian $\exp(-\pi x^2)$, $x \in \mathbb{R}$.

Remarks. Note that if we set $\Phi_n = n^{-1/2} F_n$, the unitary matrix multiple of the matrix of the DFT from Equation (7), we find that $\Phi_n^4 = I$. This means that all the eigenvalues of Φ_n are fourth roots of unity, that is, $\pm 1, \pm i$. Carlitz [1959] determined the characteristic polynomial of ϕ_n as we shall see in Chapter 8. This is related to the evaluation of the Gauss sum since the trace of F_n is the Gauss sum.

Exercise.
a) Check Table I.7. It is much easier to check than the corresponding tables in the continuous case.
b) Assume that n is an odd integer. Compute the DFT of the function defined by

$$f(x) = \begin{cases} x, & 0 \leq x \leq (n-1)/2, \\ -x, & -(n-1)/2 \leq x \leq 0. \end{cases}$$

Here we identify $\{-(n-1)/2, \ldots, -1, 0, 1, \ldots, (n-1)/2\}$ with $\mathbb{Z}/n\mathbb{Z}$. Graph the results.

c) Under the same hypothesis as in Part b, compute the DFT of the function defined by $g(x) = x$, for $x \in \{0, 1, 2, \ldots, (n-1)\}$, which can be identified with $\mathbb{Z}/n\mathbb{Z}$. Graph the results.

d) Compare your results with the analogous continuous problem, where $f(x) = |x|$ and $g(x) = x$ are considered as periodic functions on some interval.

Exercise. Use Matlab to find the DFT mod 18 of the vector v defined as follows:

$c = 3/20$;
for $i = 1 : 9$.
$v(i) = \exp(-c^*(i-1)\hat{\ }2); \; v(19-i) = v(i);$ end.

The command to get the FFT is $w = fft(f, 18)$. Next check to see if the inverse transform leads back to the original vector via $u = ifft(w, 18)$. Then, to see the graphs, use the commands

$$\text{plot(real}(v)), \text{plot(real}(w)).$$

If you don't put in "real," the plot becomes junk as the computer does not realize it is looking at real numbers.

Exercise. **Translation and Dilation Properties of the DFT.**

a) Let $a \in \mathbb{Z}/n\mathbb{Z}$. Define the *translation* (or shift) *operator* T_a on a function $f : \mathbb{Z}/n\mathbb{Z} \to \mathbb{C}$ by $T_a f(x) = f(x - a)$, for all $x \in \mathbb{Z}/n\mathbb{Z}$. Show that $T_a f = \delta_a * f$ and thus

$$\mathcal{F}(T_a f)(y) = \exp(-2\pi i a y/n) \mathcal{F}f(y). \tag{12}$$

b) Let a be an element of the unit group $(\mathbb{Z}/n\mathbb{Z})^*$. Define the *dilation* (or scaling) *operator* $D_a f(x) = f(ax)$, for all $x \in \mathbb{Z}/n\mathbb{Z}$. Show that

$$\mathcal{F}D_a f(y) = D_{a^{-1}} \mathcal{F}f(y). \tag{13}$$

Chapter 3

Graphs of $\mathbb{Z}/n\mathbb{Z}$, Adjacency Operators, Eigenvalues

The far-reaching power of the theory of groups resides in its
revelation of identity behind apparent dissimilarity.

E. T. Bell [1989, p. 185]

Here we consider various *Cayley graphs* attached to the additive group $G = \mathbb{Z}/n\mathbb{Z}$ with symmetric set of generators S (meaning $s \in S$ implies $s^{-1} \in S$). That is, we draw a graph with vertices the elements of the additive group $G = \mathbb{Z}/n\mathbb{Z}$. Then we draw an edge between two vertices v and $v + s$, for $s \in S$. We will call this graph $X(G, S)$. Actually we will draw only one undirected edge between $v = w - s$ and $v + s = w$ rather than two directed edges. See Figure I.11. Usually we will assume that the identity of our group is not in the edge set S to avoid having loops at each vertex.

And we will not give edges colors corresponding to the different elements of S. Moreover, we can also look at $X(G, S)$ when S is not a set of generators. This simply means that the graph is not connected.

Our main example will be $S = \{\pm 1 \pmod n\}$, which for $n = 9$ gives the cycle graph shown in Figure I.12.

Another example of a set S is the *shell*

$$S(r) = \{\pm r \pmod n\}. \tag{1}$$

This example and the next come up in studies of finite analogues of Radon transforms (see J. Fill [1989] and E. Velasquez [1991]).

Exercise. For what values of r is $S(r)$ a symmetric set of generators of the additive group $\mathbb{Z}/n\mathbb{Z}$?

46

Figure I.11. Identify edges corresponding to s and $-s$ in a Cayley graph.

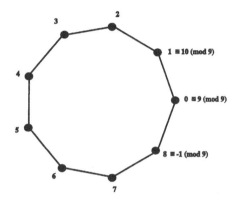

Figure I.12. The Cayley graph $X(\mathbb{Z}/9\mathbb{Z}, \{\pm 1 \ (\mathrm{mod}\ 9)\})$.

Answer. For r such that r and n are relatively prime, that is, r and n have no common divisors but ± 1:

$$\text{g.c.d.}(r, n) = 1.$$

Note that $X(\mathbb{Z}/n\mathbb{Z}, S(1))$ is graph-isomorphic[†] to $X(\mathbb{Z}/n\mathbb{Z}, S(r))$ if $(r, n) = 1$. This is because we can define a map $T : \mathbb{Z}/n\mathbb{Z} \to \mathbb{Z}/n\mathbb{Z}$ that is 1-1 and onto and such that if the vertices v, $v \pm 1$ are connected by an edge in the first graph, then the vertices Tv, $T(v \pm 1)$ are connected by an edge in the second graph. The map is $T(x \ \mathrm{mod}\ n) = rx \ (\mathrm{mod}\ n)$. It is easily checked that this map is 1-1 and thus onto (by the Dirichlet pigeonhole principle). And $T(x \pm 1) = r(x \pm 1) = rx \pm r$ so that $T(x)$ and $T(x \pm 1)$ are connected by an edge of the second graph.

Therefore the graph in Figure I.12 is isomorphic to that of Figure I.13. Nevertheless, the picture that we draw for the second graph seems more interesting than that for the first.

Still another example of a generating set S for $\mathbb{Z}/n\mathbb{Z}$ is the *ball*

$$B(r) = \{-r, -r + 1, \dots, r - 1, r \ (\mathrm{mod}\ n)\}, \quad \text{for } r \geq 1. \tag{2}$$

See Figure I.14 for an example.

[†] Two graphs X and Y are isomorphic iff there is a 1-1, onto map $T : X \to Y$ such that if x and x' are adjacent vertices in X, then Tx and Tx' are adjacent in Y.

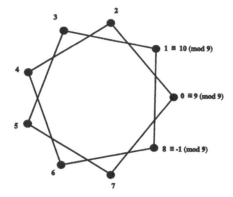

Figure I.13. $X(\mathbb{Z}/9\mathbb{Z}, S(2))$.

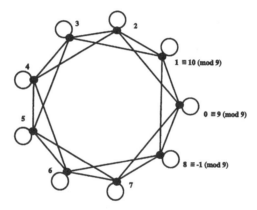

Figure I.14. $X(\mathbb{Z}/9\mathbb{Z}, B(2))$.

Basic Graph Theory

The origins of graph theory are humble, even frivolous.
Biggs, Lloyd, and Wilson [1986]

Our motivating problem is to construct a model for an efficient communication network: it could be the telephone network of a country, of the wiring of a big parallel computer, or even the neuronal system of a human body. The individuals wishing to communicate through the network will be represented by nodes, also called vertices of a graph, and the communication cables (i.e., the phone lines, the electrical cables or the axon–synapse connections) will be represented by arcs, also called edges of a graph.

Bien [1989]

The references for graph theory include N. Biggs [1974], B. Bollobás [1979], A. E. Brouwer, A. M. Cohen, and A. Neumaier [1989], Fan Chung [1996], D. Cvetković, M. Doob, and H. Sachs [1980], J. Gallian [1990], R. Graham et al. [1995], I. Grossman and W. Magnus [1992], J. Seidel [1989], and M. Starzak [1989].

We want to use Fourier analysis to study graphs. Historically, graphs first arose in puzzles such as the Königsburg bridge puzzle (see Biggs, Lloyd, and Wilson [1986]). But they also arose in the work of Kirchoff on electric networks. Any book on electrical networks will discuss Kirchoff's laws. See Doyle and Snell [1984] for a nice discussion. Many interesting examples from the history of graph theory can be found in Biggs, Lloyd, and Wilson [1986].

Of course, as noted by Bien above, there are many other sorts of networks. Many examples come from the finite element method, which places networks of nodes on structures such as bridges to analyze the stability of the structure. Another example comes from chemistry: networks of atoms in a molecule such as that which forms benzene or a buckyball. . . . Many questions from physics or engineering or chemistry can be translated into basic graph theoretic questions (see Chapter 13 or Starzak [1989]). We will, of course, be mainly interested in graphs with symmetry coming from a group action.

Basic Graph Theory

The *degree* of a vertex of a graph X is the number of edges coming out of the vertex. We assume our graphs are undirected. We say the graph is k-*regular* if all vertices have the same degree k. Our Cayley graphs are regular of degree the order of the generating set S. So, for example, the cycle graph of Figure I.12 has degree 2.

The *distance* between two vertices x, y in X is the minimum number of edges in a path connecting x to y. If our graph is connected, such a path will exist. We will try to stick to the study of connected graphs X. Our Cayley graphs are connected because the set S is assumed to be a set of generators of G.

The maximum distance over all pairs x, y of vertices in X is the *diameter* of X. The diameter of the Cayley graph $X(\mathbb{Z}/9\mathbb{Z}, \{\pm 1 \pmod 9\})$ in Figure I.12 is 4. It does not at first appear to be easy to compute the diameter directly for large graphs. However, one can make use of the fact proved in the exercise after the definition of adjacency matrix in the next section.

The *girth* of a graph X is the length of its smallest cycle or circuit. The girth may be viewed as an analogue of the length of the shortest closed geodesic on a Riemannian manifold. The girth of the Cayley graph $X(\mathbb{Z}/9\mathbb{Z}, \{\pm 1 \pmod 9\})$ is 9. The girth of $X(\mathbb{Z}/9\mathbb{Z}, \{\pm 2, \pm 3 \pmod 7\})$ is 3.

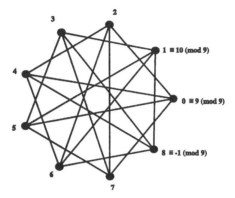

Figure I.15. $X(\mathbb{Z}/9\mathbb{Z}, \{\pm 2, \pm 4 \,(\text{mod } 9)\})$.

Exercise. Find the degrees, diameters, and girths of the graphs in Figures I.13–I.17.

Adjacency Matrices and Operators

Much information about a graph is stored in its *adjacency matrix A*. Suppose that the vertices of the graph are $\{v_1 \ldots, v_n\}$. Then A is an $n \times n$ matrix whose i, j entry is 1 if v_i is connected to v_j and 0 otherwise.

The adjacency matrix of the Cayley graph $X(\mathbb{Z}/7\mathbb{Z}, \{\pm 2, \pm 3 \,(\text{mod } 7)\})$ is given by the following matrix writing the vertices as $\{0, 1, 2, 3, 4, 5, 6\}$:

$$\begin{pmatrix} 0 & 0 & 1 & 1 & 1 & 1 & 0 \\ 0 & 0 & 0 & 1 & 1 & 1 & 1 \\ 1 & 0 & 0 & 0 & 1 & 1 & 1 \\ 1 & 1 & 0 & 0 & 0 & 1 & 1 \\ 1 & 1 & 1 & 0 & 0 & 0 & 1 \\ 1 & 1 & 1 & 1 & 0 & 0 & 0 \\ 0 & 1 & 1 & 1 & 1 & 0 & 0 \end{pmatrix}.$$

This is an example of a *circulant matrix*. Each row is a shift of the row above. More precisely, $a_{ij} = a_{i-1, j-1} = a_{1, j-i+1}$, where the indices are computed modulo n. We will see circulants again in Equation (11) later on.

Exercise. Let X be a graph with diameter d and adjacency matrix A.

a) Show that the i, j entry of the matrix A^r is the number of length r walks going from the ith vertex to the jth vertex in the graph.

b) Show that the diameter of X is the smallest integer r such that all entries of $(I + A)^r$ are nonzero.

c) Show that if X is a connected graph with diameter d, then A has at least $d + 1$ distinct eigenvalues.

Hints.

b) See Fan Chung [1996, p. 42] and Theorem 1 of Chapter 4.

c) See N. Biggs [1974, p. 10].

Remarks. It will sometimes (but not always) simplify our calculations if we take a coordinate-free approach and refuse to write down matrices and vectors. Such an approach has been very popular among pure mathematicians. We could try to blame this on the group of French mathematicians writing under the pen name Nicolas Bourbaki but perhaps they were just representative of the times. The coordinate-free approach might be viewed as the opposite of the approach which emphasizes complicated formulas full of subscripts and even sub-subscripts. In the coordinate-free language, we consider the adjacency operator defined below rather than the adjacency matrix.

Definition. Given a graph X, the *adjacency operator* A acts on functions $f : X \to \mathbb{C}$ by

$$Af(x) = \sum_{y \text{ adjacent to } x} f(y), \quad \text{for any vertex } x \text{ in } X. \tag{3}$$

As usual, we write $L^2(X)$ for the set of all functions $f : X \to \mathbb{C}$. Note again that if $X = \{p_1, \ldots, p_n\}$ (where p_i is the ith vertex of the graph X), we can identify $f \in L^2(X)$ with the column vector $(f(p_1), \ldots, f(p_n))$. So the operator Af really does correspond to the action of the adjacency matrix on vectors.

The *combinatorial Laplacian* of the k-regular graph X is $\Delta = A - kI$. See Chapter 7 for some motivation for this definition. There are other normalizations of the combinatorial Laplacian (see Fan Chung [1996] or Chung and Yau [1995], Lubotzky [1994, p. 44], and Chapter 7).

For a Cayley graph $X(\mathbb{Z}/n\mathbb{Z}, S)$, the *adjacency operator is a convolution operator*:

$$Af(X) = \sum_{s \in S} f(x + s) = (\delta_s * f)(x), \tag{4}$$

where as usual $*$ denotes convolution on the additive group $\mathbb{Z}/n\mathbb{Z}$ and $\delta_s(x)$ is the function that is 1 if $x \in S$ and 0 otherwise.

The Spectral Theorem for Self-Adjoint Operators

Given a linear transformation or operator T mapping the vector space $V = L^2(X)$ into itself, one wants to find a basis of V to make the matrix of T as simple as possible. For then it will be easier to solve problems involving T such as the problem of finding $\lim_{n \to \infty} T^n$. We can make the matrix of T diagonal under certain conditions.

We have defined an inner product $\langle v, w \rangle$ on our vector space $V = L^2(X)$ as

$$\langle f, g \rangle = \sum_{x \in X} f(x)\overline{g(x)}, \quad \text{for } f, g : X \to \mathbb{C}.$$

With respect to this inner product our adjacency operator A is a *self-adjoint operator*. This means $\langle Av, w \rangle = \langle v, Aw \rangle$ for all $v, w \in V$.

Exercise. Prove this.
Hint. By the definitions,

$$\langle Af, g \rangle = \sum_{x \in X} (Af)(x)\overline{g(x)} = \sum_{x \in X} \overline{g(x)} \sum_{y \text{ adjacent to } x} f(y).$$

So you just have to reverse the sums and note that x is adjacent to y iff y is adjacent to x.

Note that it is obvious that the adjacency matrix is symmetric. However, the operator version of this fact looks more complicated – a case when the coordinate-free version of things looks less clear.

The *spectral theorem* says that for a self adjoint operator $T : V \to V$, we can choose an orthonormal basis of V to make the matrix of T diagonal. Most linear algebra books have a proof. See, for example, Strang [1976]. More precisely, there is a basis of V consisting of vectors $v_1, \ldots, v_n$ such that

$$\langle v_i, v_j \rangle = \begin{cases} 1, & i = j, \\ 0, & \text{otherwise} \end{cases} \tag{5}$$

and $Tv_i = \lambda_i v_i$. This says that v_i is an *eigenvector* of T corresponding to the *eigenvalue* λ_i. So if $v \in V$, it has the Fourier expansion

$$v = \sum_{j=1}^{n} \langle v_i, v_j \rangle v_j \quad \text{and then} \quad Tv = \sum_{j=1}^{n} \lambda_j \langle v, v_j \rangle v_j. \tag{6}$$

This is just the statement that the matrix of T with respect to the basis $v_1, \ldots, v_n$ is diagonal with jth diagonal entry λ_j.

Another interpretation for real symmetric matrices T such as the adjacency matrix of a graph is that there is an orthogonal matrix $U = (v_1, \ldots, v_n)$ so that

$$U\,{}'U = I \quad \text{and} \quad U^{-1}TU = D,$$

where D is diagonal with jth diagonal entry λ_j, for

$$T(v_1, \ldots, v_n) = (v_1, \ldots, v_n)D = (\lambda_1 v_1, \ldots, \lambda_n v_n).$$

You can rewrite the result as

$$T = UDU^{-1} = UD\,{}'U.$$

We discuss the spectral theorem at some length in Terras [1988, Vol. II, pp. 23ff (after formula (1.22)]. The decomposition $T = UDU^{-1} = UD\,{}'U$, where U is orthogonal and D real diagonal, is often called the *Schur decomposition*. See T. Coleman and C. VanLoan [1988, p. 250]. Matlab finds the Schur decomposition with the command $[U, D] = \text{schur}(T)$. We are assuming T is real symmetric.

Eigenvalues are so very useful that computers are full of routines to approximate them (Matlab, Mathematica . . .). If a matrix is not self-adjoint, it may not be possible to diagonalize it by a change of basis. For example, consider the matrix

$$\begin{pmatrix} 1 & 1 \\ 0 & 1 \end{pmatrix}.$$

However, there will still be eigenvalues (both one in this case) and ways to simplify. In Mathematica, the command Eigensystem[T] produces a list of eigenvalues and eigenvectors of the square matrix T. Matlab has a similar command.

Older books may use the words characteristic value or proper value instead of eigenvalue. See Courant and Hilbert [1953, p. 17] where they also replace the eigenvalue by its reciprocal. (Still, they give a nice discussion.)

For a general complex $n \times n$ matrix A, the eigenvalues λ of A are roots of the *characteristic polynomial* $\det(A - \lambda I) = 0$. However, it is not necessarily true that there is a basis of V for which the matrix of A is diagonal. For this, the matrix A must be normal (i.e., A commutes with its adjoint $\,{}'\bar{A} = A^*$).

The set of all the eigenvalues of the self-adjoint operator $T : V \to V$ is called the *spectrum* of T. The theory of spectra of operators (or linear transformations or matrices) is called *spectral theory*. Much of my book (Terras [1985]) concerns spectral theory for the Laplace operator on various kinds of infinite-dimensional vector spaces. Here we are concerned with finite-dimensional analogues (except in Chapter 24).

There is also another related use of the word spectrum. It comes from chemistry and physics: spectroscopy and the analysis of data. Elements are identified by their spectral lines. Hidden periodicities are sought in rainfall data. Basically this is a search for sinusoidal components of a signal. This requires the Fourier or discrete Fourier transform. And some spectral properties of molecules can be deduced from knowledge of spectra of adjacency operators of graphs. See Chapter 13, Cvetković, Doob, and Sachs [1980], Starzak [1989], or Sternberg [1994] for more information. Fan Chung and S. Sternberg [1992, 1993] discuss an intriguing recent example. One needs Fourier analysis on the alternating group A_5 to give a complete explanation of the spectral lines of the buckyball, which is a newly discovered stable carbon molecule with the form of a soccerball. Since A_5 is not a commutative group, the Fourier transform is more complicated than that for $\mathbb{Z}/n\mathbb{Z}$. We will consider noncommutative groups in Part II.

How do we plan to use the spectral theorem stated in Equations (5) and (6)? We will replace the adjacency matrix by a diagonal matrix of its eigenvalues. Then it will be easy to find the limit of A^n, as $n \to \infty$. We also want to find out what the eigenvalues say about the graph's properties.

Spectra of Graphs

Note that the largest eigenvalue of the adjacency operator of any k-regular graph X is k since the constant function is always an eigenfunction. The following facts can also be proved.

Theorem 1. **General Facts about Spectra of Regular Graphs.** Assume that X is a k-regular graph with adjacency operator A.

1) The degree k is an eigenvalue of A. If λ is any eigenvalue of A, then $|\lambda| \le k$.
2) The graph X is connected iff k has multiplicity 1 as an eigenvalue of A.
3) A graph is said to be *bipartite* if you can separate the vertices into two disjoint sets V_1 and V_2 so that each set V_i contains no adjacent vertices. The k-regular graph X is bipartite iff $-k$ is an eigenvalue of the adjacency operator.

Proof.

1. Clearly k is an eigenvalue corresponding to the constant eigenfunction. Suppose $Af = \lambda f$, for $f \in L^2(X)$, and the maximum of $|f(x)|$ occurs at $x = a$.

Then

$$|\lambda||f(a)| = |\lambda f(a)| = |Af(a)| = \left| \sum_{y \text{ adjacent to } a} f(y) \right| \le k|f(a)|,$$

since there are k vertices y adjacent to vertex a. It follows that $|\lambda| \le k$.

2. Suppose $Af = kf$, for $f : X \to \mathbb{C}$. We need to show that f is constant. Again suppose that the maximum of $|f(x)|$ occurs at $x = a$. We can assume that $f(a) > 0$. As in the preceding proof, we have

$$kf(a) = Af(a) = \sum_{y \text{ adjacent to } a} f(y) \le kf(a).$$

To have equality we must have no cancellation in the sum and $f(y) = f(a)$, for all y adjacent to a. Then if the graph is connected, it follows that f must be constant on X.

The proof of the converse is left as an exercise.
Hint. If X is not connected, the adjacency operator can be block diagonalized with the blocks corresponding to the connected components of X.

3. Exercise. Prove part 3. ∎

Examples.
1) $X(\mathbb{Z}/6\mathbb{Z}, \{\pm 1 \pmod 6\})$ is bipartite.
2) $X(\mathbb{Z}/5\mathbb{Z}, \{\pm 1 \pmod 5\})$ is not bipartite.

Remarks on Expansion Constants of Graphs and Ramanujan Graphs

To minimize the transmission time [in our communication network], what is clearly needed is that every subset of vertices has a lot of distinct neighbors.... On the other hand, the total length of cables needed to wire a network is also a quantity we would like to minimize for several reasons.

Bien [1989] writes

For the reasons Bien has given in the quote above, one is interested in maximizing expansion constants of graphs.

Definition. A graph X has *expansion constant* c if for every set Y of vertices of X such that $|Y| \le \frac{n}{2}$ (where X has n vertices), we have

$$|\partial Y| \ge c|Y|, \tag{7}$$

where the boundary ∂Y is defined to be

$$\partial Y = \{b \in X - Y \mid b \text{ is adjacent to some } a \in Y\}.$$

A graph with large expansion constant is called an *expander graph*.

The definition of expansion constant is not well agreed upon. We are following P. Sarnak [1990]. But compare the definitions in F. Bien [1989], Fan Chung [1989, 1991], A. Lubotzky [1988], and J. Friedman [1993]. The expansion constant is analogous to the Cheeger constant of a Riemannian manifold. Lubotzky [1988, p. 2] notes that expander graphs can be used to produce superconcentrators. See also Maria Klawe [1984].

One can show that if λ_1 is the second largest eigenvalue of the adjacency operator A of a regular graph of degree k, then the expansion constant c satisfies

$$c \geq \frac{1}{2}\left(1 - \frac{\lambda_1}{k}\right). \tag{8}$$

This motivates us to look for graphs with small λ_1. For a proof of this inequality, see the proof of Theorem 2 in Chapter 4.

Definition. We say that a k-regular graph is *Ramanujan* if for all eigenvalues λ of the adjacency operator of the graph such that $|\lambda| \neq k$, we have

$$|\lambda| \leq 2\sqrt{k - 1}. \tag{9}$$

This definition was made by Lubotzky, Phillips, and Sarnak [1988]. The Cayley graph $X(\mathbb{Z}/n\mathbb{Z}, \{\pm 1\})$ will soon be seen to be Ramanujan. It is a consequence of Theorem 2 below.

Connected with the good expansion behavior of Ramanujan graphs is the fact that random walks on (nonbipartite) Ramanujan graphs converge rapidly to uniform. We will prove the result about random walks in Chapter 6. Lubotzky [1994, p. 41] notes:

They [the eigenvalues of the adjacency operator] control the speed of convergence of the random walk on X to the uniform distribution. To ensure fast convergence one wants the nontrivial eigenvalues of δ [the adjacency operator] to be as small as possible in absolute values. It turns out that for an infinite family of k-regular graphs, the best possible bound we can hope for is $2\sqrt{k-1}$. This leads to the important notion of a Ramanujan graph. . . . Such graphs are very good expanders.

See Theorem 3 of Chapter 4 for a precise statement of the last few remarks.

Another reason that the Ramanujan graphs are of interest is that these are exactly the k-regular graphs whose Ihara zeta function satisfies the analogue of the Riemann hypothesis. See Chapter 24 (also Bass [1992], Hashimoto [1989, 1990, 1992], Stark and Terras [1996], and Sunada [1986, 1988]).

Lubotzky [1994, p. 41] notes that there is a close analogy between the spectral theory of graphs and Riemannian manifolds. In fact he says: "This connection between manifolds and graphs gives more than just an analogy." Theorems from continuous geometry suggest results in graph theory and vice versa. See also F. Chung [1996].

The name "Ramanujan" refers to an Indian mathematician S. Ramanujan [1887–1919]. He was self-trained and is famous for filling notebooks with complicated identities of interest to number theorists, combinatorists, and even physicists. Some of his work involves bounds on coefficients of modular forms (i.e., functions holomorphic on the complex upper half plane with an invariance property under transformation by 2×2 matrices with integer entries and determinant 1). See Terras [1985, Vol. I, Ch. 3]. One might view Ramanujan and his notebooks full of elaborate formulas with many subscripts as the opposite of N. Bourbaki and his many volumes of elegant coordinate-free mathematics.

Hardy [1940] calls Ramanujan "the most romantic figure in the recent history of mathematics. It was his insight into algebraical formulas, transformations of infinite series, and so forth, that was most amazing." Selberg [1989, Vol. I, pp. 699–700] notes: "The most important lesson that one could draw from Ramanujan's story about the educational system is that allowances should be made for the unusual perhaps lopsidedly gifted child with very strong interests in one direction." Selberg also says: "a felicitous but unproved conjecture may be of much more consequence for mathematics than the proof of many a respectable theorem." Certainly Ramanujan's conjecture was the inspiration for the work of many modern mathematicians such as Selberg, Weil, and Deligne.

In the 1980s Margulis [1984] and independently Lubotzky, Phillips, and Sarnak [1988] found infinite families of Ramanujan graphs of fixed degree. Since Lubotzky, Phillips, and Sarnak made use of the Ramanujan conjecture on the size of Fourier coefficients of modular forms, they named these graphs Ramanujan graphs. See also Sarnak [1990] and Lubotzky [1994]. The Ramanujan conjecture was proved by Deligne using all of the machinery of modern algebraic geometry. One can get away with using some weaker results proved earlier by Eichler for the graph-theoretic applications. See Sarnak [1990].

It is easy to find Ramanujan graphs of large degree. For example, the complete graph on n vertices (all possible edges) is a Ramanujan graph of degree $n - 1$. Why? The problem is to find infinite families of Ramanujan graphs of fixed degree. When the degree is 2, the finite circle graphs do the job as the following theorem proves. We will not really have much to say about this problem. The smallest degree for which this problem is open is degree 7. Instead we will be content to consider Ramanujan graphs whose degrees approach infinity with the number of vertices. See the next two chapters and Chapter 19.

Spectra of Cayley Graphs on $\mathbb{Z}/n\mathbb{Z}$

Theorem 2. Consider the Cayley graph $X(\mathbb{Z}/n\mathbb{Z}, S)$. The eigenvalues of the adjacency operator of this graph are

$$\mathscr{F}\delta_s(a) = \sum_{s \in S} \exp\left(\frac{-2\pi i a s}{n}\right).$$

So, for example if $S = \{\pm 1 \pmod{n}\}$, the eigenvalues are

$$2\cos\left(\frac{2\pi a}{n}\right), \quad a = 0, 1, 2, \ldots, n - 1.$$

Proof. First note from Equation (3) that the adjacency operator for a Cayley graph $X(G, S)$, $G = \mathbb{Z}/n\mathbb{Z}$, is a convolution operator:

$$Af = \delta_s * f,$$

where

$$\delta_s(x) = \begin{cases} 1, & x \in S, \\ 0, & \text{otherwise.} \end{cases}$$

Now take the discrete Fourier transform and find that

$$\mathscr{F}Af(a) = \mathscr{F}(\delta_s * f)(a) = \mathscr{F}\delta_s(a) \cdot \mathscr{F}f(a).$$

Setting $h = \mathscr{F}f$, we find

$$[(\mathscr{F}A\mathscr{F}^{-1}(h)](a) = \mathscr{F}\delta_s(a) \cdot h(a).$$

This means that we have diagonalized the operator A. The eigenvalues of A are the numbers $\mathscr{F}\delta_s(a)$, $a \in \mathbb{Z}/n\mathbb{Z}$. ∎

Morals.

1) The discrete Fourier transform diagonalizes the adjacency operators of the Cayley graphs of the cyclic group $\mathbb{Z}/n\mathbb{Z}$.

2) We don't need Mathematica to find the eigenvalues of $X(\mathbb{Z}/n\mathbb{Z}, S)$. We just need to compute some exponential sums!

3) The finite circle graph is 2-regular and it is Ramanujan if its eigenvalues λ with $|\lambda| \neq 2$ satisfy $|\lambda| \leq 2\sqrt{2-1} = 2$. This is clearly true since $2|\cos(2\pi \frac{j}{n})| \leq 2$ for all j. So the finite circle graph is Ramanujan.

4) The matrix version of Theorem 2 can be stated as follows, where F_n is the matrix of the DFT, i.e., $F_n = (e^{2\pi i a b})_{0 \leq a,b \leq n-1}$ and $\Phi_n = n^{-1/2}F_n$:

$$\Phi_n A \Phi_n^{-1} = \begin{pmatrix} \hat{\delta}_s(0) & \cdots & 0 \\ \vdots & \ddots & \vdots \\ 0 & \cdots & \hat{\delta}_s(n-1) \end{pmatrix}.$$

Exercise. Find another proof for Theorem 2 above by noticing that the function

$$e_a(x) = \exp\left(\frac{2\pi i a x}{n}\right), \quad \text{for } a, x \in \mathbb{Z}/n\mathbb{Z},$$

is an eigenfunction of the adjacency operator A for the Cayley graph $X(\mathbb{Z}/n\mathbb{Z}, S)$ corresponding to the eigenvalue $\mathscr{F}\delta_s(a)$.

Exercise. Consider the linear map $T_a : L^2(\mathbb{Z}/n\mathbb{Z}) \rightarrow L^2(\mathbb{Z}/n\mathbb{Z})$ given by

$$T_a(f) = \delta_a * f, \quad \text{for some } a \in \mathbb{Z}/n\mathbb{Z}.$$

Let W_a denote the matrix of T_a using the basis of delta functions for $L^2(\mathbb{Z}/n\mathbb{Z})$ in the usual order. Show that

$$W_a = W_1^a,$$

where $W = W_1$ is the *shift matrix* given by

$$W = \begin{pmatrix} 0 & 0 & 0 & \cdots & 0 & 1 \\ 1 & 0 & 0 & \cdots & 0 & 0 \\ 0 & 1 & 0 & \cdots & 0 & 0 \\ \vdots & \vdots & \vdots & \cdots & \vdots & \vdots \\ 0 & 0 & 0 & \cdots & 0 & 0 \\ 0 & 0 & 0 & \cdots & 1 & 0 \end{pmatrix}. \tag{10}$$

Finally, show that if F_n is the matrix of the discrete Fourier transform, that is,

$$F_n = (e^{2\pi i a b/n})_{0 \leq a,b \leq n},$$

then $\Phi_n = n^{-1/2} F_n$ is a unitary matrix and $\Phi_n W_n \Phi_n^{-1}$ is diagonal with $(j+1)$st entry $e^{-2\pi i aj/n}$.

Hint. Recall the preceding proof and that $\delta_1 * \delta_a = \delta_{a+1}$.

Exercise. The shift matrix W in the preceding problem was shown to be diagonalized by the discrete Fourier transform. More generally, a *circulant matrix* has the form

$$C = \begin{pmatrix} c_1 & c_2 & c_3 & \cdots & c_n \\ c_n & c_1 & c_2 & \cdots & c_{n-1} \\ \vdots & \vdots & \vdots & \ddots & \vdots \\ c_3 & c_4 & c_5 & \cdots & c_2 \\ c_2 & c_3 & c_4 & \cdots & c_1 \end{pmatrix} = (c_{j-i+1})_{1 \le i, j \le n}, \quad \text{with subscripts (mod } n).$$

(11)

Show that this matrix is also diagonalized by the discrete Fourier transform on $\mathbb{Z}/n\mathbb{Z}$, since it is

$$C = c_1 I + c_2{}' W + c_3{}' W^2 + \cdots + c_n{}' W^{n-1} = p(W),$$

where $p(x)$ is the polynomial

$$p(x) = c_1 + c_2 x + c_3 x^2 + \cdots + c_n x^{n-1}.$$

what are the eigenvalues of C?

Note. See Davis [1979] for more information on circulants and their applications.

Exercise. Let C denote the circulant matrix of the preceding exercise. Show that

$$\det C = \prod_{j=1}^{n} \hat{c}(j),$$

where

$$\hat{c}(j) = \sum_{k=1}^{n} c(k) e^{-2\pi i jk/n}.$$

Note. This Exercise on group determinants has many applications in number theory, for example, to the computation of discriminants of cyclic extensions of number fields. Dedekind wrote Frobenius in the late 1890s asking him to generalize this to noncommutative groups. This letter inspired much of the work of Frobenius on group representations. See Part 2, particularly the exercise at the end of Chapter 15. See also Curtis [1979].

Finite Fields and Winnie Li's Graphs

Winnie Li [1992] found some interesting examples of Cayley graphs using a little finite field theory. We take our discussion of this subject from L. Dornhoff and F. Hohn [1978] and W. Gilbert [1976, pp. 243–254]. Or see J. Gallian [1990], Jessie MacWilliams and N. Sloane [1988], or C. Small [1991].

Suppose that $\mathbb{F}_q$ denotes a finite field with q elements. Sometimes this is called a Galois field. Then q must be a prime power ($q = p^r$, where p is a prime and $r \geq 1$) and $\mathbb{F}_p \subset \mathbb{F}_q$. The prime p is called the *characteristic* of $\mathbb{F}_q$.

Exercise. Prove that a finite field must have order the power of a prime.
Hint. Consider the map $T : \mathbb{Z} \to \mathbb{F}_q$, defined by

$$T(n) = \begin{cases} 1 + \cdots + 1(n \text{ times}), & \text{for } n > 0, \\ 0, & \text{for } n = 0, \\ -1 - \cdots - 1(|n| \text{ times}), & \text{for } n < 0, \end{cases}$$

where 1 denotes the multiplicative identity in $\mathbb{F}_q$. Show that T is a ring homomorphism, that is, $T(n + m) = T(n) + T(m)$ and $T(nm) = T(n)T(m)$.

Use the fundamental ring homomorphism theorem and the fact that every ideal in $\mathbb{Z}$ is principal to conclude that $\mathbb{Z}/a\mathbb{Z}$ is isomorphic to a subring of $\mathbb{F}_q$ for some $a \in \mathbb{Z}$. Since $\mathbb{F}_q$ is a field, a must be prime.

To see that $\mathbb{F}_q$ has $a = p^r$ elements, note that $\mathbb{F}_q$ is a vector space over $\mathbb{F}_p$.

Of course $\mathbb{F}_p$ is just $\mathbb{Z}/p\mathbb{Z}$. But $\mathbb{F}_q$, for $q = p^r$, $r > 1$, is definitely not $\mathbb{Z}/q\mathbb{Z}$. To construct $\mathbb{F}_q$, you need to find a polynomial $f(x)$ of degree r and irreducible in $\mathbb{F}_p[x]$. Then if α is a root of $f(x)$, the field obtained by adjoining α to $\mathbb{F}_p$ is $\mathbb{F}_q = \mathbb{F}_p(\alpha) \cong \mathbb{F}_p[x]/(f(x))$. Here $(f(x))$ denotes the principal ideal consisting of all multiples $g(x)f(x)$, for $g(x) \in \mathbb{F}_p[x]$.

Example. $\mathbb{F}_4 \cong \mathbb{F}_2[x]/(x^2 + x + 1) = \mathbb{F}_2(\alpha) = \{0, 1, \alpha, \alpha + 1\}$, where $\alpha^2 + \alpha + 1 = 0$. To see that $x^2 + x + 1$ is irreducible over $\mathbb{F}_2$, you just need to try to divide it by all lower degree polynomials in $\mathbb{F}_2$, i.e., x, $x + 1$. Note that $x^2 + 1 = (x + 1)^2$ is reducible.

Every element of $\mathbb{F}_2(\alpha)$ has the form $a_1\alpha + a_0$, where $a_i \in \mathbb{F}_2$. That is, the set $\{1, \alpha\}$ is a vector space basis of $\mathbb{F}_4$ over $\mathbb{F}_2$. Thus $\mathbb{F}_2(\alpha)$ does have four elements. It is a field because it can be identified with the quotient ring $\mathbb{F}_2[x]/(x^2 + x + 1)$

Table I.8. *Addition in* $\mathbb{F}_2(\alpha)$, *where*
$$\alpha^2 + \alpha + 1 = 0$$

+	0	1	α	$\alpha + 1$
0	0	1	α	$\alpha + 1$
1	1	0	$\alpha + 1$	α
α	α	$\alpha + 1$	0	1
$\alpha + 1$	$\alpha + 1$	α	1	0

Table I.9. *Multiplication in*
$\mathbb{F}_2(\alpha)$, *where* $\alpha^2 + \alpha + 1 = 0$

$\cdot$	0	1	α	$\alpha + 1$
0	0	0	0	0
1	0	1	α	$\alpha + 1$
α	0	α	$\alpha + 1$	1
$\alpha + 1$	0	$\alpha + 1$	1	α

and the polynomial $f(x) = x^2 + x + 1$ is irreducible. The proof is the same as that of Theorem 1 in Chapter 1. If $b(x)$ is not a multiple of $f(x)$, the coset $b + (f(x))$ has a multiplicative inverse, for $b(x)$ and $f(x)$ are relatively prime polynomials. Thus (by the polynomial version of the Euclidean algorithm) there are $s, t \in \mathbb{F}_2[x]$ with $sb + tf = 1$. So the coset of $s(x)$ gives the desired inverse.

We can write down the addition and multiplication tables for our field with four elements as shown in Tables I.8 and I.9.

Exercise. Find the addition and multiplication tables for the finite fields with 8, 9, and 16 elements.

Theorem 3. **Basic Facts about Finite Fields.**
1) A field K has p^r elements iff it is the splitting field of $x^{p^r} - x$ over $\mathbb{F}_p$. The *splitting field* is the smallest field containing K such that the given polynomial factors completely into linear factors. Thus for any prime p and $r \geq 1$, there is a field with p^r elements and it is unique up to isomorphism.
2) The multiplicative group $K^* = K - \{0\}$ of a finite field K is cyclic.

Proof Sketch.
1) The multiplicative group $K^* = K - \{0\}$ has $p^r - 1$ elements and so all nonzero elements of K satisfy the polynomial $x^{p^r-1} - 1 = 0$. Multiply by x to get a polynomial satisfied by all elements of K.

For the converse, you need to see that if K is a splitting field of $x^{p^r} - x$ over $\mathbb{F}_p$, then it has p^r elements. This requires you to show that $x^{p^r-1} - 1$ has no double roots and that the set of roots of $x^{p^r} - x$ is indeed a field.

2) For a proof, see Dornhoff and Hohn [1978, p. 369], for example. You can use a similar argument to the one we gave for $(\mathbb{Z}/p\mathbb{Z})^*$, $p = $ prime in Theorem 4 of Chapter 1. ∎

Exercise. Fill in the details in the preceding proof.

As for $\mathbb{Z}/p\mathbb{Z}$, we say that an element α of $\mathbb{F}_q$ is a *primitive element* if it generates the multiplicative group of $\mathbb{F}_q$. The field $\mathbb{F}_q$ contains $\phi(q - 1)$ primitive elements, where ϕ is the Euler phi function. If the primitive element α is a root of the irreducible monic polynomial $f(x)$, we say $f(x)$ is a *primitive polynomial*. There is no easy way to find primitive polynomials. Consequently, extensive tables exist.

You can use the primitive elements to develop a table of logs for the finite field.

Example. Suppose α is a root in $\mathbb{F}_9$ of the polynomial $x^2 + x + 2$. Write out the table of powers of α, to see whether α is a primitive element.

Each element of $\mathbb{F}_9$ has the form $a_0 + a_1\alpha$, where $a_i \in \mathbb{F}_3$. And since $\alpha^2 + \alpha + 2 = 0$, we have $\alpha^2 = 2\alpha + 1$. Thus

$$\alpha(a_0 + a_1\alpha) = a_0\alpha + a_1(2\alpha + 1) = (a_0 + 2a_1)\alpha + a_1.$$

So with $\alpha^i = a_0 + a_1\alpha$, we find Table I.10.

Table I.10. *Elements of* $\mathbb{F}_9(\alpha)^*, \alpha^2 + \alpha + 2 = 0.$ $\alpha^i = a_0 + a_1\alpha, a_j$ in $\mathbb{F}_3$

i	a_0	a_1
0	1	0
1	0	1
2	1	2
3	2	2
4	2	0
5	0	2
6	2	1
7	1	1
8	1	0

We get the $(i + 1)$th row from the ith row by sending

$$(a_0, a_1) \rightarrow (a_1, a_0 + 2a_1).$$

The table means that

$$\alpha^0 = 1, \quad \alpha^1 = \alpha, \quad \alpha^2 = 1 + 2\alpha, \quad \alpha^3 = 2 + 2\alpha,$$
$$\alpha^4 = 2, \quad \alpha^5 = 2\alpha, \quad \alpha^6 = 2 + \alpha, \quad \alpha^7 = 1 + \alpha, \quad \alpha^8 = 1.$$

So α is indeed a primitive element.

Exercise. Given that $x^5 + x^2 + 1$ is a primitive polynomial over $\mathbb{F}_2$ and $x^3 + 2x + 1$ is a primitive polynomial over $\mathbb{F}_3$, construct log tables for $\mathbb{F}_{2^5}$ and $\mathbb{F}_{3^3}$.

You can mechanize this process by what is called a *feedback shift register.* See Dornhoff and Hohn [1978, p. 385]. We did that in the example above.

Definition. A *field automorphism* $\sigma : K \rightarrow K$ is a 1-1, onto map so that

$$\sigma(x + y) = \sigma(x) + \sigma(y) \quad \text{and} \quad \sigma(xy) = \sigma(x)\sigma(y).$$

Definition. The *Galois group* $\mathrm{Gal}(K/F)$ of the field extension K/F is the set of all field automorphisms $\sigma : K \rightarrow K$ such that $\sigma(x) = x$ for all $x \in F$. The group operation is composition of functions.

We also need to know the following results.

*Theorem 4. **Basic Facts about Automorphisms of Finite Fields.***

Fact 1) If p is a prime then $\mathbb{F}_{p^r} \subset \mathbb{F}_{p^s}$ iff r divides s.

Fact 2) Every field automorphism of $K = \mathbb{F}_{p^s}$ over $F = \mathbb{F}_p$ is a power of the *Frobenius map* $\psi(x) = x^p$, for $x \in \mathbb{F}_p$. So $\mathrm{Gal}(\mathbb{F}_{p^s}/\mathbb{F}_p)$ is a cyclic group of order s, generated by the Frobenius automorphism.

Similarly $\mathrm{Gal}(\mathbb{F}_{p^s}/\mathbb{F}_{p^r})$ is cyclic of order s/r, generated by $\psi^r(x) = x^{p^r}$.

Proof.

1) Suppose $\mathbb{F}_{p^r} \subset \mathbb{F}_{p^s}$. Since the larger field is a vector space over the smaller,

$$p^s = (p^r)^d = p^{rd}, \quad \text{where } d \text{ is the dimension of } \mathbb{F}_{p^s} \text{ over } \mathbb{F}_{p^r}.$$

Conversely if r divides s, $(p^r - 1)$ divides $(p^s - 1)$, then $(x^{p^r-1} - 1)$ divides $(x^{p^s-1} - 1)$. So the splitting field of $(x^{p^r} - x)$ is contained in the splitting field of $(x^{p^s} - x)$.

2) You can find a proof in Dornhoff and Hohn [1978, p. 375]. ∎

Exercise. List all the subfields of $\mathbb{F}_{2^{30}}$.

Definition. The *trace* and *norm* of the extension K/F are defined as follows:

$$\mathrm{Tr}_{K/F}(\alpha) = \sum_{\sigma \in \mathrm{Gal}(K/F)} \sigma(\alpha), \tag{12}$$

$$\mathrm{N}_{K/F}(\alpha) = \prod_{\sigma \in \mathrm{Gal}(K/F)} \sigma(\alpha), \quad \text{for all } \alpha \in K. \tag{13}$$

Exercise.
a) Suppose $\alpha \in K$. Show that $\mathrm{Tr}_{K/F}(\alpha)$ and $\mathrm{N}_{K/F}(\alpha)$ are in F (the base field).
b) Show that the norm maps K onto F.
c) Show that $\mathrm{Tr}(\alpha + \beta) = \mathrm{Tr}(\alpha) + \mathrm{Tr}(\beta)$ and $\mathrm{N}(\alpha\beta) = \mathrm{N}(\alpha)\mathrm{N}(\beta)$.

We could study the main theorems of Galois theory for finite fields but we don't seem to need these things at the moment. The fact that the Galois group is cyclic simplifies everything tremendously.

Finally, after all these preliminaries, we look at a special case of graphs defined by Winnie Li [1992]. Suppose that F is a finite field of q elements and F_n is a degree n extension field of F. Let Ξ_n be the kernel of the norm map from F_n to F, that is,

$$\Xi_n = \{\alpha \in F_n \,|\, \mathrm{N}_{F_n/F}\alpha = 1\}.$$

Note that the kernel contains $d_n = (q^n - 1)/(q - 1)$ elements, for

$$\mathrm{N}_{F_n/F}(\alpha) = \alpha \cdot \alpha^q \cdot \alpha^{q^2} \cdot \cdots \cdot \alpha^{q^{n-1}} = \alpha^{\frac{q^n-1}{q-1}}.$$

Assume that n is even so that Ξ_n is a symmetric set of generators of F_n. In fact we will usually take $n = 2$.

Definition. Winnie Li's graph is the Cayley graph $X(F_n, \Xi_n)$ with vertices the elements of F_n and edges of vertex x given by vertices $x + s$, $s \in \Xi_n$. These are d_n-regular graphs with q^n vertices.

Example 1. Consider $F = \mathbb{F}_2$ and $K = \mathbb{F}_4 = \mathbb{F}_2(\alpha) = \{0, 1, \alpha, \alpha^2\}$, with $\alpha^2 + \alpha + 1 = 0$. Then

$$\mathrm{N}\beta = \beta^{1+2} = \beta^3 = 1 \quad \text{holds for} \quad \beta = 1, \alpha, \alpha^2.$$

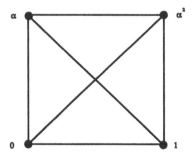

Figure I.16. $X(\mathbb{F}_4, \Xi_2)$ is the tetrahedron or the complete graph on four vertices.

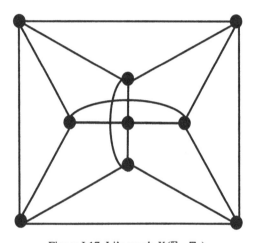

Figure I.17. Li's graph $X(\mathbb{F}_9, \Xi_9)$.

So our set $\Xi_4 = \{1, \alpha, \alpha^2\}$. We draw the graph $X(\mathbb{F}_4, \{1, \alpha, \alpha^2\})$ in Figure I.16. It is K_4, the complete graph on four vertices – a tetrahedron. That is, you draw all possible edges!

Example 2. Consider $K = \mathbb{F}_9 = \mathbb{F}_3(\alpha)$, where $\alpha^2 + \alpha + 2 = 0$. Now

$$N\beta = \beta^{1+3} = \beta^4 = 1 \quad \text{implies} \quad \beta = \alpha^{(3-1)j} = \alpha^{2j}, \quad j = 0, 1, 2, 3.$$

So we just take the even powers of α for Ξ and obtain

$$\Xi_9 = \{1, \alpha^2, \alpha^4, \alpha^6\} = \{1, 1 + 2\alpha, 2, 2 + \alpha\}.$$

We get the graph shown in Figure I.17.

Exercise.
(a) Draw W. Li's graph for $\mathbb{F}_{25} = \mathbb{F}_5(\alpha)$, where α satisfies $\alpha^2 + \alpha + 2 = 0$.
(b) Draw W. Li's graph for $\mathbb{F}_{27} = \mathbb{F}_3(\alpha)$, where $\alpha^3 + 2\alpha + 1 = 0$.

Hint. Mathematica includes a package loaded with the command:

$$\ll \text{DiscreteMath'Combinatorica'}$$

which allows you to type in the adjacency matrix m and the desired vertices v and then you get the graph with

$$\text{ShowLabeledGraph[Graph[m,v]]}.$$

One can often obtain a better picture by using the commands

$$g = \text{Graph[m,v]}$$
$$g1 = \text{SpringEmbedding[g]}.$$

This produces a graph that looks better by viewing it as a system of springs and minimizing some sort of energy. You can also use the command

$$g2 = \text{ShakeGraph[g]}.$$

Then

$$\text{ShowGraph[g] or ShowLabeledGraph[g]}$$

actually draws the picture.

The Beginnings of a Trace Formula

One of our goals in Part I is to develop the Poisson summation formula, and in Part 2 (Chapters 22 and 23) this will become a finite analogue of Selberg's trace formula. See Terras [1985, Ch. 3] for a continuous version. When the trace formula is developed for compact or finite-volume Riemannian manifolds, it may appear somewhat difficult. However, the discrete version is appealingly elementary. It begins with a self-adjoint matrix M (i.e., $M = {}^t\bar{M}$) and the spectral theorem from Equations (5) and (6). Write $M = (m_{ij})_{1 \le i, j \le n}$. Then the spectral theorem says $M = \bar{U}^{-1} DU$, where $I = {}^t\bar{U}U$ and D is a diagonal matrix with diagonal entries λ_j, the eigenvalues of M. The *original trace formula* says:

$$\text{Tr}\,(M) = \sum_{j=1}^{n} m_{jj} = \sum_{j=1}^{n} \lambda_j. \tag{14}$$

Now we can view M as an operator on functions $f : \mathbb{Z}/n\mathbb{Z} \to \mathbb{C}$ by writing M as a map

$$M : G \times G \to \mathbb{C}, \quad G = \mathbb{Z}/n\mathbb{Z}$$

given by $m_{kj} = M(k, j)$. Then define

$$L_M f(j) = \sum_{k=1}^{n} M(j, k) f(k).$$

Now Equation (14) is just the usual formula for the trace of a self-adjoint integral operator, namely

$$\text{Tr}\,(L_M) = \sum_{j=1}^{n} M(k, k) = \sum_{j=1}^{n} \lambda_j, \quad \text{where } \lambda_j = j\text{th eigenvalue of } L_M. \tag{15}$$

This is analogous in the continuous case to the trace formula for a Hilbert–Schmidt operator. See Terras [1985, Chapter 2] or Courant and Hilbert [1958]. Luckily we do not have to worry about convergence in the matrix case.

To begin to understand the Selberg trace formula, we need more structure on our operators. That is, we need to think of $G = \mathbb{Z}/n\mathbb{Z}$ as a group under addition and not just a set with n elements. We have $M : G \times G \to \mathbb{C}$ acting via L_M as above on functions $f : G \to \mathbb{C}$. There is a group action of $a \in G$ on functions f via the translation operator:

$$T_a f(x) = f(x - a). \tag{16}$$

We want to consider only operators L_M such that L_M commutes with the group action; that is,

$$L_M T_a = T_a L_M, \quad \text{for all } a \in G. \tag{17}$$

This requires $M(x, y) = k(x - y)$ for some $k : G \to \mathbb{C}$. Thus this requires that the matrix M be a circulant matrix and the operator L_M is convolution by k:

$$L_M f(x) = L_k f(x) = (k * f)(x) = \sum_{y \in G} k(x - y) f(y). \tag{18}$$

In order to produce a trace formula, we need to know the eigenvalues of the operator (18). It is easy to find them, since an orthogonal set of eigenfunctions of (18) is the set of characters from Chapter 2:

$$e_a(X) = \exp\left(\frac{2\pi i a x}{n}\right), \quad a, x \in G = \mathbb{Z}/n\mathbb{Z}. \tag{19}$$

We know from Chapter 2 that the e_a, $a \in G = \mathbb{Z}/n\mathbb{Z}$, form a complete orthonormal set in $L^2(\mathbb{Z}/n\mathbb{Z})$. They are eigenfunctions of L_k by the following computation:

$$(L_k e_a)(x) = \sum_{y \in G} k(x - y)e_a(y) = \sum_{u \in G} k(u)e_a(x - u)$$
$$= \sum_{u \in G} k(u)e_a(x)e_a(-u) = e_a(x)\langle k, e_a \rangle = \hat{k}(a)e_a(x).$$

So we have proved

$$(L_k e_a)(x) = \hat{k}(a)e_a(x). \tag{20}$$

The *trace formula* (15) for convolution operators becomes

$$\text{Tr}\,(L_k) = \sum_{y \in \mathbb{Z}/n\mathbb{Z}} k(y - y) = n\,k(0) = \sum_{y \in \mathbb{Z}/n\mathbb{Z}} \hat{k}(y).$$

This is simply the *Fourier inversion formula at the origin*.

You can derive Fourier inversion at an arbitrary point a from this by replacing k by the translated version $T_a k$, defined by (16). Then

$$nT_a k(0) = \sum_{y \in \mathbb{Z}/n\mathbb{Z}} \mathscr{F}(T_a k)(y).$$

Now use the *translation property of the DFT* (see Equation (12) of Chapter 2):

$$\mathscr{F}(T_a k)(y) = e_{-a}(y)\,\mathscr{F}k(y). \tag{21}$$

Putting all this together we have

$$n\,k(-a) = \sum_{y \in \mathbb{Z}/n\mathbb{Z}} e_{-a}(y)\hat{k}(y),$$

which is the inversion formula for the DFT, which we proved in Chapter 2.

Thus, in this simple situation, the Selberg trace formula is just Fourier inversion. To get closer to the version of the trace formula envisioned by Selberg, we will need to have a subgroup H of the group G. We will do this in the chapter on the Poisson summation formula (Chapter 12).

Chapter 4

Four Questions about Cayley Graphs

> Outrage, disgust, the characterization of group theory as a
> plague or as a dragon to be slain – this is not an atypical
> physicist's reaction in the 1930s–50s to the use of group
> theory in physics.
>
> Sternberg [1994, p. xi]

In this section, we hope to make use of a little group theory (in the guise of
Theorem 2 from the last chapter) and a lot of linear algebra, to answer a few
questions about the Cayley graphs we introduced in the last chapter. We will
also consider some of the main results needed in the work on expander graphs
(the last three theorems in this chapter). Hopefully no outrage or dragons will
be encountered.

Suppose A is the adjacency operator of a connected regular (undirected)
graph X of degree k (without multiple edges). Let spec(A) denote the spectrum
of A, that is, the set of all eigenvalues of A. Let d be the diameter of X and g
be the girth. The last two concepts were defined in the last chapter.

Question 1. Is X Ramanujan, that is, if $\lambda \in \text{spec}(A)$, $|\lambda| \neq k$, does λ satisfy

$$|\lambda| \leq 2\sqrt{k-1}?$$

Question 2. Is $0 \in \text{spec}(A)$ or, equivalently, is A invertible?
Question 3. Can we bound the diameter d?
Question 4. Can we bound the girth g?

Finite Circle Graph

Example 1. $X(\mathbb{Z}/n\mathbb{Z}, \{\pm 1 (\mathrm{mod}\ n)\})$, where $n \geq 3$.

This is a graph with n vertices each of degree 2. We know from Theorem 2 of Chapter 3 that if A is the adjacency operator for this graph,

$$\mathrm{spectrum}(A) = \left\{ 2\cos\left(\frac{2\pi a}{n}\right) \,\middle|\, a = 0, 1, \ldots, n-1 \right\}.$$

Now let's answer our four questions.

Question 1. Is X Ramanujan?

Yes. The graph is Ramanujan since

$$\left| 2\cos\left(\frac{2\pi a}{n}\right) \right| \leq 2 = 2\sqrt{2-1}.$$

But note that if n is even $-2 \in \mathrm{spec}(A)$, as is seen by setting $a = n/2$. The graph is bipartite in this case. This causes some problems with random walks, as we will see in Chapter 6.

Question 2. $0 \in \mathrm{spectrum}(A)$ iff n is divisible by 4.

Then we can solve

$$\frac{2\pi a}{n} = \frac{\pi}{2} \quad \text{with } a = \frac{n}{4}.$$

Question 3. The diameter is given by

$$d = \begin{cases} (n-1)/2, & \text{if } n \text{ is odd,} \\ n/2, & \text{if } n \text{ is even.} \end{cases}$$

Question 4. The girth is n.

Remarks on the Connection with Finite Radon Transforms

Note that the sets $S(r) = \{\pm r(\mathrm{mod}\ n)\}$ give Cayley graphs $X_r = X(\mathbb{Z}/n\mathbb{Z}, S(r))$ which are isomorphic to X_1 (the graph of Example 1) when r and n are relatively prime. Fill [1989] calls the adjacency operators for $X(\mathbb{Z}/n\mathbb{Z}, S(r))$ Radon transforms on shells and considers the case when g.c.d.$(r, n) \neq 1$ also.

We will also take the next two examples from Fill's paper. The main question preoccupying the Radon transformer is Question 2. For if the adjacency matrix is invertible, then one can reconstruct f from Af, its finite Radon transform.

Even if A is not invertible, it is possible to make use of the Moore–Penrose generalized inverse.

The classical Radon transform is an integral over planes in $\mathbb{R}^3$. It was studied by Radon in 1917 and is useful today in medicine (CAT scanners), radio astronomy, and seismology. See Terras [1983, Ch. 2]. Recently many papers on finite analogues have appeared (see Bolker [1987], Fill [1989], and Elinor Velasquez [1991, 1997]). Our treatment of the subject has been greatly influenced by Velasquez.

Much of the motivation for the study of these finite analogues comes from statistics, for example, for the analysis of circular data such as measurements of wind directions at a given location, time series from economics, and frequency of distribution of first digits in long lists of large numbers. See Batschelet [1971], Mardia [1972], and Diaconis [1988] for many examples.

Graphs Based on a Ball in $\mathbb{Z}/n\mathbb{Z}$

Example 2. $X(\mathbb{Z}/n\mathbb{Z}, B(r))$, where $B(r) = \{0, \pm 1, \ldots, \pm r \pmod{n}\}$. Here $n \geq 3$.

Remarks on an Application to Random Number Generators
The case $r = 1$ is studied in a paper by Chung, Diaconis, and Graham [1987]. The speed of convergence of the random walk on the graph $X(\mathbb{Z}/n\mathbb{Z}, B(1))$ has implications for random number generators. We will have more to say about this in Chapter 6. One nice aspect of this graph is that when $r = 1$, -3 is never an eigenvalue of the adjacency operator. That is, the graph is not bipartite.

The graph $X = X(\mathbb{Z}/n\mathbb{Z}, B(r))$ has n vertices of degree $2r + 1$. The eigenvalues of the adjacency operator A are

$$\lambda_a = \sum_{k=-r}^{r} \exp\left(\frac{2\pi i a k}{n}\right) = 1 + 2\cos\left(\frac{2\pi a}{n}\right) + \cdots + 2\cos\left(\frac{2\pi a r}{n}\right). \quad (1)$$

Question 1. Is $X = X(\mathbb{Z}/n\mathbb{Z}, B(r))$ Ramanujan?
  Not in general. We see from formula (1) that

$$\lambda_1 \to 1 + 2r, \quad \text{as } n \to \infty, r \text{ fixed.}$$

But $1 + 2r > 2\sqrt{2r}$, if $r \geq 1$. So the graphs are not Ramanujan for large values of n. What happens for smaller n and for r large $= n/2$ say? You can use Mathematica to study the question (as an exercise).

Question 2. Is $0 \in \mathrm{spec}(A)$?

To answer this question, we need another formula for our eigenvalue λ_a. Setting $w = e^{2\pi i/n}$ and assuming that n does not divide a, we get

$$\lambda_a = \sum_{k=-r}^{r} w^{ak} = w^{-ra} \sum_{j=0}^{2r} w^{aj} = w^{-ra} \frac{w^{a(2r+1)} - 1}{w^a - 1}$$

$$= w^{-a(r+1/2)} \frac{w^{a(2r+1)} - 1}{w^{a/2} - w^{-a/2}} = \frac{w^{a(r+1/2)} - w^{-a(r+1/2)}}{w^{a/2} - w^{-a/2}}.$$

Therefore

$$\lambda_a = \frac{\sin(\pi a(2r + 1)/n)}{\sin(\pi a/n)}. \tag{2}$$

Here when the denominator vanishes, you have to use l'Hôpital's rule to interpret the result. So now we can answer Question 2. We find that

$$0 \notin \mathrm{spec}(A) \quad \mathrm{iff} \quad \mathrm{g.c.d.}(n, 2r + 1) = 1.$$

Question 3. Can we find a bound for the diameter d?

We leave this as an *exercise*. The case $r = 1$ follows from the preceding example. In general this seems complicated. But we will be able to estimate the diameter if we have a good bound on $|\lambda|$ for $\lambda \in \mathrm{spec}(A), |\lambda| \neq k$ using Theorem 1 below.

Question 4. Can we find a bound for the girth g?

Since the graph has loops, $g = 1$.

Here is a final example from Fill [1989].

Exercise. Consider the Cayley graph $X(\mathbb{Z}/p\mathbb{Z}, S)$, for prime p, S any proper subset of $\mathbb{Z}/p\mathbb{Z}$, $S \neq \{0\}$. Show that 0 is not in the spectrum of the adjacency operator.

Hint. Set $w = e^{2\pi i/p}$. We know by Theorem 2 of Chapter 3 that the eigenvalues of A are

$$\lambda_a = \sum_{s \in S} w^{sa} = p_s(w^a), \quad a = 0, 1, 2, \ldots, p - 1,$$

where $p_s(x)$ is a polynomial of degree less than p. Note that $p_s(x)$ can have no roots among the pth roots of unity, for, by Eisenstein's irreducibility criterion, we know that

$$\frac{x^p - 1}{x - 1} = x^{p-1} + \cdots + x + 1$$

is an irreducible polynomial with w as root.

Winnie Li's Graphs

Example 3. Let's consider only the special case of Li's graphs $X = X(\mathbb{F}_{p^2}, \Xi)$, where

$$\Xi = \text{kernel of norm} = \{\alpha \in \mathbb{F}_{p^2} \mid N\alpha = \alpha\alpha^p = 1\}.$$

We saw some examples at the end of Chapter 3. The graphs have p^2 vertices (the elements of the additive group of the field with p^2 elements). The degree of each vertex is the order of S, which is $p + 1$. To see this, you have to use the fact that the multiplicative group of a finite field is cyclic. Suppose γ generates this multiplicative group: $\mathbb{F}_{p^2}^* = \langle \gamma \rangle$. Then, since we are in a finite field, the equation $\alpha^{p+1} = 1$ has at most $p + 1$ roots. To see that it has exactly $p + 1$ roots, note that the roots are

$$\alpha = \gamma^{(p-1)j}, \quad j = 0, 1, 2, \ldots, p,$$

because $\gamma^{p^2-1} = 1$, since $p^2 - 1$ is the order of $\mathbb{F}_{p^2}^*$.

To answer our questions, we need to convince ourselves that the analogue of Theorem 2 of Chapter 3 holds for the noncyclic additive group of the finite field with p^2 elements. To do this, you must figure out what the discrete Fourier transform should be for this finite abelian group $\mathbb{F}_{p^2}$ under addition. This means that we must know the complete list of characters, that is, the group homomorphisms

$$\psi : \mathbb{F}_{p^2} \to \mathbb{T} = \{z \in \mathbb{C} \mid |z| = 1\}.$$

Here $\mathbb{F}_{p^2}$ is a group under $+$ and $\mathbb{T}$ is a group under *. It turns out that the characters are, for $a \in \mathbb{F}_{p^2}$,

$$\psi_a(x) = \exp\left(\frac{2\pi i \operatorname{Tr}(ax)}{p}\right), \quad \text{where } \operatorname{Tr}(x) = x + x^p, x \in \mathbb{F}_{p^2}.$$

Exercise.

i) Prove that the functions ψ_a, $a \in \mathbb{F}_{p^2}$, form a complete orthogonal set of functions on $L^2(\mathbb{F}_{p^2})$.

ii) Show that the function ψ_a, $a \in \mathbb{F}_{p^2}$, is an eigenfunction of the adjacency operator for Winnie Li's graphs $X(\mathbb{F}_{p^2}, \Xi)$ corresponding to the eigenvalue

$$\lambda_a = \sum_{s \in S} \psi_a(S) = \sum_{\substack{s \in \mathbb{F}_{p^2} \\ Ns = 1}} \exp\left(\frac{2\pi i (as + (as)^p)}{p}\right).$$

Hint. You need to show that $\mathrm{Tr} : \mathbb{F}_{p^2} \to \mathbb{F}_p$ is an onto mapping. Then you need to generalize formula (8) of Chapter 2. Then recall the exercise after Theorem 2 of Chapter 3.

Thus we see that the eigenvalue λ_a of the preceding exercise is a *Kloosterman sum*

$$\lambda_a = \sum_{\substack{u \in \mathbb{F}_{p^2} \\ Nu = Na}} \exp\left(\frac{2\pi i (u + u^{-1})}{p}\right), \quad \text{setting } u = as. \tag{3}$$

Deligne [1977, p. 219] estimated these sums for $a \neq 0$ by

$$|\lambda_a| \leq 2\sqrt{p}.$$

Li [1992] finds a simpler proof which, makes use of work of Weil [1979, Vol. I, pp. 399–508] and Weil [1973, Appendix V]. We cannot discuss the proof here as it requires some knowledge of zeta functions of curves over finite fields.

Now we can think about the four graph theory questions.

Question 1. Are the graphs Ramanujan?

The answer is yes, using the exponential sum estimate of Deligne and W. Li.

Question 2. Can the eigenvalues vanish?

Exercise. Answer Question 2 for Winnie Li's graphs.

Hint. See Katz [1980].

Question 3. We can obtain estimates for the diameter from Fan Chung's Theorem 1 below. We obtain

$$d \leq \frac{\log(p^2 - 1)}{\log((p+1)/(2\sqrt{p}))} + 1.$$

As p goes to infinity the bound on d approaches 5.

The finiteness of the diameter shows that the graph is connected.

Exercise. Find another proof that the graph $X(\mathbb{F}_{p^2}, \text{ker norm})$ is connected by finding a set of generators of the additive group of $\mathbb{F}_{p^2}$ each of which have norm 1.

Hint. Since the kernel of the norm has $p + 1$ elements, it must contain some element outside of $\mathbb{F}_p$.

Question 4. Li [1992] shows that the girth is 3 unless $p \equiv 1 \pmod{3}$ when it is 4.

Remarks on Kloosterman Sums

The sums that give the eigenvalues of the adjacency operator for Li's graphs are *Kloosterman sums*, which are famous exponential sums, at least to number theorists. In fact, they have seemed so important that some number theorists have spoken of Kloostermania (see Huxley [1985]).

The original *Kloosterman sum* is

$$K_p(a, b) = \sum_{x \in \mathbb{F}_p^*} \exp\left(\frac{2\pi i (ax + bx^{-1})}{p}\right). \tag{4}$$

It appears in Fourier coefficients of modular forms and thus estimates of the sums (4) played an important role in work giving partial results on the Ramanujan conjecture (now theorem) on $\tau(n)$. See Sarnak [1990, Chapter 1] and Selberg [1989, pp. 506–520].

Just as Gauss sums may be viewed as finite analogues of the gamma function, the sum $K_p(a, b)$ may be viewed as a finite analog of the K-Bessel function defined by

$$K_s(a) = \frac{1}{2} \int_0^\infty t^{s-1} e^{-a(t+t^{-1})/2} dt, \quad \text{for } \operatorname{Re} a > 0. \tag{5}$$

Many authors have considered finite analogues of classical special functions. See, for example, Evans [1986].

Recently Kloosterman sums for fields of characteristic 2 have been applied to error-correcting codes. See G. Lachaud [1989, 1990]. The Kloosterman sum also gives the number of points on certain elliptic curves. See G. Lachaud and J. Wolfmann [1987]. Another reference is C. Moreno [1991]. W. Schmidt [1976, p. 46] considers estimates for the original Kloosterman sum. And sums similar to those of Winnie Li appear in the theory of the representations of the group of 2×2 nonsingular matrices over $\mathbb{F}_p$ (see Piatetski–Shapiro [1983]).

We will see Kloosterman sums again in the next chapter on finite euclidean graphs as well as in Part 2 when we look at noneuclidean finite upper half plane graphs.

Fan Chung's Diameter Estimate

Theorem 1. (Fan Chung). Estimating the Diameter in Terms of an Eigenvalue Bound. Let X be a k-regular nonbipartite graph with n vertices and adjacency operator A. Define

$$\mu = \max\{|\lambda| \mid \lambda \in \text{spec}(A), \ |\lambda| \neq k\}.$$

Then the diameter of X satisfies

$$d \leq \frac{\log(n-1)}{\log k - \log \mu} + 1.$$

Proof. Chung [1989, 1996] and Quenell [1994]. First recall that if A is the adjacency matrix of X, $(A^m)_{a,b}$ is the number of paths of length m from a to b. Pick $a, b \in X$ such that the distance between a and b is $d =$ the diameter of X. Then $(A^d)_{a,b} \neq 0$ whereas $(A^{d-1})_{a,b} = 0$. So set $m = d - 1$. In the proof below,

$$(A^m)_{b,a} = (A^m \delta_a)(b),$$

where

$$\delta_a(x) = \begin{cases} 1 & \text{if } x = a, \\ 0 & \text{otherwise.} \end{cases}$$

Next use the spectral theorem for the operator A on $L^2(X)$ (see formulas (5) and (6) in Chapter 3) to obtain a complete orthonormal set ϕ_j of eigenfunctions of A, with $\phi_1 = $ constant and $A\phi_2 = \pm\mu\phi_2$:

$$A^m f = \sum_{j=1}^{n} \langle f, \phi_j \rangle \lambda_j^m \phi_j, \quad \text{for any } f \in L^2(X).$$

From our choices of a, b and $m = d - 1$, we have

$$0 = (A^m \delta_a)(b) = \sum_{j=1}^{n} \langle \delta_a, \phi_j \rangle \lambda_j^m \phi_j(b).$$

Use the fact that $\phi_1(x) = 1/\sqrt{n}$ and $\lambda_1 = k = $ the largest eigenvalue of A in absolute value to see that

$$0 \geq \frac{k^m}{n} - \mu^m \sum_{j=2}^{n} |\phi_j(a)|\,|\phi_j(b)|.$$

Thus by the Cauchy–Schwarz inequality

$$0 \geq \frac{k^m}{n} - \mu^m \left(\sum_{j=2}^{n} |\phi_j(a)|^2 \right)^{1/2} \left(\sum_{j=2}^{n} |\phi_j(b)|^2 \right)^{1/2}.$$

Now since the $(\phi_j(a))_{j \in Y, a \in X}$, $Y = \{1, 2, \ldots, n\}$ form an orthogonal matrix, we have

$$\sum_{j=1}^{n} |\phi_j(b)|^2 = 1, \quad \text{for all } b \in X.$$

Thus with $m = d - 1$, we obtain the estimate

$$0 \geq \frac{k^m}{n} - \mu^m \sqrt{1 - (\phi_1(a))^2}\sqrt{1 - (\phi_1(b))^2} = \frac{k^m}{n} - \mu^m \left(1 - \frac{1}{n}\right).$$

This quickly implies the theorem. ∎

Exercise. Suppose X is a bipartite k-regular graph with n vertices. Show that the diameter d of X satisfies

$$d \leq \frac{\log(n-2) - \log 2}{\log k - \log \mu} + 2.$$

Hint. (See Quenell [1994].) Imitate the proof of the theorem using the fact that for a connected k-regular bipartite graph $-k$ is also an eigenvalue of multiplicity 1. Why is it legal to assume our graphs are connected?

Fan Chung ([1989] and [1996]) also gives a brief history of expander graphs and some other examples of expander graphs that are Cayley graphs for the multiplicative groups of finite fields. At first it was shown that a "random" regular graph would be a good expander, but it was not known how to find a good expander. And Chung [1989] considers graphs bases on the multiplicative group of a finite fields. The graphs are shown to be Ramanujan using N. Katz's estimates of the exponential sums involved.

Paley Graphs

Exercise. Let p be an odd prime. Consider the Cayley graph $X(\mathbb{Z}/p\mathbb{Z}, S_\square)$, where

$$S_\square = \{x^2 \mid x \in (\mathbb{Z}/p\mathbb{Z})^*\}.$$

a) Show that $p \equiv 1 (\bmod\, 4)$ implies that $S_\square$ is symmetric, that is, $y \in S_\square$ implies $-y \in S_\square$. If $p \equiv 1 (\bmod\, 4)$, the graph $Z(\mathbb{Z}/p\mathbb{Z}, S_\square)$ is called a *Paley graph*.

b) Show that if A is the adjacency operator of $X(\mathbb{Z}/p\mathbb{Z}, S_\square)$, $\lambda \in \mathrm{spec}(A)$, $|\lambda| \neq \frac{p-1}{2}$, implies $|\lambda| \leq \frac{1+\sqrt{p}}{2}$.

c) Answer the four questions for the graph with $p \equiv 1 (\bmod\, 4)$. It will help to know the facts about Gauss sums proved in Chapter 8.

d) Compute a bound for the expansion constant of the graph (see Equation (8) of Chapter 3).

Hints.

a) $p \equiv 1 (\bmod\, 4)$ implies that -1 is a square mod p.

b) Note that $|S_\square| = (p-1)/2$. You find that the eigenvalues are very close to being Gauss sums defined in formula (11) of Chapter 2. You will need to know that Gauss sums mod p have absolute value $\sqrt{p}$. We will consider Gauss sums in more detail in Chapter 8.

c) Answers to the four questions for Paley graphs:

Question 1: false for $p = 3$; true for $p \geq 5$.

Question 2: false.

Question 3: $d \leq 4$ for $p \geq 11$.

Question 4: $g = 5$ for $p = 5$ and g is 3 or 4 for $p \geq 7$.

Use a result in Bollobás (1, p. 68) for the girth. It says that if the graph X has n vertices, degree k, and girth g, then

$$n = \begin{cases} 1 + \frac{k}{k-2}\left((k-1)^{(g-1)/2} - 1\right), & g \text{ odd}, \\ \frac{2}{k-2}\left((k-1)^{g/2} - 1\right), & g \text{ even}. \end{cases} \tag{6}$$

d) Use Theorem 2 below.

Note. Chung [1996, p. 94] says: "The Paley graph is a favorite and frequently cited example in extremal graph theory, no doubt due to its many nice properties, the existence of which are guaranteed by eigenvalue bounds." See also Chung [1991] and Bollobás [1985]. Chung also considers *sum graphs*, which are defined to have vertex set $\mathbb{Z}/p\mathbb{Z}$ with two vertices a and b joined by an edge if $a + b$ is a square mod p. It is also possible to consider the analogue of Paley graphs with $\mathbb{Z}/p\mathbb{Z}$ replaced by a finite field $\mathbb{F}_q$. See Bollobás [1985, p. 315].

The Expansion Constant

The following Theorem says that Ramanujan graphs give good expanders.

Theorem 2. Suppose X is a k-regular graph with n vertices. Then X has expansion constant c defined by

$$|\partial Y| \geq c|Y| \quad \text{for } Y \subset X \text{ with } |Y| \leq \frac{n}{2},$$

as in formula (7) of Chapter 3. Then we can take

$$c = \frac{1}{2}\left(1 - \frac{\lambda_1}{k}\right),$$

where λ_1 is the second largest eigenvalue of the adjacency operator of X.

Proof. (From a preprint of Sarnak [1990] as in Terras [1991].)

By the maximum principle characterization of eigenvalues (see Courant and Hilbert [1953, pp. 23–27], Horn and Johnson [1991, p. 178], or Strang [1976]) we can write λ_1 as the maximum of the Rayleigh quotient:

$$\lambda_1 = \max_{\substack{g \in L^2(X) \\ \langle g,1 \rangle = 0}} \frac{\langle Ag, g \rangle}{\langle g, g \rangle}.$$

Let W be any subset of X such that $|W| \leq n/2$ and let $f = \delta_W$, that is, f is 1 on W and 0 off W. Take W so that $|\partial W| = c|W|$. Let $g(x) = f(x) - \bar{f}$, where

$$\bar{f} = \frac{\langle f, 1 \rangle}{n} = \text{mean value of } f \text{ on } X = \frac{|W|}{n}.$$

Note that $\langle g, 1 \rangle = 0$. We can obtain a lower bound for λ_1 by finding a lower bound for $\langle Ag, g \rangle$, where A is the adjacency operator of X. First we have

$$\langle Ag, g \rangle = \langle f, Af \rangle - kn\bar{f}^2.$$

Then

$$2\langle Af, f \rangle = 2 \sum_{\substack{x,y \text{ adjacent}}} f(x)f(y)$$

$$= - \sum_{\substack{x,y \text{ adjacent} \\ \geq 2k|W|-2k|\partial W|,}} (f(x) - f(y))^2 + 2k \sum_{x \in X} (f(x))^2,$$

where

$$\partial W = \{x \notin W \mid x \text{ is adjacent to a vertex in } W\}.$$

Therefore

$$\lambda_1 \langle g, g \rangle \geq \langle Ag, g \rangle \geq k|W| - k|\partial W| - kn\bar{f}^2.$$

Use $|W| \leq n/2$ and the fact that $\langle g, g \rangle = |W| - |W|^2/n$ to finish the proof. ∎

The following theorem says that the Ramanujan bound is optimal for an infinite sequence of k-regular graphs with number of vertices going to infinity.

Theorem 3 (Alon and Boppana). Let $X_{m,k}$ run through a sequence of k-regular graphs with $m \geq n$ vertices. Then

$$\lim_{n \to \infty} \left\{ \inf_{m \geq n} \lambda_1(X_{m,k}) \right\} \geq 2\sqrt{k-1}.$$

For a proof of Theorem 3, see Sarnak [1990, p. 71] or Terras [1991]. It is important to note that the degree of the graphs $X_{m,k}$ is fixed. In the next chapter we will be considering graphs whose degrees go to infinity along with the number of vertices.

Exercise. Check Theorem 1 for the finite circle graph $X(\mathbb{Z}/n\mathbb{Z}, \{\pm 1\})$, for n odd.

Exercise. (From p. 455 of Angel, Trimble, Shook, and Terras [1995].) Show that if A is the adjacency matrix of a k-regular nonbipartite graph with n vertices and if μ is the second largest $|\lambda|$ for λ in the spectrum of A, then we have the inequality

$$\mu \geq \sqrt{\frac{kn - k^2}{n - 1}}.$$

Hint. $\text{Tr}(A^2) = kn$, since A is symmetric and the ith row of A has k ones and $(n-k)$ zeros. So the product of the ith row of A and the ith column of A is k.

Other Questions.

1. One can ask an analogue of a question of Kac [1966]. Can you hear the shape of a graph? That is, must two graphs with the same eigenvalues (having the same multiplicities) be isomorphic? In Chapter 22 we will use Cayley graphs

of nonabelian finite groups to find examples of graphs whose shapes cannot
be heard.

2. Do our Cayley graphs have Hamiltonian cycles? This means a cycle in the
graph that passes through every vertex exactly once before returning to the
starting vertex. In 1859 Hamilton invented a puzzle, which was to find such
a path on a regular dodecahedron. See Gallian [1990, p. 451–462]. It turns
out that undirected Cayley graphs of finite abelian groups are known to have
Hamiltonian cycles. See the article of L. Babai in Graham et al. [1995, Vol.
II, pp. 1472–3]. It appears to be hard to find Cayley graphs that do not have
Hamiltonian cycles. In fact, according to Babai, only four connected vertex-
transitive graphs (with more than two vertices) without Hamiltonian cycles
are known.

Chapter 5

Finite Euclidean Graphs and Three Questions about Their Spectra

> The discovery of new types of geometries – including non-Euclidean, affine, projective, etc. – led, eventually, to the famous Erlangen program of Klein, which proposed that the true study of any geometry lies in an analysis of its group of motions.
>
> Sternberg [1994, p. ix]

The goal of this chapter is to discuss the distribution of eigenvalues of combinatorial Laplacians – or adjacency operators – for graphs corresponding to finite analogues of the simplest euclidean symmetric space over finite fields of odd characteristic. In a later chapter we will consider finite analogues of non-euclidean symmetric spaces. In the process we will find some new examples of Ramanujan graphs. And, as we know from the last chapter, knowledge of the spectra of graphs leads to knowledge about graph invariants such as diameter, expansion constants, and girths.

In this chapter we consider results from Medrano, Myers, Stark, and Terras [1996]. It is easy to find a finite analogue of n-dimensional euclidean space $\mathbb{R}^n$ (considered in Chapter 1 of Terras [1985]) by replacing $\mathbb{R}$ with the finite field $\mathbb{F}_q$ with $q = p^r$ elements, *where p is an odd prime*.

Definition. Finite euclidean n-space is $\mathbb{F}_q^n$, which consists of column vectors x with jth entries x_j in $\mathbb{F}_q$.

Definition. The *"distance"* between two column vectors x and y in $\mathbb{F}_q^n$ is defined by

$$d(x, y) = \sum_{k=1}^{n} (x_j - y_j)^2. \tag{1}$$

Why did we put quotes around the word "distance?" Since $d(x, y)$ has values in the finite field, this distance is not a metric in the sense of the triangle inequality and analysis but rather a metric in the sense of quadratic forms and algebra (see Lang [1984]). It does have finite analogues of the group invariance properties of the euclidean metric in $\mathbb{R}^n$, for clearly,

$$d(x + u, y + u) = d(x, y) \quad \text{for all } u \in \mathbb{F}_q^n \tag{2}$$

and

$$d(kx, ky) = d(x, y) \quad \text{for all } k \in O(n, \mathbb{F}_q). \tag{3}$$

Here the orthogonal group $O(n, \mathbb{F}_q)$ consists of $n \times n$ matrices k with entries in $\mathbb{F}_q$ such that ${}^t k k = I$, where $I =$ the $n \times n$ identity matrix.

Definition. The *euclidean graph* $E_q(n, a)$ associated to $a \in \mathbb{F}_q$ has vertices the elements of $\mathbb{F}_q^n$ and edges between vertices x, y if $d(x, y) = a$.
   The graph $E_q(n, a)$ is a Cayley graph $X(\mathbb{F}_q^n, S_q(n, a))$ where

$$S_q(a) = S_q(n, a) = \left\{ x \in \mathbb{F}_q^n \mid d(x, 0) = x_1^2 + \cdots + x_n^2 = a \right\}. \tag{4}$$

It is not hard to show (see the exercise below, Borevitch and Shafarevitch [1966, Thm. 2, p. 14], Carlitz [1953], Ireland and Rosen [1982, Ch. 8], or Small [1991, pp. 86–91, 145–6]) that our euclidean graphs $E_q(n, a)$, with $a \neq 0$, are connected and regular with degree $|S_q(n, a)| = q^{n-1} + O(q^{(n-1)/2})$. A few examples drawn by Mathematica are shown in Figures I.18 and I.19.

Definition. The *finite euclidean group* G consists of $2n \times 2n$ matrices with block form

$$g = \begin{pmatrix} k & u \\ 0 & 1 \end{pmatrix}, \quad \text{where } k \in O(n, \mathbb{F}_q) \text{ and } u \text{ a column vector in } \mathbb{F}_q^n.$$

Figure I.18. The euclidean graph $E_5(2, 1)$.

Figure I.19. The euclidean graph $E_7(2, 1)$.

Here 0 denotes a row vector of n zeros. Then g acts on $x \in \mathbb{F}_q^n$ by

$$x \mapsto kx + u = \begin{pmatrix} k & u \\ 0 & 1 \end{pmatrix} \begin{pmatrix} x \\ 1 \end{pmatrix}.$$

Note that this action preserves the distance $d(x, y)$ defined in (1).

Now define K to be the subgroup of G of matrices with $u = 0$. Then $G/K \cong \mathbb{F}_q^n$. This space is a *symmetric space* since we have a commutative algebra $L^2(\mathbb{F}_q^n)$ of functions $f : \mathbb{F}_q^n \to \mathbb{C}$ with multiplication defined by *convolution*:

$$(f * g)(x) = \sum_{y \in \mathbb{F}_q^n} f(y)g(x - y).$$

We will say more about nonabelian groups such as G and symmetric spaces G/K in Chapter 19.

Note also that in the exceptional case $(q, n, a) = (q, 2, 0)$ with -1 a non-square in $\mathbb{F}_q$, we have $S_q(n, 0) = \{0\}$. Otherwise we can ask if the K-orbits in $\mathbb{F}_q^n - \{0\}$ are the sets $S_q(n, a)$, for $a \neq 0$, or $S_q(n, 0) - \{0\}$, for $a = 0$. This is true by Witt's theorem (see Lang [1984]).

For a symmetric space we need an analogue of the Laplace operator or the G-invariant operators. Recall that the *adjacency operator* A of a graph $E_q(n, a)$ acts on functions $f \in L^2(\mathbb{F}_q^n)$ by

$$Af(x) = \sum_{\substack{y \in \mathbb{F}_q^n \\ d(x,y) = a}} f(y) = \sum_{\substack{u \in \mathbb{F}_q^n \\ d(u,0) = a}} f(x + u) = \delta_{S_q(n,a)} * f(x), \qquad (5)$$

where the distance $d(x, y)$ is defined by (1) and the set $S_q(n, a)$ is defined by (4). Here for a set S

$$\delta_S(x) = \begin{cases} 1, & x \in S, \\ 0, & x \notin S. \end{cases}$$

Three Questions about Euclidean Graphs

Question 1

Consider a connected regular graph X of degree k with adjacency operator A. Define

$$\mu = \max\{|\lambda| \mid \lambda \text{ an eigenvalue of } A, |\lambda| \neq k\}. \qquad (6)$$

Recall from Chapter 3 that a graph is *Ramanujan* if

$$\mu \leq 2\sqrt{k - 1}. \qquad (7)$$

Are our graphs Ramanujan? Can 0 be an eigenvalue of A?

Question 2

What can one say about the distribution of the nontrivial eigenvalues of A – the spectrum? That is, we look at the histogram of the spectrum of A minus the degree. This means divide up the interval $[-\mu, \mu]$ into equal subintervals I and count the number $b(I)$ of eigenvalues in I. Then plot rectangles of base I and height $b(I)$.

Matlab has a command to compute the histogram. In Mathematica you need the statistics packages as well as the graphics packages. An explanation of how to do histograms can be found in Shaw and Tigg [1994, pp. 122–4].

Once you can do these histograms, you can ask what happens as q goes to infinity. Do the eigenvalues spread out in a solid line? What about the multiplicities? Does the histogram of eigenvalues approach a semicircle?

Question 3

What can one say about the "level curves" of the eigenfunctions?

Discussion of the Three Questions

We explained in the last chapter some of the motivation for seeking Ramanujan graphs.

Question 2 arises in several recent works (see Friedman [1993], Lafferty and Rockmore [1992], McKay [1981]). Lafferty and Rockmore [1992] consider Cayley graphs attached to groups G of 2×2 matrices over finite fields with q elements and subsets $S \subset G$ with $|S| = 4$. They have asked whether you see a continuous band of eigenvalues in the interval $[-\mu, \mu]$ as $q \to \infty$, with μ as in formula (6). For the Lafferty–Rockmore graphs there are gaps in the spectra.

If the spectra do seem to give a solid band between $-\mu$ and μ, it is natural to ask for even more detailed knowledge of the spectra. One can ask whether the eigenvalues approach the distribution of a random real symmetric matrix as q goes to infinity (see Mehta [1967]). This is a question asked by Adolphson [1989], Katz [1980,1988], or Bollobás [1985, p. 351]. That is: Do the eigenvalues approach the *Wigner semicircle distribution* (also known as the *Sato–Tate distribution*) as $q \to \infty$? This means that if a set $B \subset [-2, 2]$, we want

$$\frac{1}{q-1} \#\{\lambda \mid 2\lambda/\mu \in B\} \sim \frac{1}{2\pi} \int_{x \in B} \sqrt{4 - x^2} \, dx, \quad \text{as} \quad q \to \infty. \tag{8}$$

Here $q - 1$ is the number of eigenvalues given by Kloosterman sums in Equation (12) below. We are neglecting the multiplicity of the eigenvalues. Take $x = 2 \cos \theta$ to obtain the Sato–Tate distribution as in Adolphson [1989] and Katz [1988]. See also McKay [1981] and Sarnak [1990].

As part of Question 2, one can also ask: What is the level spacing distribution? See Sarnak [1996, 1997]. This means that you must order the eigenvalues of the adjacency operator

$$\lambda_1 \leq \lambda_2 \leq \lambda_3 \leq \cdots \leq \lambda_{m-1} \leq \lambda_m,$$

and normalize so that the mean of the level spacings $\lambda_i - \lambda_{i+1}$ is 1. Then plot the histogram of the differences $\lambda_{i+1} - \lambda_i$. The question is: Does the result look like e^{-x}?

Question 3 is a finite euclidean analogue of questions considered in mathematical physics (see Courant and Hilbert [1953, p. 302] and D. L. Powers [1987]). Other references are Gutzwiller [1990], Hejhal and Rackner [1992], and Sarnak [1995].

In the case of eigenfunctions of

$$\Delta = \frac{\partial^2}{\partial x^2} + \frac{\partial^2}{\partial y^2},$$

which are 0 on the boundary of a domain D in the real plane $\mathbb{R}^2$, we are asking for the points of a vibrating drum with a given height above the plane of the resting drum. For a circular drum, the contour lines of radial eigenfunctions are circles. For real-life drums with dark dust on them, one does see these circles (and rays). See Figure I.20. The *nodal lines* of an eigenfunction v are the (x, y) in the plane such that $v(x, y) = 0$.

We can think of the eigenfunctions of the combinatorial Laplacian on the finite euclidean graphs as finite spherical functions. See Chapter 10 and Part 2

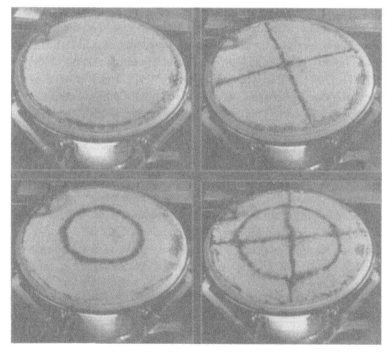

Figure I.20. Cover of D. L. Powers [1987] showing the nodal lines for an actual vibrating drum.

for more information on spherical functions. For continuous symmetric spaces G/K such as $\mathbb{R}^n$, the spherical functions are the K-invariant eigenfunctions of the Laplacian (normalized to be 1 at the origin). Because the eigenfunctions of the adjacency operators of the finite euclidean graphs are functions of the distance defined by (1) of the point to the origin, the contour plot can be obtained by computing the function $d(x, 0), x = (x_1, x_2)$, with $x_j \in \{1, 2, \ldots, p-1\}$. Then on a $p \times p$ grid we color the square at the point (x_1, x_2) according the value of $d(x, 0)$. The Mathematica command List Density Plot gives Figures I.21 and I.22. As q approaches infinity we seem to be seeing convergence to a pretty pattern of circles much like a Fresnel diffraction pattern. See Goetgheluck [1993]. The result does not appear to be chaotic at all.

This euclidean result should be compared with the analogous noneuclidean pictures in Part 2, which look very chaotic.

Answers to the Three Questions

An answer to Question 1 is given in Medrano, Myers, Stark, and Terras [1996]. Let us summarize the results here. First recall from Chapter 3 that the *trace* from $\mathbb{F}_q$ down to $\mathbb{F}_p$ where $q = p^r$ is given by

$$\text{Tr}(x) = x + x^p + \cdots + x^{p^{r-1}}, \quad \text{for } x \in \mathbb{F}_q.$$

Use the notation

$$e(x) = \exp\left(\frac{2\pi i \, \text{Tr}(x)}{p}\right), \quad \text{for } x \in \mathbb{F}_q. \tag{9}$$

We will write our vectors $x \in \mathbb{F}_q^n$ as column vectors and ${}^t x$ is the transpose of x. Then as in Theorem 2 of Chapter 3, for $b \in \mathbb{F}_q^n$, the exponentials

$$e_b(x) = \exp\left(\frac{2\pi i \, \text{Tr}({}^t x b)}{p}\right) \tag{10}$$

form a complete orthogonal set of eigenfunctions of the adjacency operator for the euclidean graph $E_q(n, a)$. The eigenvalue corresponding to e_b is

$$\lambda_b = \sum_{d(x,0)=a} e_b(x), \quad \text{for } b \in \mathbb{F}_q. \tag{11}$$

Exercise. Prove the preceding statements.

Hint. Recall the proof of the orthogonality relations of the characters of $\mathbb{Z}/n\mathbb{Z}$ in Chapter 2. The same proof works for any finite abelian group. Then recall the exercise after Theorem 2 in Chapter 3.

After some computation, we find in Medrano et al. [1996] that, for $b \neq 0$, the eigenvalues λ_b of formula (11) can be rewritten as generalized Kloosterman sums. See the exercise at the end of this chapter.

Definition. Suppose that κ is a character of the multiplicative group $\mathbb{F}_q^*$ and let $a, b \in \mathbb{F}_q^*$. Then the *generalized Kloosterman sum* is

$$K(\kappa \mid a, b) = \sum_{t \in \mathbb{F}_q^*} \kappa(t) e\left(-\left(at + \frac{b}{t}\right)\right). \tag{12}$$

As we mentioned in the last section, Kloosterman sums are finite analogues of Bessel functions. See the discussion around Equation (4) in Chapter 4. Kloosterman sums also occur as Fourier coefficients of the modular forms known as Poincaré series (see Sarnak [1990, Ch. 1]).

The eigenvalues (not equal to the degree) of the adjacency operator of $E_q(n, a)$ have the form

$$\lambda_{2c} = \frac{1}{q} G_1^n K(\varepsilon^n \mid a, d(c, 0)). \tag{13}$$

Here G_1 is a *Gauss sum* defined by

$$G_1 = \sum_{t \in \mathbb{F}_q} \varepsilon(t) e(t), \tag{14}$$

and ε is the quadratic character of $\mathbb{F}_q^*$:

$$\varepsilon(t) = \begin{cases} 1, & t \in \mathbb{F}_q^*, \, t = u^2, \quad \text{for some } u \in \mathbb{F}_q^*, \\ 0, & t = 0, \\ -1, & \text{otherwise.} \end{cases} \tag{15}$$

The Gauss sum is a finite analogue of the gamma function as you can see by recalling Euler's formula

$$\Gamma(s) = \int_0^\infty t^s \exp\{-t\} \frac{dt}{t},$$

which is the integral of a multiplicative character times an additive character. We will discuss Gauss sums more in Chapter 8. In particular, we will see why formula (14) and formula (11) of Chapter 2 can both be called Gauss sums. Also see the exercise below.

By formula (13) the eigenvalue λ_c is also a radial eigenfunction of the combinatorial Laplacian as a function of c. This may appear to be rather strange but it is a general phenomenon for symmetric spaces. We will say more about this in Chapter 10 and in Part 2. Anyway, once you believe this, it is no surprise that we have found finite analogues of Bessel functions to be the eigenfunctions of the Laplacian for our euclidean graphs. See Terras [1985, Vol. I, Ch. 2].

The Kloosterman sum (12) was bounded by A. Weil [1979, Vol. I, pp. 386–9] using his proof of the Riemann hypothesis for zeta functions of curves over finite fields. See Schmidt [1976] for a more elementary approach. This leads to a partial answer to Question 1 about our euclidean graphs.

Theorem 1. Finite Euclidean Graphs are Asymptotically Ramanujan. Let λ_b be an eigenvalue of the adjacency operator A of the graph $E_q(n, a)$ corresponding to the eigenfunction defined by (10). Then for $b \neq 0$, we have the bound

$$|\lambda_b| \leq 2q^{(n-1)/2}.$$

Proof. Use formula (13) and the result of Weil mentioned above. ∎

The degrees of these graphs for $a \neq 0$ are

$$|S_q(n, a)| = q^{n-1} + O\left(q^{(n-1)/2}\right).$$

The error may be 0, positive, or negative. It always is asymptotic to a lower power of q as $q \to \infty$. This says that the graphs are asymptotically Ramanujan as $q \to \infty$. But one can still ask whether the graphs are exactly Ramanujan. To answer this question, one needs the exact formula for the degree. When $a \neq 0$, it is

$$|S_q(n, a)| = q^{n-1} + \begin{cases} \varepsilon((-1)^{(n-1)/2}a)\, q^{(n-1)/2}, & \text{for } n \text{ odd,} \\ -\varepsilon((-1)^{n/2})\, q^{(n-2)/2}, & \text{for } n \text{ even.} \end{cases} \tag{16}$$

Here $\varepsilon(a)$ is the quadratic character of $\mathbb{F}_q$, which is 0 at 0, 1 on squares, and -1 on nonsquares as in (15).

For odd dimensions n, we can tell whether our graphs are Ramanujan easily, since the Kloosterman sums are essentially cosines. This is a result of Salié [1932]. A proof can be found in Carlitz [1953]. See the exercise below. So we find that *for odd dimensions n*, if $a \cdot d(b, 0) \neq 0$, with ε as in (15):

$$\lambda_{2b} = \begin{cases} 2G_1^{n-1}\varepsilon(d(b, 0)) \cos(4\pi \operatorname{Tr}(c)/p), & \text{if } a \cdot d(b, 0) = c^2, \\ 0, & \text{if } a \cdot d(b, 0) \text{ is not a square.} \end{cases} \tag{17}$$

If the dimension n is odd and $a \cdot d(b, 0) = 0$, with $b \neq 0$, we have

$$\lambda_{2b} = \begin{cases} q\varepsilon(-a), & \text{if } d(b, 0) = 0, \\ q\varepsilon(-d(b, 0)), & \text{if } a = 0, \ d(b, 0) \neq 0. \end{cases} \tag{18}$$

From formula (17), it is not hard to see that if $p \equiv 3 \pmod 4$, for $p > 158$, the graphs $E_p(3, 1)$ are *not Ramanujan*. Recalling formula (16), we see that the graphs $E_p(3, 1)$, for $p \equiv 1 \pmod 4$, are Ramanujan. It is interesting to note that the graphs $E_p(3, 1)$ give examples of similarly constructed graphs whose Ihara zeta functions either do or do not satisfy the Riemann hypothesis according as $p \equiv 1$ or $3 \pmod 4$. See Stark and Terras [1996].

For $E_p(2, 1)$, we have Ramanujan graphs if $p \equiv 3 \pmod 4$. But when $p = 17$ and 53, the graphs $E_p(2, 1)$ are not Ramanujan.

It follows from formula (17) that for odd dimensions n, the eigenvalues can indeed be 0. In even dimensions, however, Katz [1980] proves that the Kloosterman sums do not vanish.

In answer to Question 2 one needs to understand the distribution of the Kloosterman sums (12). We include histograms of these sums for the graphs $E_{1021}(2, 1)$ and $E_{1019}(3, 1)$ as Figure I.21. Note that the two histograms look very different. In fact, Katz [1988] proves that the distribution of Kloosterman sums $K(1 \mid a, 1)$ approaches the semicircle distribution given in formula (8), as $q \to \infty$. See also Adolphson [1989]. This is the distribution one expects for a large random graph (see McKay [1981]). However, McKay studies graphs of fixed degree. We will say more about this question for other graphs later.

Now we come to Question 3. What are the "level curves" of the eigenfunctions of the adjacency operators for our euclidean graphs? These are the sets $S_q(n, c)$ for fixed q and n, as c varies over $\mathbb{F}_q$. For $n = 2$, such sets can be viewed

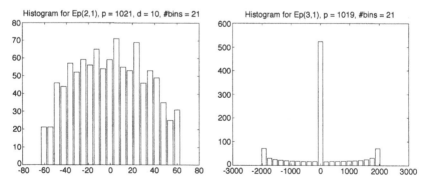

Figure I.21. Histogram of Kloosterman sums giving the spectrum for $E_{1021}(2, 1)$ on the left and $E_{1019}(3, 1)$ on the right. Figure drawn using Matlab.

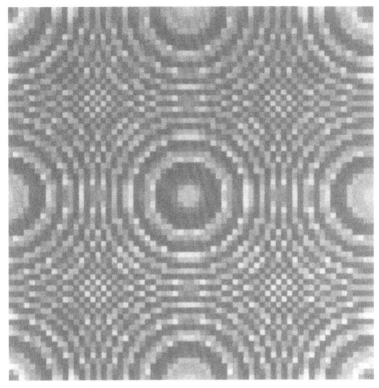

Figure I.22. A "radial" or K-invariant eigenfunction of the adjacency operator on a euclidean graph $E_q(n, a)$ has essentially the same level curves as the distance function $d(x, 0), x \in F_q^n$. Here we show a $q \times q$ grid with point x given a color determined by the value of the distance function $d(x, 0)$. Here $q = 67$.

as finite analogues of circles. Figures I.22 and I.23 give these curves in the case that $n = 2$ by associating a color or grey level to $c \in \mathbb{F}_q$, when q is a prime. The pictures look best in color and are obtained by a simple Mathematica command:

ListDensityPlot[Table[Mod[x^2 + y^2,p],{x,1,p},{y,1,p}],

Mesh-> False,ColorFunction-> Hue].

These pictures look very much like Fresnel diffraction patterns. See Goetgheluck [1993] for a discussion of similar figures.

Finally we should note that it is proved in Medrano et al. [1996] that given q and n, there are only two distinct graphs $E_q(n, a)$ for even n and at most three distinct graphs for odd n. This is in contrast to the finite upper half plane case in the next chapter when there seem to be $q - 2$ different graphs for each finite field $\mathbb{F}_q$.

Figure I.23. This is the analogue of Figure I.22 for $q = 127$.

A. Rudvalis noted that we should look at more general distances than (1) such as

$$d_c(x, y) = \sum_{j=1}^{n} c_j (x_j - y_j)^2 \quad \text{for } c_j \in \mathbb{F}_q.$$

With given n one can produce more Ramanujan graphs by varying c. Perla Myers [1995] has considered these more general euclidean graphs. The figures analogous to Figures I.22 and I.23 for $c = (1, -1)$ or $(1, 5)$ are quite interesting. Here the level "curves" are finite analogues of hyperbolas or ellipses. Myers also has some beautiful animations of these figures as the prime goes to infinity.

The theory of association schemes associated to finite symmetric spaces G/K is also of interest in the study of the finite euclidean space. See Bannai and Ito [1984], Bannai [1990], and Kwok [1992]. We do not say more about this here as this chapter is restricted to abelian groups.

Archie Medrano [1997] has considered super-euclidean graphs where the distance in (1) is replaced by

$$x_1^{2c} + \cdots + x_n^{2c}, \quad \text{for } c = 1, 2, 3, \ldots, p - 1.$$

The cases $2c = p - 1$ and $2c = p - 1/2$ are particularly interesting. It is also interesting to view analogues of Figures I.22 and I.23 for this super-euclidean distance.

Exercise.

a) Prove formula (13).
b) Prove formula (17).
c) Use the method of part a to obtain a formula for $|S_p(n, a)|$ as defined by formula (4).

Hints. See Medrano et al. [1996] and Carlitz [1953].

(a) Look at

$$B_r(b) = \sum_{x \in \mathbb{F}_q^n} e(2^t b \cdot x + r^t x \cdot x).$$

Show that for $b \neq 0$,

$$q \lambda_{2b} = \sum_{r \in \mathbb{F}_q} B_r(b) e(-ar) \quad \text{and} \quad B_r(b) = (\varepsilon(r) G_1)^n e\left(\frac{-^t b \cdot b}{r}\right),$$

where $\varepsilon(r)$ is defined by (15) and G_r is the *Gauss sum*

$$G_r = \sum_{y \in \mathbb{F}_q} e(ry^2) = \sum_{x \in \mathbb{F}_q} (1 + \varepsilon(x)) e(rx) = \varepsilon(r) G_1.$$

We will have more to say about Gauss sums in Chapter 8.

Next you need to know the evaluation of Gauss sums for $q = p^s$:

$$G_1 = \begin{cases} (-1)^{s-1} \sqrt{q}, & \text{if } p \equiv 1 \ (\text{mod } 4), \\ -(-i)^s \sqrt{q}, & \text{if } p \equiv 3 \ (\text{mod } 4). \end{cases}$$

For $s = 1$, this result is due to Gauss. We will sketch a proof of that case in Chapter 8. See also Ireland and Rosen [1982, p. 75]. The general case comes from a result of Davenport and Hasse. See Ireland and Rosen [1982, p. 162] or Small [1991, pp. 17, 123, 176]. See also Berndt and Evans [1981].

(b) For n odd, $d \neq 0$,

$$K(\varepsilon \mid a, d) = \varepsilon(d) K(\varepsilon \mid ad, 1).$$

Now show that

$$\sum_{x \in \mathbb{F}_q} e(bx^2 - 2x) = \varepsilon(b) G_1 e(-1/b).$$

So for $a \neq 0$, we have

$$G_1 \sum_{\substack{b \neq 0 \\ b \in \mathbb{F}_q}} \varepsilon(b) e(-ab - 1/b) = \sum_{x \in \mathbb{F}_q} e(-2x) \sum_{b \in \mathbb{F}_q} e(b(x^2 - a))$$

$$= \begin{cases} 0, & a \text{ not a square,} \\ q\{e(-2c) + e(2c)\}, & \text{for } a = c^2. \end{cases}$$

(c) Set $b = 0$ in the argument given in Part a.

Exercise. Consider what happens to the euclidean graphs if you replace $\mathbb{F}_q$ with $\mathbb{Z}/q\mathbb{Z}$.
Hint. See Medrano et al. [1996]. The case $q = 2^n$ can be found in Michelle De Deo [1998]. It is more complicated. Euclidean graphs over finite fields of characteristic two have not been studied up to this point.

Exercise. Arrange the eigenvalues (without multiplicity) of a finite upper half plane graph in increasing order $\lambda_1 \leq \lambda_2 \leq \cdots \leq \lambda_n$. Then construct a histogram of differences $\lambda_j - \lambda_{j-1}$. Does the graph look like e^{-x} (Poisson) as the number of vertices in the graph goes to infinity? This is really a question about Kloosterman sums.

Note. This is the *level spacing question* of Sarnak, though we should actually normalize the eigenvalues to have mean spacing 1. That is, we want the number of eigenvalues in a sufficiently small interval of length L to be approximately L. See Sarnak [1996]. The word "level" refers to the relationship between eigenvalues of differential operators and energy levels for physical systems. It has been found empirically for many systems of numbers arising in eigenvalue problems or as zeros of zeta functions that there are only two answers to the level spacing question. One answer is e^{-x}. The other is called the Gaudin distribution. We will not describe it here.

Chapter 6

Random Walks on Cayley Graphs

Anyone who considers arithmetical methods of producing
random digits is, of course, in a state of sin.

J. von Neumann

In the first part of this chapter we obtain limit theorems for the simplest random
walks on $\mathbb{Z}/n\mathbb{Z}$, for n odd, using the DFT and Markov chains. In the second part
we redo some of the first part, replacing Markov chains with sums of random
variables. We begin with the example of random number generators.

References for this chapter include: Fan Chung [1996], F. Chung, P. Diaconis,
and R. Graham [1987], P. Diaconis [1988], P. Diaconis and M. Shashahani
[1986], P. Diaconis and D. Stroock [1987], P. Doyle and J. L. Snell [1984],
W. Feller [1968], R. Guy [1984, Vol. 3, Section K45], J. G. Kemeny and
J. L. Snell [1960], W. LeVeque [1974, Vol. 3, Section K45], K. Rosen [1993,
Section 8.7], J. T. Sandefur [1990], J. L. Snell [1975], and M. Schroeder [1986,
Chapter 27]. See also Chapters 17 and 18 for more information on random
number generators.

Random Number Generators

There are many reasons why programs such as Mathematica and Matlab are
capable of giving us a random number at the drop of a hat or the push of a
key. Applications include computer simulations, sampling, testing of computer
algorithms, decision making, Monte Carlo methods for numerical integration,
and fault detection.

In the "good" old days, people obtained random numbers from tables. For example, Knuth [1981, p. 2] notes that "in 1955 the RAND Corporation
published a widely used table of a million random digits." With the advent of

computers, people sought better methods. Von Neumann suggested a method in 1946, which Knuth [1981, p. 3] notes "has actually proved to be a comparatively poor source of random numbers. The danger is that the sequence tends to get into a rut, a short cycle of repeating elements."

Again quoting Knuth [1981, p. 9]: "By far the most popular random number generators in use today are special cases of the following scheme, introduced by D. H. Lehmer [1951] in 1949." Start with a modulus m, a multiplier a, an increment b, and an initial value X_0. Create a sequence X_n, $n = 0, 1, 2, 3, \ldots$ of *pseudorandom numbers* recursively via

$$X_{n+1} \equiv aX_n + b(\mathrm{mod}\ m),$$

with $m = 2^{31} - 1$ or $m = 2^{32}$ being popular choices. A detailed discussion can be found in Knuth [1981]. Lehmer took $b = 0$.

Random Walks on the Finite Circle

The random number generator above with $a = 1$ and b chosen at random itself from ± 1, with equal probabilities of $+1$ and -1, can be viewed as a random walk on the finite circle. Consider a creature moving on a Cayley graph $X(\mathbb{Z}/n\mathbb{Z}, S)$ with $S = \{\pm 1(\mathrm{mod}\ n)\}$. The simplest case is that the creature moves from vertex x to $x + 1$ or $x - 1$ (mod n) with equal probability. This is an example of a random walk on $\mathbb{Z}/n\mathbb{Z}$ and it gives a Markov chain as explained below. There are applications not only to random number generators but also to the modeling of diffusion, genetics, economics, etc.

To be concrete let's assume the walker starts out at the origin at time 0. Then at time 1 it is either at 1 or -1 with equal probability. And at time 2 it has probability $1/4$ of being at 2 or -2 and probability $1/2$ of being back at 0 (always assuming n is sufficiently large that all the points are distinct.) We can set Mathematica to work on this problem using the concept of Markov chain – an idea introduced by Markov in 1907.

Let A be the adjacency matrix of the graph $X = X(\mathbb{Z}/n\mathbb{Z}, S)$. Let k be the degree of the graph, that is, $k = |S|$. We get a *Markov chain* from A as follows: We have a set of states (the vertices of our graph). At time t, the process (walker) goes from ith state to jth state with probability p_{ij} given by $1/k$ if vertex i is adjacent to vertex j and 0 otherwise. The *Markov transition matrix* is

$$T = (p_{ij})_{1 \le i, j \le n} = \frac{1}{k}A, \tag{1}$$

where A is the adjacency matrix of the graph. We may also view A and T as operators on functions on X.

Suppose that $p^{(t)}(x)$ is a *probability density* on $x \in X$, at time t, that is,

$$p^{(t)}(x) \geq 0 \quad \text{for all } x \in X, \ t \geq 0 \text{ and } \sum_{x \in X} p^{(t)}(x) = 1. \tag{2}$$

Here $p^{(t)}(x)$ represents the probability that our random walker is at vertex x of the graph at time t. So we have

$$p^{(t+1)}(x) = T p^{(t)}(x) = T^{t+1} p^{(0)}(x). \tag{3}$$

Example. The Random Walk on the Circle. Take $m = 11$. Even moduli are bad, as we explain later. The Markov transition matrix T is

$$T = \frac{1}{2} \begin{pmatrix} 0 & 1 & 0 & 0 & 0 & 0 & 0 & 0 & 0 & 0 & 1 \\ 1 & 0 & 1 & 0 & 0 & 0 & 0 & 0 & 0 & 0 & 0 \\ 0 & 1 & 0 & 1 & 0 & 0 & 0 & 0 & 0 & 0 & 0 \\ 0 & 0 & 1 & 0 & 1 & 0 & 0 & 0 & 0 & 0 & 0 \\ 0 & 0 & 0 & 1 & 0 & 1 & 0 & 0 & 0 & 0 & 0 \\ 0 & 0 & 0 & 0 & 1 & 0 & 1 & 0 & 0 & 0 & 0 \\ 0 & 0 & 0 & 0 & 0 & 1 & 0 & 1 & 0 & 0 & 0 \\ 0 & 0 & 0 & 0 & 0 & 0 & 1 & 0 & 1 & 0 & 0 \\ 0 & 0 & 0 & 0 & 0 & 0 & 0 & 1 & 0 & 1 & 0 \\ 0 & 0 & 0 & 0 & 0 & 0 & 0 & 0 & 1 & 0 & 1 \\ 1 & 0 & 0 & 0 & 0 & 0 & 0 & 0 & 0 & 1 & 0 \end{pmatrix}$$

and we take

$$p^{(0)} = \begin{pmatrix} 1 \\ 0 \\ 0 \\ 0 \\ 0 \\ 0 \\ 0 \\ 0 \\ 0 \\ 0 \\ 0 \end{pmatrix}.$$

So we obtain the following table of vectors $p^{(t)} = {}^t(v_0, v_1, v_2, \ldots, v_{10})$, $t = 1, 2, \ldots$ (see Table I.11).

Table I.11. *Probability vectors for simple random walk on the finite circle at times* $t = 0, 1, 2, 3, 4, 5$

t	v_0	v_1	v_2	v_3	v_4	v_5	v_6	v_7	v_8	v_9	v_{10}
0	1	0	0	0	0	0	0	0	0	0	0
1	0	0.5	0	0	0	0	0	0	0	0	0.5
2	0.5	0	0.25	0	0	0	0	0	0	0.25	0
3	0	0.375	0	0.125	0	0	0	0	0.125	0	0.375
4	0.375	0	0.25	0	0.0625	0	0	0.0625	0	0.25	0
5	0	0.3125	0	0.15625	0	0.03125	0.03125	0	0.15625	0	0.3125

Looking at Table I.11 is not too enlightening. It is also tedious to compute. In fact, we should make Mathematica do the work and plot the points as in Figure I.24. The plots are ListPlots of the vectors at $t = 10$, 50, and 100. The fact that the probability distribution looks uniform at $t = 100$ fits in with Theorem 2 below. See Example 1 after that theorem, which says it should take about 11^2 steps for the probability distribution to look uniform.

Exercise. Produce a figure similar to Figure I.24 when 11 is replaced by 10. Then change the random walk to that which either stays at x or goes from x to $x \pm 1$ – each with probability $1/3$ – and make a figure of probability distributions at various times for this second walk, as well.

Questions About Random Walks

- Question 1. After how many steps is the probability distribution of the creature close to random?
- Question 2. How many steps must we take before we can be sure that the walker hits every vertex with some positive probability?

These are the sort of questions considered by Diaconis [1988]. We will look only at Question 1. An answer is given for $X(\mathbb{Z}/n\mathbb{Z}, \{\pm 1\})$, n odd, in Example 1 after Theorem 2 below.

As we said above, these questions are related to random number generators based on the formula

$$X_{j+1} \equiv aX_j + b \,(\text{mod } n).$$

The sequence X_j is not actually random, although it can have many properties of

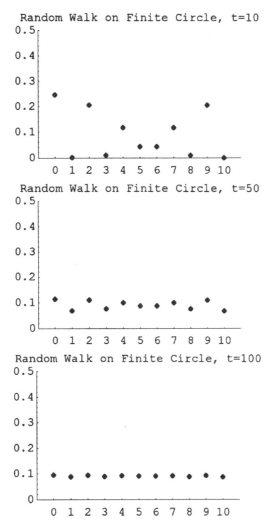

Figure I.24. Plots of the probability distributions of a simple random walk on the finite circle $\mathbb{Z}/11\mathbb{Z}$ at times $t = 10, 50$, and 100.

a random sequence. To increase randomness, one can shuffle several generators. For example, if a_j and b_j come from some other random number generator, one can form

$$X_{j+1} \equiv a_j X_j + b_j \pmod{n}.$$

If we hold $a_j \equiv 1 \pmod{n}$ and let b_j take values $\pm 1 \pmod{n}$, each with probability $1/2$, we have the original random walk above. If we take a fixed value of

$a \equiv a_j \not\equiv 1 \pmod{n}$, we can also analyze the process using the DFT on $\mathbb{Z}/n\mathbb{Z}$. For $a = 2$ and b chosen from $0, \pm 1$, each with probability $1/3$, Chung, Diaconis, and Graham [1987] show that it takes

$$c \log n \, \log \log n \text{ steps}$$

for this walk to become random. They also show that for almost all odd n, $1.02 \log_2 n$ steps suffice. See Chapter 19 for more information on such random number generators.

Next let's suppose we know that our Cayley graph $X(\mathbb{Z}/n\mathbb{Z}, S)$ is *not bipartite*. This means that the spectrum $(T) = (1/k)$ spectrum (A) is

$$\lambda_1 = 1 > \lambda_2 \geq \cdots \geq \lambda_n > -1. \tag{4}$$

Since X is connected, the eigenvalue $\lambda_1 = 1$ has multiplicity 1. The graph X is bipartite iff -1 is an eigenvalue. See Chapter 3. In the following theorem we show that as time goes to infinity the probability distribution for the random walker approaches the uniform distribution meaning that the walker has essentially equal probability of being anywhere on the graph, after sufficient time.

Theorem 1. Random Walks Approach the Uniform Distribution. Suppose X is a connected nonbipartite regular graph of degree k with n vertices. Let A be the adjacency operator of X and $T = (1/k)A$, the Markov transition operator. Then for every probability function on X, that is, $p(x) \geq 0$, for all $x \in X$ and $\sum_{x \in X} p(x) = 1$, we have

$$\lim_{t \to \infty} T^t p = u,$$

where the *uniform distribution* is

$$u(x) = \frac{1}{n}, \quad \text{for all } x \in X.$$

Proof. By the spectral theorem for T, there is a complete orthonormal set $\phi_1, \ldots, \phi_n$ in $L^2(X)$ consisting of eigenfunctions of T, that is,

$$T\phi_j = \lambda_j \phi_j, \quad j = 1, \ldots, n, \qquad \langle \phi_i, \phi_j \rangle = \begin{cases} 1, & i = j, \\ 0, & \text{otherwise.} \end{cases}$$

Then any $f \in L^2(X)$ has the Fourier expansion

$$f(x) = \sum_{i=1}^{n} \langle f, \phi_i \rangle \phi_i(x), \qquad Tf(x) = \sum_{i=1}^{n} \langle f, \phi_i \rangle \lambda_i \, \phi_i(x).$$

It follows that for any probability function p, we have

$$T^t p(x) = \sum_{i=1}^{n} \langle p, \phi_i \rangle \lambda_i^t \phi_i(x). \tag{5}$$

Now make use of (4). Since $\lambda_1 = 1$ and $|\lambda_j| < 1, \; j \neq 1$, we see that

$$\lim_{t \to \infty} \lambda_i^t = \begin{cases} 0, & i \neq 1, \\ 1, & i = 1. \end{cases}$$

So only one term remains in the sum (5) after we send t to infinity. Therefore

$$\lim_{t \to \infty} T^t p = \langle p, \phi_1 \rangle \phi_1 = \left\langle p, \frac{1}{\sqrt{n}} \right\rangle \frac{1}{\sqrt{n}} = \frac{1}{n} \langle p, 1 \rangle = \frac{1}{n} = u.$$

Here we have used the definition of ϕ_1 and the fact that p is a probability function. ∎

Exercise. Suppose that X is a graph satisfying the hypotheses of the preceding theorem and suppose A and T are as in the theorem. Show that

$$\lim_{m \to \infty} T^m = \frac{1}{n} \begin{pmatrix} 1 & 1 & 1 & \cdots & 1 \\ 1 & 1 & 1 & \cdots & 1 \\ \vdots & \vdots & \vdots & \cdots & \vdots \\ 1 & 1 & 1 & \cdots & 1 \\ 1 & 1 & 1 & \cdots & 1 \end{pmatrix}.$$

It follows that this limit theorem holds for $X = X(\mathbb{Z}/n\mathbb{Z}, \{\pm 1\})$, for n odd. What happens for n even?

Now we want to use our knowledge of the spectra of Cayley graphs to obtain more precise information about the convergence of the random walk to the uniform distribution. In particular we want to answer Question 1 above:

Question 1. After how many steps is the probability distribution of the walker close to random? To make this question precise, we need to know what we mean by "close." It is known that all norms on $\mathbb{R}^n$ are equivalent, that is, there is a constant $c > 0$ independent of x so that for two norms $\|x\|_1, \|x\|_2$

$$\frac{1}{c}\|x\|_1 \leq \|x\|_2 \leq c\|x\|_1, \quad \text{for all } x \in X$$

(see Lang [1983, p. 123]). However, the constant c would usually involve the order of G and could be humongous. Thus the random walker would certainly want to know which norm was being used.

Exercise. Comparison of $\|x\|_1, \|x\|_2 \|x\|_\infty$ *Norms on* $\mathbb{R}^n$. Let X be our graph, and let f, g be two real-valued functions on X. Define

$$\|f\|_2 = \langle f, f \rangle^{1/2}, \quad \text{where } \langle f, g \rangle = \sum_{x \in X} f(x)g(x), \quad \|f\|_1 = \sum_{x \in X} |f(x)|, \quad (6)$$
$$\|f\|_\infty = \max\{|f(x)| \mid x \in X\}.$$

a) Show that

$$\|f\|_2 \leq \|f\|_1 \leq |X|^{1/2}\|f\|_2. \tag{7}$$

b) Show that

$$\|f\|_\infty \leq \|f\|_2 \leq |X|^{1/2}\|f\|_\infty \quad \text{and} \quad \|f\|_\infty \leq \|f\|_1 \leq |X|\,\|f\|_\infty. \tag{8}$$

Hints. a) The proof uses the Cauchy–Schwarz inequality:

$$|\langle f, g \rangle| \leq \|f\|_2 \cdot \|g\|_2. \tag{9}$$

Also, you can assume f is not identically 0. Then you can divide f by $\|f\|_1$.

Remarks. $(1/2)\|f - g\|_1$ is what Diaconis calls *the variation distance* between f and g. See Diaconis [1988, especially pp. 21–3]. For example, on p. 25, Diaconis gives an argument to show that the L^1-norm is preferable to the L^2-norm. The latter can be very small for large groups, for example, for a $p(x)$ uniform on half the group and 0 on the rest of the group. So we will use the L^1-norm in the next theorem.

Theorem 2. A More Precise Result on Convergence of the Random Walk to the Uniform Distribution. Suppose we have a connected nonbipartite k-regular graph X with n vertices. Let the Markov transition operator on X be $T = (1/k)A$, where $A =$ the adjacency operator. Define

$$\beta = \max\{|\lambda| \mid \lambda \in \text{spectrum } (T), \lambda \neq 1\} < 1.$$

Then for every probability function p on X, that is, $p(x) \geq 0$, all $x \in X$ and $\sum_{x \in X} p(x) = 1$, we have

$$\|T^m p - u\|_1 \leq \sqrt{n} \, \beta^m.$$

Here $u(x)$ is the uniform probability distribution, that is,

$$u(x) = \frac{1}{n} \quad \text{for all } x \in X.$$

Corollary. If the graph in Theorem 2 is Ramanujan, that is, $\beta \leq \frac{2}{k}\sqrt{k-1}$, then

$$\|T^m p - u\|_1 \leq \sqrt{n} \left(\frac{2}{k}\sqrt{k-1} \right)^m.$$

Proof of Theorem. We saw in the exercise above that

$$\|T^m p - u\|_1 \leq \sqrt{n} \, \|T^m p - u\|_2.$$

Once more we use the existence of a complete orthonormal set of eigenfunctions of T, which we call $\phi_1, \ldots, \phi_n$. Then from Equation (5) above we get

$$\|T^m p - u\|_2^2 = \left\| \sum_{j=2}^{n} \langle p, \phi_j \rangle \lambda_j^m \phi_j \right\|_2^2 = \sum_{j=2}^{n} |\langle p, \phi_j \rangle|^2 \, |\lambda_j|^{2m}.$$

To see that u cancels with the first term of the sum, use the following facts:

$$\lambda_1 = 1, \quad \phi_1(x) = n^{-1/2}, \quad \langle p, 1 \rangle = 1, \quad u(x) = \frac{1}{n}, \quad \text{for all } x \in X.$$

Now, by the definition of β, we have

$$\|T^m p - u\|_2^2 \leq \beta^{2m} \sum_{j=2}^{n} |\langle p, \phi_j \rangle|^2 \leq \beta^{2m} \sum_{j=1}^{n} |\langle p, \phi_j \rangle|^2$$
$$= \beta^{2m} \|p\|_2^2.$$

Here we used *Parseval's equality* in a normed vector space, which says:

$$\sum_{j=1}^{n} |\langle p, \phi_j \rangle|^2 = \|p\|_2^2.$$

But now

$$\|p\|_2^2 = \sum_{x \in X} |p(x)|^2 \le \sum_{x \in X} |p(x)| = 1.$$

Our result follows from putting all our inequalities together. ∎

Example 1. Finite Circle Graph

$X(\mathbb{Z}/n\mathbb{Z}, \{\pm 1\})$, where n is odd (to prevent the graph from being bipartite).

Exercise. Consider the finite circle graph $X(\mathbb{Z}/n\mathbb{Z}, \{\pm 1\})$, where n is odd.

a) Show that for this graph the value of β in Theorem 2 is given by $\beta = |\cos(\frac{\pi}{n})|$.
b) Use the inequality

$$\cos x \le \exp\left(\frac{-x^2}{2}\right), \quad \text{for } 0 \le x \le \frac{\pi}{2},$$

to see that

$$\beta \le \exp\left(\frac{-\pi^2}{2n^2}\right).$$

c) Conclude that, for the Markov transition matrix T of the simple random walk on $X(\mathbb{Z}/n\mathbb{Z}, \{\pm 1\})$,

$$\|T^m p - u\|_1 \le \sqrt{n}\, \exp\left(\frac{-m\pi^2}{2n^2}\right).$$

The factor of $\sqrt{n}$ in front of the exponential can be eliminated using Lemma 2 (the upper bound lemma) at the end of this chapter.

Hint. See Diaconis [1988, p. 26]. Recall that the eigenvalues of the Markov matrix T are $\cos(\frac{2\pi a}{n})$, $a = 0, 1, \ldots, n-1$. For odd n, we get a maximum absolute value at $a = (n-1)/2$.

Exercise. Assume that $X = X(\mathbb{Z}/n\mathbb{Z}, \{\pm 1\})$, with n even.

a) What happens to $T^m p$ as $m \to \infty$, if n is even?
b) Can you make $T^m p$ converge to a uniform distribution by replacing $S = \{\pm 1\}$ with $S = \{0, \pm 1\}$, that is, by allowing the graph to have loops. This means that the walker at x has $1/3$ probability each of remaining in place, moving to $x + 1$, or moving to $x - 1$.

Answer.

a) The random walk is periodic. There is no limit. For -1 is an eigenvalue of T.

Example 2. Winnie Li's Graph

Consider $X = X(\mathbb{F}_{p^2}, \text{ker norm})$. These graphs are Ramanujan with $n = p^2$, $k = p + 1$, $\beta = 2\sqrt{p}/(p+1)$. So for every probability function v on X, we have

$$\|T^m v - u\|_1 \leq \sqrt{n}\,\beta^m = p\left(\frac{2\sqrt{p}}{p+1}\right)^m.$$

Therefore, if we take $m = 3$, we find that

$$\|T^3 v - u\|_1 \leq \frac{8}{\sqrt{p+1}}.$$

For large primes p, this is quite small. The random walk looks uniform after three steps. That is pretty darn fast. Of course, the degree of the graph is increasing with p. One might prefer Ramanujan graphs with fixed degree as the number of vertices goes to infinity, such as those of Lubotzky, Phillips, and Sarnak [1988].

Example 3. Finite Euclidean Graphs

Exercise. Imitate the last example to see how many steps are required for a random walker to be lost on the finite euclidean graphs of the last chapter.

Example 4. The Complete Graph

Exercise. Consider the complete graph on n vertices, that is, connect each vertex to every other vertex.

a) Is it Ramanujan?
b) How large should m be for $\|T^m - u\|_1 < \sqrt{n}/(n-1)$?

Answers. (a) Yes. (b) $m = 1$.

Moral. After only one step, the random walker is essentially lost (even though it cannot possibly be back at its origin in one step as there are no loops)! Of course, this would not be a good graph if you want to minimize the number of connections.

Exercise on the Variation Distance.

a) Suppose that p and q are probabilities on a graph X. Show that, for any $V \subset X$, if we define

$$p(V) = \sum_{x \in V} p(x),$$

we have

$$\frac{1}{2} \| p - q \|_1 = \max_{V \subset X} |p(V) - q(V)|.$$

b) If $f : X \to \mathbb{R}$, define the *expected value* of f under p to be

$$E_p(f) = \langle f, p \rangle = \sum_{x \in X} f(x) p(x).$$

Prove that

$$\frac{1}{2} \| p - q \|_1$$
$$= \frac{1}{2} \max \{ |E_p(f) - E_q(f)| \mid f : X \to \mathbb{R} \text{ with } \| f \|_\infty = 1 \}.$$

Hints. a) Look at the set $W \subset X$ on which the maximum value of $|p(V) - q(V)|$ occurs. We can assume that $p(x) \geq q(x)$ for all $x \in W$. Also

$$W = \{ x \in X \mid p(x) \geq q(x) \}.$$

Use the fact that

$$\sum_{x \in X} p(x) = 1 = \sum_{x \in X} q(x).$$

b) To maximize $|E_p(f) - E_q(f)|$, take f to be 1 on the set W from the preceding hint and -1 on $X - W$.

Exercise.

a) Kemeny and Snell [1960] call a Markov chain *regular* if T^m has no nonzero entries for some m. Show that if our k-regular graph is connected and non-bipartite, the corresponding Markov chain is regular.
b) The value of m obtained in part a provides an answer to Question 2 above. This was the question: How many steps must we take before we can be sure that the walker hits every vertex with some positive probability? Can we use Chung's theorem (which was Theorem 1 of Chapter 4) to find such an m?
c) Consider the example of the Markov chain for the Cayley graph $X(\mathbb{Z}/n\mathbb{Z}, \{\pm 1\})$. Find a value of m such that T^m has no nonzero entries, assuming n is *odd*.

In the next paragraphs we give an alternate discussion of Theorem 1 which replaces the notion of Markov chain with that of sum or product of random variables. We will see that the discrete Fourier transform gives a quick proof of the desired limit theorem.

Limits of Sums of Random Variables

Let G be a finite abelian group such as $\mathbb{Z}/n\mathbb{Z}$ (with operation denoted by $+$). Suppose that $p : G \to [0, 1]$ such that $\sum_{x \in G} p(x) = 1$. Then we say that p is a *probability* (distribution). An example is the *uniform distribution* $p_U(x) = u(x) = 1/|X|$, for all $x \in G$. A *random variable* with values in X is a (measurable) function $X : S \to G$, where S is some sample space (a measure space with a measure P such that $P(S) = 1$). We have a *density* or *distribution* associated to X via

$$p_X(r) = P\{\omega \in S \mid X(\omega) = r\}, \quad \text{for } r \in G,$$

where P is a probability measure on S. We can forget about our sample space S and the measure P if we like and consider just the probability distribution $p = p_X$. There are many examples in Snell [1975], for example.

Given two random variables X and Y on G, the *joint distribution* of X and Y is

$$p_{X \times Y}(r, s) = P\{\omega \in S \mid X(\omega) = r \text{ and } Y(\omega) = s\}.$$

We say that X and Y are *independent random variables* if

$$p_{X \times Y}(r, s) = p_X(r) \cdot p_Y(s), \quad \text{for all } r, s \in G.$$

Now suppose that two random variables X and Y on G are independent. The *sum* of X and Y is defined to be $X + Y$ with

$$p_{X+Y}(r) = P\{\omega \in S \mid X(\omega) + Y(\omega) = r\}.$$

Here $X(\omega) + Y(\omega)$ denotes the sum in the group G.

Now after all these definitions we can give the probabilist's interpretation of convolution. The probability distribution associated to the sum of two independent random variables is the *convolution* of the probability distributions, that is,

$$p_{X+Y}(r) = P\{\omega \in S \mid X(\omega) + Y(\omega) = r\} = \sum_{\substack{u,v \in G \\ u+v=r}} p_X(u) p_Y(v)$$

$$= \sum_{u \in G} p_X(u) p_Y(r - u) = (p_X * p_Y)(r).$$

So we have proved the following theorem.

Theorem 3. Probabilistic Meaning of Convolution. The distribution of the sum of *independent* random variables X and Y on G is the convolution of p_X and p_Y, that is,

$$p_{X+Y}(r) = (p_X * p_Y)(r).$$

One of the *central* questions in probability and statistics is the following: Suppose that we have a sequence $\{X_v\}_{v \geq 1}$ of independent random variables on G, each having the same probability distribution $p(r) = p_{X_v}(r), r \in G$. What is the limiting distribution

$$\lim_{m \to \infty} (X_1 + \cdots + X_m)? \tag{10}$$

By Theorem 3 this translates immediately into the question: What is

$$\lim_{m \to \infty} \overbrace{(p * \cdots * p)}^{m}? \tag{11}$$

Taking the discrete Fourier transform (^ or DFT) on the finite abelian group G translates this problem to

$$\lim_{m \to \infty} (\hat{p})^m = ? \tag{12}$$

Here $\hat{p}$ is the DFT of p.

Limits (11) and (12) are equivalent by the following lemma.

Lemma 1. Suppose that q_m is a sequence of functions on G. Then we have the following continuity result for the discrete Fourier transform^:

$$\lim_{m \to \infty} \hat{q}_m(x) = \hat{t}(x) \quad \text{for each } x \in G \quad \text{iff}$$

$$\lim_{m \to \infty} q_m(x) = t(x) \quad \text{for each } x \in G.$$

Proof. Note that because our set G is finite, pointwise convergence of functions is the same as $\|\;\|_\infty$-convergence (uniform convergence). By an exercise above, we know that convergence in $\|\;\|_\infty$-norm is equivalent to convergence in $\|\;\|_2$-norm on the finite-dimensional vector space $L^2(G)$. We know that the DFT is an $\|\;\|_2$-norm isometry. This means that $\|\;\|_2$-convergence of a sequence of functions is equivalent to $\|\;\|_2$-convergence of the sequence of Fourier transforms. ∎

Example. The Finite Circle Again. Consider the random walk on the additive group $G = \mathbb{Z}/n\mathbb{Z}$ with

$$p(x) = p_{x_v}(x) = \begin{cases} 1/2, & \text{if } x \equiv \pm 1 (\text{mod } n), \\ 0, & \text{otherwise.} \end{cases}$$

We want to find the limit (10):

$$\lim_{m \to \infty} (X_1 + \cdots + X_m).$$

This translates to finding the limit (12):

$$\lim_{m \to \infty} (\hat{p})^m.$$

We know

$$(\hat{p})(r) = \frac{1}{2}(\hat{\delta}_1 + \hat{\delta}_{-1})(r) = \cos(2\pi r/n).$$

Thus we need to know about

$$\lim_{m \to \infty} (\hat{p})^m (r) = \lim_{m \to \infty} (\cos(2\pi r/n))^m.$$

When n is even, the limit does not exist if $r = n/2$, as it is the limit of $(-1)^m$ as $m \to \infty$. So (as before) we must *assume n is odd*. For n odd, we have

$$\lim_{m \to \infty} (\cos(2\pi r/n))^m = \begin{cases} 1, & r \equiv 0 (\text{mod } n), \\ 0, & \text{otherwise.} \end{cases}$$
$$= \delta_0(r) = \hat{p}_U(r),$$

where $p_U(r) = u(x) = 1/n$, for all $r \in \mathbb{Z}/n\mathbb{Z}$, that is, $p_U = u$ is the uniform probability distribution.

In summary, we have the following proposition:

Proposition. Limit of Sums of Random Variables on the Finite Circle. Suppose n is odd. Define $p(x) = 1/2$, for $x \equiv \pm 1 \pmod{n}$, and $p(x) = 0$, otherwise. Then for every $x \in \mathbb{Z}/n\mathbb{Z}$, we have

$$\lim_{m \to \infty} \overbrace{(p * \cdots * p)}^{m} = p_U(x) = u(x) = \frac{1}{n}.$$

This says that if X_ν, $\nu = 1, 2, \ldots$ is a sequence of independent random variables with the same probability distribution $p(x)$, then the sum $X_1 + \cdots + X_m$ approaches the uniform distribution as $m \to \infty$.

Remark. See Dvoretzky and Wolfowitz [1951] for a similar result for more general random variables on $\mathbb{Z}/n\mathbb{Z}$. Diaconis [1988] gives many applications of such limit theorems, for example, to random number generators, card shuffling, ...

Gerl [1981] views such limit theorems as *central limit theorems.* See Terras [1985] for the classical central limit theorem. Kemeny and Snell [1960] would seek a result closer to the classical central limit theorem.

Next we revisit random number generators and the paper of Chung, Diaconis, and Graham [1987].

Lemma 2. The Upper Bound Lemma. (See Chung, Diaconis, and Graham [1987, p. 1150.] Suppose that p is a probability density on the Cayley graph $X(\mathbb{Z}/n\mathbb{Z}, S)$. Let u be the uniform density. Then if $\hat{p}(a)$ denotes the Fourier transform of p on $\mathbb{Z}/n\mathbb{Z}$, we have

$$\|p - u\|_1^2 \leq \sum_{a=1}^{n-1} |\hat{p}(a)|^2.$$

Proof. By the definition of $\| \ \|_1$, the Cauchy–Schwarz inequality, and the Plancherel theorem for the DFT (in Chapter 2), we set

$$\|p - u\|_1^2 = \left(\sum_{x=0}^{n-1} |p(x) - u(x)| \right)^2 \leq n \sum_{x=0}^{n-1} |p(x) - u(x)|^2$$

$$= \sum_{a=0}^{n-1} |\hat{p}(a) - \hat{u}(a)|^2.$$

But $\hat{u}(a) = 0$ unless $a \equiv 0 \pmod{n}$. And $\hat{u}(0) = \hat{p}(0) = 1$. Thus the last sum is $\sum_{a=1}^{n-1} |\hat{p}(a)|^2$ and the lemma is proved. ∎

Exercise. Use Lemma 2 and similar ideas to those needed in our earlier exercise on the finite circle graph to show that if n = prime and $S = \{\pm 1, 0\}$, defining $p = \frac{1}{3}\delta_s$ and

$$p^{*k} = p * \underbrace{\cdots}_{k} * p,$$

we have

$$\left\| p^{(k)} - u \right\|_1 \leq e^{-\beta k/n^2}.$$

This is a better result by a factor of $\sqrt{n}$.

Hint. See Chung, Diaconis, and Graham [1987], who also find a similar lower bound. This means that it takes approximately cn^2 steps for p^{*k} to converge to uniform. Therefore the random number generator

$$X_{k+1} \equiv a_k X_k + b_k \pmod{n} \text{ with } a_k = 1 \text{ and } b_k \in \{\pm 1, 0\}$$
$$\text{with probability } 1/3$$

does not converge very fast to uniform. Surprisingly there is a big improvement if $a_k = 1$ is replaced by $a_k = 2$. Then Chung, Diaconis, and Graham [1987] show that it takes $c \log n \log\log n$ steps.

Diaconis and Saloff-Coste [1995] find that if we set $a = 1$, $S_1 = \{\pm 1, \pm 2\}$, and $S_2 = \{\pm 1, \pm \lfloor \sqrt{n} \rfloor\}$, then the random number generator associated with S_1 takes cn^2 steps whereas that associated with S_2 takes only cn steps to uniform (assuming that n is odd). Greenhalgh [1987] and Hildebrand [1994] show that for most sets S of order s in $\mathbb{Z}/n\mathbb{Z}$ with s fixed and n a large prime, $cn^{2/(s-1)}$ steps are necessary and sufficient to reach uniform.

Diaconis and Saloff–Coste [1994, 1995] also note that the above behavior is typical of nilpotent groups like $\mathbb{Z}/n\mathbb{Z}$ where one expects to take d^2 steps to uniform, where d is the diameter of the Cayley graph $X(\mathbb{Z}/n\mathbb{Z}, S)$. See Chapter 18 for more information on nonabelian nilpotent groups.

Hildebrand [1993] looks at random number generators of the form

$$X_{k+1} \equiv a_k X_k + b_k \pmod{n},$$

where n is a prime under certain restrictions, and finds that $c (\log n)^2$ steps suffice unless $a_j = 1$ always or $b_j = 0$ always or both a_j and b_j can take on only one value. The restrictions on n are a bit complicated. For example, n must be relatively prime to the elements of the sets S_1 and S_2 from which the a_k and b_k are taken, respectively, as well as differences $s_1 - s_2$ of elements. There is a further condition involving the Fourier transform of $p_2 = \frac{\delta S_2}{|S_2|}$.

Chapter 7

Applications in Geometry and Analysis. Connections between Continuous and Finite Problems. Dido's Problem for Polygons

The diving bell of your head descends.
You cut the murk and peer at luminous razorthin creatures
   who peer back
creatures with eyes and ears sticking out of their backsides
lit up like skyscrapers or planes taking off.
M. Piercy [1990, p. 17] (from "Concerning the Mathematician")

Connections between Continuous and Finite Problems

Recall that in Chapter 2 we said that the DFT on $\mathbb{Z}/n\mathbb{Z}$ is an approximation to the usual Fourier transform on the real line $\mathbb{R}$. In this chapter, our aim is to say a little more about the connection between finite and continuous analysis. This is generally treated in books on finite difference methods to solve partial (and ordinary) differential equations. See, for example, Cvetković et al. [1980, Section 8.4], Forsythe and Wasaw [1960], Garabedian [1964, pp. 475 ff.], Greenspan [1973, 1974], Haberman [1983], Powers [1987], Strang [1976, Section 6.5], and Strikwerda [1989]. Another reference is Fan Chung [1996].

Spectra of Laplacians

Let's just consider the simplest possible case. How does the combinatorial Laplacian on the circle graph $X(\mathbb{Z}/n\mathbb{Z}, \{\pm 1\})$ approximate the operator d^2/dx^2 on the real line? Let $f : \mathbb{R} \to \mathbb{C}$. Suppose we know $f(p_j) = f(j)$ at n equally spaced points p_j, $j = 0, 1, 2, \ldots, n-1$. Let the spacing be $\delta = L/n$. The first difference operator

$$D_1 f(j) = \frac{f(j+1) - f(j)}{\delta} \quad \text{approaches } f'(p_j) \text{ as } \delta \text{ approaches } 0. \quad (1)$$

And

$$D_2 f(j) = D_1 \left(\frac{f(j+1) - f(j)}{\delta} \right)$$

$$= \frac{f(j+2) - f(j+1) - f(j+1) + f(j)}{\delta^2}.$$

Thus

$$D_2 f(j) = \frac{f(j+2) + f(j) - 2f(j+1)}{\delta^2}$$

$$= \delta^{-2}(A - 2I)f(j+1) = \delta^{-2}\Delta_c, \qquad (2)$$

where A is the adjacency operator of the graph consisting of the n points in a line and Δ is the combinatorial Laplacian

$$\Delta_c = A - 2I.$$

If we assume that the function satisfies periodic boundary conditions, A is the adjacency operator for the finite circle graph $X(\mathbb{Z}/n\mathbb{Z}, \{\pm 1 (\mathrm{mod}\, n)\})$. As δ approaches 0, $D_2 f(j)$ approaches $f''(p_j)$. Thus the finite difference operator approaches the second derivative.

So we are interested in the operator $n^2(A - 2I) = n^2 \Delta_c$, where I is the identity operator, A is the adjacency operator of the circle graph, and Δ_c is the combinatorial Laplacian. Then $n^2 \Delta_c$ is a difference operator approximating the second derivative operator on the continuous circle.

The eigenvalues of $n^2 \Delta_c$ are, for $a \in \mathbb{Z}/n\mathbb{Z}$,

$$n^2 \left\{ 2\cos \left(\frac{2\pi a}{n} \right) - 2 \right\} = -4n^2 \sin^2 \left(\frac{\pi a}{n} \right)$$

$$= -4\pi^2 a^2 + \text{terms involving powers of } \frac{a}{n}. \qquad (3)$$

And the operator d^2/dx^2 on $\mathbb{R}/\mathbb{Z}$ has as eigenfunctions $\exp(2\pi i a x)$, with eigenvalues $-4\pi^2 a^2$, for $a \in \mathbb{Z}$. Thus the eigenvalues λ_a of $n^2 \Delta_c$, with $|a/n|$ small, do approach those of d^2/dx^2 as $n \to \infty$.

Turing [1938] says that one should perhaps only expect such things for compact abelian groups like the circle group $\mathbb{R}/\mathbb{Z}$. See also Hodges [1983, pp. 129–30]. Greenspan [1973] would say that the discrete circle is at least as good a physical model as the real circle. See also Nambu [1987], who considers dynamical systems over finite fields. Another reference is Martin, Odlyzko, and Wolfram [1984].

Analogies between the spectral theory of the Laplacian on Riemannian manifolds and graphs have been of great interest lately. See Brooks [1986, 1991, 1998], Buser [1988, 1992], Gordon, Webb, and Wolpert [1992], Lubotzky [1994], and Sunada [1986, 1988].

The Heat Equation

Fan Chung [1996, p. 145] says: "All information about a graph is, of course, contained in its heat kernel."

One of the most important kernels for various applications in such diverse fields as statistics and differential geometry is the heat kernel. On the real line, the heat kernel (or fundamental solution of the heat equation $\partial u / \partial t = a^2(\partial^2 u / \partial x^2)$ is a Gaussian $G_{2a^2t}(x)$, where

$$G_t(x) = (2\pi t)^{-1/2} \exp\left(\frac{-x^2}{2t}\right), \quad \text{for } t > 0.$$

See Terras [1985, Vol. I, pp. 4, 13]. On the circle $\mathbb{R}/\mathbb{Z}$, the heat kernel is a theta function, which is the sum of integer translates of $G_t(x)$. See Terras [1985, Vol. I, p. 39]. See Chapter 23 for a sketch of an application of the heat kernel on the Poincaré upper half plane H to eigenvalues of the Laplacian on a Riemann surface $\Gamma \setminus H$. References for the continuous heat equation are: Buser [1988], Chavel [1984], Dym and McKean [1964], Körner [1988], Powers [1987], and Terras [1985].

Here we consider two different sorts of finite analogues of the heat kernel. We ultimately want to compare the finite heat kernels with the continuous one. References for the discrete heat kernels include those mentioned in the chapter on the Laplacian plus Kac [1966], Chung [1996], Chung and Yau [1995], Martin, Odlyzko, and Wolfram [1984], and Velasquez [1991].

Finite Heat Kernel #1

We begin with a discussion motivated by Kac [1947]. It is similar to our discussion of Markov chains in the last chapter. Start with $p_j = j\delta$ equally spaced points on the line, $j = 0, 1, \ldots, n-1$, with spacing $\delta = 1/n$, as well as equally spaced times $t_i \in \mathbb{R}$ with spacing τ. Let

$$u(b, j) = u(b\delta, j\tau) = \begin{cases} \text{probability that particle is at} \\ \text{position } b\delta \text{ at time } j\tau \text{ if at} \\ \text{the beginning it was at 0.} \end{cases} \tag{4}$$

Assume that we have a free particle equally likely to move right or left. Then $u(b, j)$ will satisfy

$$u(b, j + 1) = \frac{1}{2}(u(b - 1, j) + u(b + 1, j)), \tag{5}$$

or, equivalently,

$$\frac{u(b, j + 1) - u(b, j)}{\tau} = \frac{\delta^2}{2\tau}\left(\frac{\Delta_c u}{\delta^2}\right), \tag{6}$$

where Δ_c is the combinatorial Laplacian in (2). So, in the limit as $\delta \to 0$, we get the *heat equation*

$$\frac{\partial u}{\partial t} = D\frac{\partial^2 u}{\partial x^2}, \quad \text{where } D = \frac{\delta^2}{2\tau}. \tag{7}$$

Kac [1947, p. 372] notes that this "is responsible for the conclusion that the velocity of the Brownian particle is infinite." In our model, the ratio δ/τ plays the role of the instantaneous velocity and it approaches infinity as $\delta \to 0$ (if D is constant).

For a stable numerical algorithm to approximate the solution of (7), we need the quantity

$$\frac{\tau}{\delta^2} = \frac{1}{2D} \leq \frac{1}{2},$$

that is, $D \geq 1$. One reference is Garabedian [1964, pp. 475 ff]. The Lax equivalence theorem says that for consistent finite difference approximations of time-dependent linear partial differential equations that are well posed, the numerical scheme converges iff it is stable. See Strang [1986, p. 574] or Strikwerda [1989]. Strang calls the Lax equivalence theorem "the fundamental theorem of numerical analysis."

Thus to solve (6) we can use (5). To make a well-posed problem, we need an *initial condition*:

$$u(b, 0) = f(b), \quad b \in \mathbb{Z}/n\mathbb{Z}, \tag{8}$$

for some given function f on $\mathbb{Z}/n\mathbb{Z}$. We can solve (5) and (8) using the DFT:

$$\mathscr{F}(u(b, j)) = \sum_{a \in \mathbb{Z}/n\mathbb{Z}} u(a, j) \exp\left(\frac{-2\pi i a b}{n}\right). \tag{9}$$

So the DFT of (5) with respect to the b variable is

$$\mathscr{F}u(b, j + 1) = \frac{1}{2}\mathscr{F}Au(b, j),$$

where A is the adjacency operator on $X(\mathbb{Z}/n\mathbb{Z}, \{\pm 1\})$ acting on the b variable. This implies

$$\mathscr{F}u(b, j + 1) = \frac{1}{2}\mathscr{F}\big(\delta_{\{\pm 1\}} * u(b, j)\big),$$

where convolution is over the b variable. By properties of the DFT from Chapter 2, we have

$$\mathscr{F}u(b, j + 1) = \frac{1}{2}\big(\mathscr{F}\delta_{\{\pm 1\}}\big)\mathscr{F}u(b, j) = \cos\left(\frac{2\pi b}{n}\right)\mathscr{F}u(b, j),$$

since

$$\big(\mathscr{F}\delta_{\{\pm 1\}}\big)(b) = \sum_{s=\pm 1}\exp\left(\frac{2\pi isb}{n}\right) = 2\cos\left(\frac{2\pi b}{n}\right).$$

So

$$\mathscr{F}u(b, j + 1) = \left(\cos\left(\frac{2\pi b}{n}\right)\right)^{j}\mathscr{F}f(b). \tag{10}$$

To find the *fundamental solution* of (5) and (8), take $f = \delta_0$, the function that is 1 at 0 and 0 everywhere else. We obtain the *fundamental solution of the finite heat equation* on $X(\mathbb{Z}/n\mathbb{Z}, \{\pm 1\})$, for $a \in \mathbb{Z}/n\mathbb{Z}$, $j \in \mathbb{Z}$:

$$p_j(a) = \frac{1}{n}\sum_{b \in \mathbb{Z}/n\mathbb{Z}}\left(\cos\left(\frac{2\pi b}{n}\right)\right)^{j}\exp\left(\frac{2\pi iab}{n}\right), \tag{11}$$

The solution of (5) and (8) is

$$p_j * f(b) = u(b, j). \tag{12}$$

Note that $p_j = \hat{\omega}_t$, where

$$\omega_t(b) = \frac{1}{n}\left(\cos\left(\frac{2\pi b}{n}\right)\right)^{j}.$$

Let us compare this with the fundamental solution of the continuous heat equation on the circle $\mathbb{R}/\mathbb{Z}$ (see Terras [1985, Chapter 1]):

$$\frac{\partial u}{\partial t} = D\frac{\partial^2 u}{\partial x^2}, \quad \text{where } x \in \mathbb{R}/\mathbb{Z}, t > 0; u(x, 0) = f(x). \tag{13}$$

The fundamental solution for (13) is

$$p_t(x) = \sum_{b \in \mathbb{Z}} \exp(-4\pi^2 Db^2 t) \exp(2\pi i bx). \tag{14}$$

The bth Fourier coefficient of $p_t(x)$ corresponding to $b \in \mathbb{Z}$ is

$$\exp(-4\pi^2 Db^2 t) \cong 1 - \frac{2\pi^2 b^2 j}{n^2}, \quad \text{if } b/n \to 0,$$

where we used $D = (1/n)^2/(2\tau), t = \tau j$.

Exercise. (See Kac [1947].) Imitate the preceding discussion for a particle in a field of constant force and in the presence of reflecting barriers at 0 and n. Assume

(a) the probability of a move to the right is $q = \frac{1}{2} - \beta\delta > 0$;
(b) the probability of a move to the left is $p = \frac{1}{2} + \beta\delta$;
(c) when the particle reaches 0, it must at the next step move right or when it reaches n, it must at the next step move left.

Finite Heat Kernel #2

Our second approach to the fundamental solution of the heat equation on a graph comes from Fan Chung [1996], Chung and Yau [1995], and Elinor Velasquez [preprint]. Let Δ_c be the combinatorial Laplacian for $X(\mathbb{Z}/n\mathbb{Z}, \{\pm 1\})$. Let ϕ_j be a complete orthonormal set of eigenfunctions of Δ_c with corresponding eigenvalues λ_j.

The heat kernel comes from the heat matrix $H_t = \exp(t\Delta_c)$, where $\Delta_c = A - 2I$, if A is the adjacency matrix of the finite circle graph. Using the basis of eigenfunctions, one can express the heat kernel as

$$H_t(x, y) = \sum_{a \in \mathbb{Z}/n\mathbb{Z}} \exp(\lambda_a t)\phi_a(x)\overline{\phi_a(y)} = k_t(x - y), \quad \text{where}$$

$$k_t(x) = \sum_{j=1}^{n} e^{\lambda_j t}\phi_j(x). \tag{15}$$

Then the solution of

$$\frac{\partial u}{\partial t} = \Delta_c u, \quad u(x,0) = f(x)$$

is $k_t * f$. In our case $\Delta_c = \frac{1}{2}(A - 2I)$ and the eigenfunctions are $\phi_a(x) = \exp(\frac{2\pi i j}{n})$, with corresponding eigenvalues

$$\lambda_j = \cos\left(\frac{2\pi j}{n}\right) - 1 = -2\sin^2\left(\frac{\pi j}{n}\right), \quad \text{for } j \in \mathbb{Z}/n\mathbb{Z}.$$

Therefore the heat kernel for the finite circle $X(\mathbb{Z}/n\mathbb{Z}, \{\pm 1\})$ is

$$k_t(x) = \sum_{b \in \mathbb{Z}/n\mathbb{Z}} \exp\left(-2t \sin^2\left(\frac{\pi x}{n}\right)\right) \exp\left(\frac{2\pi i x b}{n}\right). \tag{16}$$

Note that $k_t = \hat{\eta}_t$, where

$$\eta_t(x) = \exp\left(-2t \sin^2\left(\frac{\pi x}{n}\right)\right), \quad \text{for } x \in \mathbb{Z}/n\mathbb{Z}.$$

Chung and Yau [1995] use this fundamental solution to obtain results on graphs that are similar to those from the spectral theory of the Laplacian on Riemannian manifolds. They also show that $(1/n)k_t(u)$ is asymptotic to $\sqrt{2}G_t(u)$, as $n \to \infty$, for t larger than u^2. The proof involves comparing the sum with an integral.

Exercise. Compare formulas (11) and (16).

Chung and Yau [1997] also obtain a trace formula close to Selberg's formula as discussed in Chapters 22–24. Selberg's formula relates eigenvalues of the Laplacian and closed paths on the graph. To do this, recall that the heat matrix is

$$H_t = \exp(t\Delta_c), \quad \text{where } \Delta_c = A - 2I,$$

if A is the adjacency matrix of the finite circle graph. Using the Taylor series for the matrix exponential, we find that

$$H_t = \exp(t(A - 2I)) = \sum_{k \geq 0} \frac{t^m}{m!}(A - 2I)^m$$

$$= \sum_{k \geq 0} \frac{t^m}{m!} \sum_{r=0}^m \binom{m}{r} A^r (-2)^{m-r}.$$

Recall that $A^r(x, y)$ is the number of paths of length r from x to y in $X(\mathbb{Z}/n\mathbb{Z}, \{\pm 1\})$. Then use (15) and take the trace to obtain a relation between eigenvalues of the Laplacian and closed paths of length r in our graph. Chung and Yau [1997] give a slightly different formulation.

Applications of the Discrete Fourier Transform in Geometry

Suppose that $\Pi = \{z(0), z(1), \ldots, z(k-1)\}$ is a closed polygon in the plane with $k \geq 2$. Let $Dz(j)$ be the midpoint of the jth side $\{z(j-1), z(j)\}$ of Π, that is, with

$$Dz(j) = \frac{1}{2}\{z(j-1) + z(j)\} = \frac{1}{2}\{(\delta_1 + \delta_0) * z\}. \tag{17}$$

Here we use the notation of Chapter 2 for convolution and the delta functions and we view $z(j)$ as a function z on $\mathbb{Z}/k\mathbb{Z}$, that is, $z(0) = z(k)$. Let the *first derived polygon* of Π be defined by

$$\Pi' = \{Dz(1), Dz(2), \ldots, Dz(k-1)\}.$$

Figure I.25 shows a sequence of polygons $\Pi, \Pi', \Pi'', \ldots$. In the picture, one sees that $\Pi^{(m)}$ approaches the *centroid* or center of gravity of Π given by

$$P = \frac{1}{k}(z(0) + z(1) + z(2) + \cdots + z(k-1)). \tag{18}$$

Just as in the last chapter we used the DFT to find the limiting behavior of a finite Markov chain, we can use the DFT to prove the theorem that the preceding figure leads us to conjecture.

Theorem 1. Derived Polygons Approach the Centroid. Consider a polygon $\Pi = \{z(0), z(1), \ldots, z(k-1)\}$ in the complex plane, that is, we can think of $z : \mathbb{Z}/k\mathbb{Z} \to \mathbb{C}$. Let $Dz(j-1)$ be the midpoint of the jth side as in Equation (16). Let Π' be the derived polygon

$$\Pi' = \{Dz(0), Dz(1), \ldots, Dz(k-1)\} \quad \text{and set } \Pi^{(r)} = \left(\Pi^{(r-1)}\right)'.$$

Then, in the limit as r approaches infinity $\Pi^{(r)}$ approaches the centroid P defined by (18).

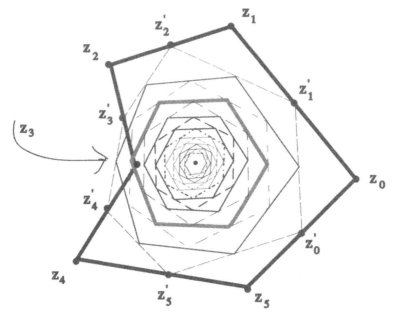

Figure I.25. Some derived polygons $\Pi^{(j)}$ of $\Pi = \{z(0), z(1), z(2), z(3), z(4), z(5)\}$, showing the approach of $\Pi^{(j)}$ to the centroid.

Proof. (From Schoneberg [1950].) Suppose the center of gravity P of Π is at the origin, that is,

$$0 = \sum_{j=0}^{k-1} z(j).$$

As we noted above,

$$Dz = \frac{1}{2}(\delta_0 + \delta_1) * z = d * z, \quad \text{where } d = \frac{1}{2}(\delta_0 + \delta_1).$$

We want to find

$$\lim_{r \to \infty} (\underbrace{d * \cdots * d}_{r} * z)(j).$$

By properties of the DFT on $\mathbb{Z}/k\mathbb{Z}$ from Chapter 2 and Theorem 2 of Chapter 3, we know that

$$\mathscr{F}(d * z)(j) = \mathscr{F}d(j)\mathscr{F}z(j) = \frac{1}{2}\left(1 + \exp\left(\frac{-2\pi i j}{k}\right)\right)\mathscr{F}z(j).$$

So to find the limit of D^r as $r \to \infty$, we need to examine

$$\lim_{r \to \infty} (\mathscr{F}d(j))^r = \lim_{r \to \infty} \left\{ \frac{1}{2} \left(1 + \exp\left(\frac{-2\pi i j}{k} \right) \right) \right\}^r.$$

Clearly this limit is

$$\lim_{r \to \infty} \exp\left(\frac{-\pi r i j}{k} \right) \left\{ \cos\left(\frac{\pi j}{k} \right) \right\}^r.$$

Since, for $j = 1, 2, \ldots, k - 1$, we have $|\cos(\pi j/k)| < 1$, we see that

$$\lim_{r \to \infty} \exp\left(\frac{-r \pi i j}{k} \right) \left\{ \cos\left(\frac{\pi j}{k} \right) \right\}^r = \begin{cases} 0, & j \not\equiv 0 \,(\mathrm{mod}\ k), \\ 1, & j \equiv 0 \,(\mathrm{mod}\ k). \end{cases}$$

As we have assumed the centroid is 0, we have $\mathscr{F}z(0) = 0$. It follows that

$$\lim_{r \to \infty} \mathscr{F}(D^r z)(j) = \lim_{r \to \infty} (\mathscr{F}d(j))^r \mathscr{F}z(j) = 0, \quad \text{for all } j \in \mathbb{Z}/k\mathbb{Z}.$$

So we see that

$$\lim_{r \to \infty} (D^r z)(j) = 0 \ \text{(the centroid)},$$

proving the theorem. ■

Exercise. In the preceding proof, explain why we can't multiply the supposed limit point by a constant multiple and still get the same result, since we can multiply the first equation in the proof by any constant.

Theorem 2. An Inequality for the Area of a Polygon. Let $\Pi = \{z(0), z(1), \ldots, z(k-1)\}$ be a closed polygon in the plane. Let A denote the oriented area of Π defined by

$$A = \frac{1}{2} \sum_{j=0}^{k-1} \mathrm{Im}(\overline{z_j}\, z_{j+1}). \tag{19}$$

Set

$$B = \sum_{j=0}^{k-1} |z(j)|^2.$$

Then we have the inequality

$$A \leq \frac{B}{2}.$$

Proof. (From Schoneberg [1950].) The Plancherel formula from Theorem 1 of Chapter 2 and the translation property of the DFT in Equation (11) of Chapter 2 imply

$$A = \frac{1}{2k} \sum_{j=0}^{k-1} |\mathcal{F}z(j)|^2 \sin\left(\frac{2\pi j}{k}\right). \tag{20}$$

Exercise.
a) Prove formula (19).
b) Then prove formula (20).

Set

$$\gamma = \max\left\{ \left.\sin\left(\frac{2\pi j}{k}\right)\right| j = 0, 1, \ldots, k-1\right\}.$$

Then, using Plancherel again, show that

$$A \leq \frac{\gamma}{2k} \sum_{j=0}^{k-1} |\mathcal{F}z(j)|^2 = \frac{\gamma}{2} \sum_{j=0}^{k-1} |z(j)|^2.$$

Since $\gamma \leq 1$, the result is proved. ■

The Isoperimetric Inequality

The *isoperimetric inequality* says that for any simply connected bounded domain in the plane with rectifiable boundary, setting A equal to the area of the domain and L to the length of its boundary, we have

$$A \leq \frac{L^2}{4\pi}. \tag{21}$$

There is equality if and only if the domain is a circle. For a proof due to Hurwitz using Fourier analysis, see Courant and Hilbert [1953, p. 97] or Körner [1988].

The inequality (21) is related to the isoperimetric problem or *Queen Dido's Problem*, named for Queen Dido of Carthage (850 B.C.). The problem asks us to find the closed plane curve of fixed length enclosing the largest area. The problem arose in Virgil's *Aeneid* when Dido made a deal to buy land on the coast of present day Tunisia. She asked the ruler for as much land as she could bound with a cow's hide. The request was granted. Then she cut the hide into strips and stretched the hide between two points on the Mediterranean coast. Thus Carthage was founded.

Unfortunately for Dido, she fell in love with Aeneus (a shipwrecked fugitive from Troy). He later abandoned her to become the founder of the Roman people. So she killed herself. This is a brief summary of part of a favorite text of many a high school Latin class and an opera by Purcell.

For the special case of polygons in the plane, a solution to Queen Dido's problem can be based on an idea of J. Steiner. See Courant and Hilbert [1953, p. 166]. For example, it is easy to see that a polygon of fixed perimeter which maximizes the area has to be convex and equilateral. Next one must show that the polygon maximizing the area can be inscribed in a circle. Steiner, a geometer who lived from 1796 to 1863, gave five proofs. Bell [1990, p. 158] notes that Steiner could not read until he was 14.

For other discrete analogues of the isoperimetric inequality, see L. Babai's article in Vol. 2 of R. Graham et al. [1995] and Fan Chung [1996, p. 165].

Theorem 3. Isoperimetric Inequality for Equilateral k-gons. Suppose that $\Pi = \{z(0), z(1), \ldots, z(k-1)\}$ is a closed polygon in the plane. That is, $z: \mathbb{Z}/k\mathbb{Z} \to \mathbb{C}$. If Π is *equilateral*, each side having length a, and if A is the area of Π, and L the length of its boundary, we have the isoperimetric inequality:

$$\frac{A}{L^2} \leq \frac{1}{4k} \cot \frac{\pi}{k} \leq \frac{1}{4\pi}.$$

Equality holds if and only if Π is regular, that is, all angles are equal.

Proof. Use Plancherel's theorem to see that $L^2/k = ka^2$ is given by

$$\frac{L^2}{k} = ka^2 = \sum_{j=0}^{k-1} |z(j+1) - z(j)|^2$$

$$= \frac{1}{k} \sum_{j=0}^{k-1} \left| \exp\left(\frac{2\pi i j}{k}\right) - 1 \right|^2 |\hat{z}(j)|^2 = \frac{4}{k} \sum_{j=0}^{k-1} \left| \sin\left(\frac{\pi j}{k}\right) \right|^2 |\hat{z}(j)|^2.$$

So

$$L^2 = 4 \sum_{j=0}^{k-1} \left| \sin\left(\frac{\pi j}{k}\right) \right|^2 |\hat{z}(j)|^2.$$

It follows from formula (20) for A that

$$4k\,A\tan\frac{\pi}{k} = 4\sum_{j=0}^{k-1}\tan\left(\frac{\pi}{k}\right)\sin\left(\frac{\pi j}{k}\right)\cos\left(\frac{\pi j}{k}\right)|\hat{z}(j)|^2.$$

Therefore

$$L^2 - 4k\,A\tan\frac{\pi}{k}$$

$$= 4\sum_{j=0}^{k-1}\sin\left(\frac{\pi j}{k}\right)\left\{\sin\left(\frac{\pi j}{k}\right) - \tan\left(\frac{\pi}{k}\right)\cos\left(\frac{\pi j}{k}\right)\right\}|\hat{z}(j)|^2.$$

To prove the isoperimetric inequality, you need to show that all the terms in the sum over j are nonnegative. We leave it to the reader to prove this.

Exercise.
a) Prove that the terms in the last sum over j are all nonnegative.
b) Show that you get equality if and only if the polygon is regular.

∎

Other Inequalities

There is another well-known inequality from analysis called *Wirtinger's inequality*. It says that (for sufficiently nice) real-valued functions $f(t)$ of period 2π with

$$\int_0^{2\pi} f(t)\,dt = 0,$$

we have

$$\int_0^{2\pi} f(t)^2\,dt \le \int_0^{2\pi} f'(t)^2\,dt.$$

We want to look at a finite analogue based on the fact that a finite difference operator such as $D_1 z(j) = z(j+1) - z(j)$ is a finite analogue of the derivative.

Theorem 4. Finite Analogue of Wirtinger's Inequality. Suppose that $z(j)$, $j = 0, 1, \ldots, k-1$, are complex numbers such that

$$\sum_{j=0}^{k-1} z(j) = 0.$$

Then we have Wirtinger's inequality:

$$\sum_{j=0}^{k-1} |z(j+1) - z(j)|^2 \geq 4\sin^2\left(\frac{\pi}{k}\right) \sum_{j=0}^{k-1} |z(j)|^2.$$

Proof. (From Davis [1979, pp. 114–19].)

We can use Plancherel's theorem. Set

$$D_1 z(j) = z(j+1) - z(j) = \{\delta_{-1} - \delta_0\} * z(j).$$

Then we are interested in the square of the L^2-norm of $D_1 z$. Using the Plancherel theorem from Theorem 1 of Chapter 2, we have

$$\|D_1 z\|_2^2 = \frac{1}{k}\| \mathscr{F} D_1 z\|_2^2 = \frac{1}{k} \sum_{j=0}^{k-1} \left| \exp\left(\frac{2\pi i j}{k}\right) - 1 \right|^2 |\mathscr{F} z(j)|^2.$$

The right-hand side is easily bounded by

$$\frac{4}{k}\sin^2\left(\frac{\pi}{k}\right) \sum_{j=0}^{k-1} |\mathscr{F} z(j)|^2.$$

Plancherel's theorem completes the proof. ∎

Remarks. The uncertainty principle for functions $f \in L^2(\mathbb{R})$ (see Terras [1985, Vol. I, Chapter 1]) is equivalent to the following inequality when $f' \in L^2(\mathbb{R})$ and $\|f\|_2^2 = \int_{\mathbb{R}} |f(t)|^2 \, dt = 1$:

$$\int_{\mathbb{R}} |t f(t)|^2 \, dt \int_{\mathbb{R}} |f'(t)|^2 \, dt \geq \frac{1}{4}.$$

We will discuss finite analogues of this inequality in Chapter 14. See also Elinor Velasquez [1998].

Chapter 8

The Quadratic Reciprocity Law

Thus, if the terms of a Gauss sum are put to proper use in what
may be called quadratic-residue phase gratings, then coherent
light, radar beams or sound waves can be very effectively
scattered.

M. R. Schroeder [1986, p. 12]

Our goal in this chapter is to prove the quadratic reciprocity law. To do so
we will make use of the DFT and Gauss sums. At the end of this chapter we
will consider the application mentioned by Schroeder above. Discussions of
the quadratic reciprocity law can be found in most number theory books, for
example, Rosen [1993] and Silverman [1997]. Gauss gave the first proof and
seven others. The proof we give is a variant of his sixth proof.

The Quadratic Reciprocity Law-First Statement

Fix two distinct odd primes p, q.

Case 1. If $\frac{p-1}{2} \frac{q-1}{2}$ is even, then the congruence

$$x^2 \equiv q \pmod{p} \text{ has a solution } x \in \mathbb{Z}$$

if and only if the congruence

$$y^2 \equiv p \pmod{q} \text{ has a solution } y \in \mathbb{Z}.$$

Case 2. If $\frac{p-1}{2} \frac{q-1}{2}$ is odd, then the congruence

$$x^2 \equiv q \pmod{p} \text{ has a solution } x \in \mathbb{Z}$$

Table I.12. *Squares mod 11*

x	1	2	3	4	5	6	7	8	9	10
x^2	1	4	9	5	3	3	5	9	4	1

if and only if the congruence

$$y^2 \equiv p \pmod{q} \text{ has } no \text{ solution } y \in \mathbb{Z}.$$

Gauss found eight proofs of this law (the first at age 18). We will give a variant of one of Gauss's proofs involving the finite Fourier transform.

Why might one care which numbers are squares mod p or *quadratic residues* mod p? Of course, mathematicians don't need a reason beyond that of mountain climbers. But there is a rather surprising application. Schroeder [1986] shows that quadratic residues can be used in the design of concert hall ceilings. See Figure I.26 near the end of the chapter.

Example. $p = 11$. The squares modulo 11 can be found by looking at Table I.12. The numbers in the lower row of the table consist of the squares modulo 11. The numbers left out of the lower row are the nonsquares modulo 11. We conclude that the numbers 1,3,4,5,9 are squares or quadratic residues mod 11. The rest of the elements of $(\mathbb{Z}/11\mathbb{Z})^*$ are nonsquares or quadratic nonresidues mod 11.

Exercise. Assume that p is an odd prime. Show that it always happens that half, that is, $(p - 1)/2$ of the elements of $(\mathbb{Z}/p\mathbb{Z})^*$ are squares.
Hint. Recall that the multiplicative group $(\mathbb{Z}/p\mathbb{Z})^*$ is cyclic generated by a primitive root g mod p. The squares in this group are the even powers of g mod p.

To reach our goal of proving the quadratic reciprocity law, we need to define the Legendre symbol. This symbol will allow us to write the two cases of the reciprocity law as one elegant formula. It is the beginning of a long line of symbols invented by the great number theorists: Jacobi, Hilbert, Artin, ... We have already seen this symbol in Chapter 5. There we called it the quadratic character.

Definition. The *Legendre symbol* for an integer a not divisible by the odd prime p is defined by

$$\left(\frac{a}{p}\right) = \begin{cases} 1, & \text{if } x^2 \equiv a \pmod{p} \text{ has a solution,} \\ -1, & \text{otherwise.} \end{cases} \tag{1}$$

Exercise. Euler's Criterion. Suppose that the odd prime p does not divide a. Show that

$$\left(\frac{a}{p}\right) \equiv a^{(p-1)/2} (\text{mod } p). \tag{2}$$

Then show that

$$\left(\frac{-1}{p}\right) = (-1)^{(p-1)/2}. \tag{3}$$

Hint. Recall the previous exercise and the fact that

$$g^{(p-1)/2} \equiv -1(\text{mod } p).$$

Then note that if both sides of a congruence modulo an odd prime p are ± 1, then the congruence must be an equality.

Exercise. Show that, as a function of the top number $a(\text{mod } p)$, the Legendre symbol $(\frac{a}{p})$ is a homomorphism from the multiplicative group $(\mathbb{Z}/p\mathbb{Z})^*$ into the group with two elements $\{\pm 1\}$.
Hint. Use Euler's criterion.

Notes. Suppose that g is a primitive root modulo the odd prime p, that is, g generates the multiplicative group

$$(\mathbb{Z}/p\mathbb{Z})^* = \{a \bmod p | p \text{ does not divide } a\}.$$

Suppose that

$$a \equiv g^e (\text{mod } p).$$

Then $e = \log_g a$ is called the *discrete logarithm base g of a*. It was once called the *index* of a mod p. The problem of computing e is called "the discrete logarithm problem." See Menezes [1993]. Of course e is only determined mod $(p-1)$. The hints to the first exercise of this chapter said that if e is the index of a square mod p, then e must be even and conversely.

A table of logarithms or indices for primes less than 1,000 was published by Jacobi in 1839. Now such tables are of use in various applications such as coding theory. See Dornhoff and Hohn [1978].

Exercise. Use Mathematica to develop a log table for $(\mathbb{Z}/p\mathbb{Z})^*$ for your favorite large p. If Mathematica is not available, take $p = 29$. Then list the squares in this group. Then list the cubes, fourth, fifth, sixth, and seventh powers.

Exercise. Suppose, as usual, that p is an odd prime. How many solutions $x \pmod{p}$ are there for the congruence $x^k \equiv a \pmod{p}$?
Hint. Use the finite logarithm to change this congruence to a linear congruence. See Rosen [1993, p. 131].

Example. From our previous example, we see that the Legendre symbol $(\frac{5}{11}) = 1$.
   You could also have figured this out by using the Euler criterion:

$$5^{(11-1)/2} = 5^5 \equiv 1 \pmod{11}.$$

Note that you should compute large powers in $\mathbb{Z}/p\mathbb{Z}$ recursively, for example,

$$5^2 \equiv 3, \ 5^3 \equiv 5 \cdot 3 \equiv 4, \ 5^4 \equiv 5 \cdot 4 \equiv -2 \pmod{11}.$$

This keeps the numbers within the size your computer can handle. Otherwise integer overflow occurs.
   An even faster method for large moduli makes use of the binary expansion of the exponent and repeated squaring. See Rosen [1993] and Silverman [1997]. So, for our previous example, you need to make a table of the numbers $5^{(2^j)} \pmod{11}$. Then

$$5^5 \equiv 5^{4+1} \equiv 5^1 5^{(2^2)} \equiv 5(5^2)^2 \equiv 5 \cdot 3^2 \equiv 5 \cdot 9 \equiv 45 \equiv 1 \pmod{11}.$$

In Mathematica you should use PowerMod(a, b, n) rather than Mod(a^b, n).
   For large p and a, it may be rather tedious to use these methods to find out if a is a square mod p. The quadratic reciprocity law will provide a method to answer this question as well as a surprising and pretty formula. Gauss called it the golden theorem. It was stated by Legendre in 1785, but Gauss gave the first proof. In fact, Gauss did not stop with one proof but gave eight. In 1963, M. Gerstenhaber [1963] gave the 152nd proof. We will content ourselves with one proof here. There is another in Terras [1985, Vol. I, Exercise on p. 195].
   Note that Mathematica can calculate the Legendre symbol. It is called JacobiSymbol[n, m], which is an extension to all odd bottom numbers in $(\frac{n}{m})$. We will define it soon.
   We now can restate the Quadratic Reciprocity Law (QRL).

$$\left(\frac{q}{p}\right)\left(\frac{p}{q}\right) = (-1)^{\frac{p-1}{2}\frac{q-1}{2}}, \quad \text{for distinct odd primes } p, q. \tag{4}$$

Using the DFT to Prove the QRL

Now we prove the quadratic reciprocity law after proving four lemmas involving the DFT of the Legendre symbol as a function of its top entry. Before our lemmas, we make a definition, for p an odd prime.

Definition.

$$h_p(x) = \begin{cases} \left(\frac{x}{p}\right), & \text{if } p \text{ does not divide } x, \\ 0, & \text{if } p \text{ divides } x \end{cases} \tag{5}$$

Lemma 1. $\hat{h}_p(-x) = h_p(x)\hat{h}_p(-1)$, where $\hat{f}$ is the DFT of f on $\mathbb{Z}/p\mathbb{Z}$.

Proof.

Case 1. p *does not divide* x.

$$\hat{h}_p(-x) = \sum_{a=0}^{p-1} h_p(a) \exp\left(\frac{2\pi i x a}{p}\right) = \sum_{a=1}^{p-1} \left(\frac{a}{p}\right) \exp\left(\frac{2\pi i x a}{p}\right).$$

Now change variables in the sum by setting $xa = b$. Since we are summing over the elements of the multiplicative group $(\mathbb{Z}/p\mathbb{Z})^*$, the sum over b is the same as that over a, just in a different order. So we find that

$$\hat{h}_p(-x) = \sum_{b=1}^{p-1} \left(\frac{bx^{-1}}{p}\right) \exp\left(\frac{2\pi i b}{p}\right) = \left(\frac{x^{-1}}{p}\right)\hat{h}_p(-1)$$
$$= h_p(x)\hat{h}_p(-1).$$

Here we used

$$h_p(x) = \left(\frac{x}{p}\right) = \left(\frac{x^{-1}}{p}\right).$$

Case 2. p *divides* x.
   In this case

$$\hat{h}_p(-x) = \sum_{a=0}^{p-1} h_p(a) = 0,$$

because there are as many squares as nonsquares in $(\mathbb{Z}/p\mathbb{Z})^*$, as we saw in an exercise above. This gives the result we want in this case, since $0 = h_p(x)$, when p divides x.

◼

Note. Lemma 1 says that $h_p(x)$ is a constant multiple of its own Fourier transform, since

$$\hat{h}_p(x) = h_p(-x)\hat{h}_p(-1) = h_p(x)h_p(-1)\hat{h}_p(-1).$$

Setting $c = h_p(-1)\hat{h}_p(-1)$, we have

$$\hat{h}_p(x) = ch_p(x).$$

This is a property it has in common with the normal density or Gauss kernel $\exp(-\pi x^2)$ on the real line. See Terras [1985].

Definition. The *Gauss sum g* is

$$g = \hat{h}_p(-1) = \sum_{a=1}^{p-1} \left(\frac{a}{p}\right) \exp\left(\frac{2\pi i a}{p}\right). \tag{6}$$

Lemma 2. If g is the Gauss sum defined by formula (6) above,

$$g^2 = (-1)^{(p-1)/2} p.$$

Proof. Take the DFT of both sides of the equation in Lemma 1. This gives, using the inversion formula for the DFT and the fact that the Fourier transform commutes with $f(x) \mapsto f(-x)$:

$$ph_p(x) = \hat{\hat{h}}_p(-x) = \hat{h}_p(x)\hat{h}_p(-1).$$

Now set $x = -1$ to obtain

$$g^2 = ph_p(-1).$$

To finish the proof, we need only recall that

$$h_p(-1) = \left(\frac{-1}{p}\right) = (-1)^{(p-1)/2}.$$

This comes from Euler's criterion above. ◼

Lemma 3. Suppose that p and q are distinct odd primes. Set

$$p^* = (-1)^{(p-1)/2} p.$$

If g is the Gauss sum in Lemma 2, then

$$g^{q-1} \equiv \left(\frac{p^*}{q}\right) (\text{mod } q).$$

Proof. The proof is left as an exercise.

Hint. Note that since q is odd, the left-hand side of the congruence is in $\mathbb{Z}$, by Lemma 2. Use Euler's criterion above. ∎

In the next lemma, we will compute in the ring $R = \mathbb{Z}(\exp(2\pi i/p))$, the ring of polynomials with integer coefficients in $w = \exp(2\pi i/p)$. So the Gauss sum is an element of R. Note that w is a primitive pth root of unity. The multiplicative group of all roots w of the equation $w^p = 1$ forms a cyclic group of order p (isomorphic with $\mathbb{Z}/p\mathbb{Z}$). A generator of this group is called a *primitive pth root of unity*.

Warning. For $p \geq 23$, this ring R is not a *unique factorization domain*. It was a famous mistake of some of those trying to prove Fermat's last theorem[†] in the last century to assume R has unique factorization into irreducibles. Dedekind's theory of ideals was developed to deal with the lack of unique factorization in such rings. Most algebraic number theory books discuss this subject. Thus we cannot assume that if q divides a product of two numbers from the ring R, it follows that q divides one or the other.

It is also possible to view $w^p = 1$ as an equation over the field $\mathbb{F}_q = \mathbb{Z}/q\mathbb{Z}$. This allows one to view the Gauss sum as an element of the finite extension field $\mathbb{F}_q[w]$. So you can view the Gauss sum as an element of a finite field. Serre [1973] takes this approach, but we will not.

Lemma 4. Suppose that p and q are distinct odd primes. Then

$$(\hat{h}_p(x))^q \equiv \hat{h}_p(qx) (\text{mod } q).$$

Proof. Here the congruence is understood to take place in the ring $R = \mathbb{Z}[e^{2\pi i/p}]$. Note that

$$(\hat{h}_p(x))^q = \left(\sum_{a=1}^{p-1} \left(\frac{a}{q}\right) \exp\left(\frac{-2\pi iax}{p}\right)\right)^q$$

$$\equiv \sum_{a=1}^{p-1} \left(\frac{a}{p}\right) \exp\left(\frac{-2\pi iqax}{p}\right) (\text{mod } q).$$

[†] Saying that there are no integer solutions for $n \geq 3$ to $x^n + y^n = z^n$ with $xyz \neq 0$.

This congruence comes from the fact that the prime q divides the binomial coefficients $\binom{q}{n}$, $n = 1, 2, \ldots, q - 1$. Thus

$$(u + v)^q \equiv u^q + v^q \,(\mathrm{mod}\, q) \quad \text{for all } u, v \in \mathbb{Z}.$$

Exercise. Prove this last statement. ∎

Finally we are ready to prove the QRL.

Theorem 1. **Quadratic Reciprocity Law.** Suppose that p and q are distinct odd primes. Then

$$\left(\frac{p}{q}\right)\left(\frac{q}{p}\right) = (-1)^{\frac{p-1}{2}\frac{q-1}{2}}.$$

Proof. Set $x = -1$ in Lemma 4 to see that if g is the Gauss sum of Lemma 2, we have

$$g^q \equiv \hat{h}_P(-q) \,(\mathrm{mod}\, q).$$

Thus, by Lemma 1, we have

$$g^q \equiv \left(\frac{q}{p}\right) g \,(\mathrm{mod}\, q).$$

Therefore,[†] multiplying by g, we get

$$g^{q+1} \equiv \left(\frac{q}{p}\right) g^2 \,(\mathrm{mod}\, q).$$

Lemma 3 gives

$$\left(\frac{p^*}{q}\right) g^2 \equiv \left(\frac{q}{p}\right) g^2 \,(\mathrm{mod}\, q).$$

Lemma 2 says

$$(-1)^{(p-1)/2} p \left(\frac{p^*}{q}\right) \equiv (-1)^{(p-1)/2} p \left(\frac{q}{p}\right) \,(\mathrm{mod}\, q).$$

[†] At this point, Serre [1973, p. 8] divides by g, arguing that we are in a finite field extension of $\mathbb{F}_q$. We do not do this because we are thinking we are in the ring $R = \mathbb{Z}[e^{2\pi i/n}] \subset \mathbb{C}$, where we do not have unique factorization. So we want to get back to a congruence in $\mathbb{Z}$. Our discussion is like that in Ireland and Rosen [1982, p. 72].

Now we have a congruence in which both sides are in $\mathbb{Z}$. So we can divide p out of this congruence since p and q are distinct primes. Thus we obtain

$$(-1)^{(p-1)/2}\left(\frac{p^*}{q}\right) \equiv (-1)^{(p-1)/2}\left(\frac{q}{p}\right)(\bmod\ q).$$

Now everything in the congruence is ± 1. Since q is odd, both sides must actually be equal. This implies that

$$\left(\frac{p^*}{q}\right) = \left(\frac{q}{p}\right),$$

which gives the quadratic reciprocity law since

$$p^* = (-1)^{(p-1)/2}p, \quad \text{using}\left(\frac{-1}{q}\right) = (-1)^{(q-1)/2}.$$

■

Example. Does there exist a solution to the congruence

$$x^2 \equiv 997(\bmod\ 991)?$$

Answer. No, as can be shown with the following calculation:

$$\left(\frac{997}{991}\right) = \left(\frac{6}{991}\right) = \left(\frac{2}{991}\right)\left(\frac{3}{991}\right).$$

Now

$$\left(\frac{3}{991}\right) = \left(\frac{991}{3}\right)(-1)^{\frac{(3-1)(991-1)}{4}} = \left(\frac{1}{3}\right)(-1) = -1.$$

We are left with the problem of showing that

$$\left(\frac{2}{991}\right) = 1, \quad \text{since}\quad 991 \equiv -1(\bmod\ 8).$$

This will be proved in an exercise below. It follows then that

$$\left(\frac{997}{991}\right) = -1$$

and the congruence above has no solution.

Exercise. Where did we use the hypothesis that p and q are primes in our proof of the quadratic reciprocity law?

Note. As we have said before, Gauss gave eight proofs of the quadratic reciprocity law. The proof above is close to his sixth. He used polynomials. Ours substituted a primitive pth root of unity for x in the polynomial. Many have looked at such proofs (e.g., Cauchy, Eisenstein, Jacobi). See Ireland and Rosen [1982] or Serre [1973]. Ireland and Rosen give our proof with a different version of Lemma 2. They also prove cubic and biquadratic and Eisenstein reciprocity laws similarly. It would perhaps be of interest to see if the DFT can be used to simplify the proof of higher reciprocity laws. These higher reciprocity laws are part of a very general result called the Artin reciprocity law. See Cassells and Fröhlich [1967] or Narkiewicz [1974].

There are proofs of the quadratic reciprocity law to suit any taste. For an elementary proof, see almost any elementary number theory book (e.g., Davenport [1983]). For an algebraic proof, see Samuel [1970]. For an analytic proof using theta functions, see Terras [1985, Vol. I]. For a plethora of references on the evaluation of Gauss sums for quadratic and higher degree congruences, see Berndt and Evans [1981] or Berndt, Evans, and Williams [1998].

Rademacher [1964, pp. 76–9] gives a proof of the QRL that uses the convolution property of the DFT to avoid Lemma 4. We base the following exercise on his method.

Exercise. Rademacher's proof of the QRL avoiding Lemma 4.

a) Show that if $a(x) = [\hat{h}_p(-x)]^k$, then

$$\hat{a}(x) = p \sum_{m_1 + \cdots + m_k \equiv x \,(\mathrm{mod}\ p)} \left(\frac{m_1 \cdots m_k}{p} \right).$$

Note that here we have only written one summation symbol even though it is a k-fold sum over $m_j \in (\mathbb{Z}/p\mathbb{Z})^*$, $j = 1, \ldots, k$.

b) On the other hand, assuming that k is odd, use Lemma 1 to show that

$$\hat{a}(x) = g^{k+1} h_p(-x).$$

c) Use Lemma 2 to see that if we set $x = q$, assuming k is an odd integer greater than or equal to 3, we have

$$g^{k-1} (-1)^{(p-1)/2} p \left(\frac{-q}{p} \right) = p \sum_{m_1 + \cdots + m_k \equiv q \,(\mathrm{mod}\ p)} \left(\frac{m_1 \cdots m_k}{p} \right).$$

d) Then set $k = q$ and obtain

$$g^{q-1} = \left(\frac{q}{p} \right) \sum_{m_1 + \cdots + m_q \equiv q \,(\mathrm{mod}\ p)} \left(\frac{m_1 \cdots m_q}{p} \right).$$

e) Finally show that

$$\sum_{m_1+\cdots+m_q \equiv q(\mathrm{mod}\ p)} \left(\frac{m_1 \cdots m_q}{p} \right) \quad \text{is congruent to 1 modulo } q.$$

Hint. The case that all the m_i are 1 gives 1 mod q. Then you must argue that the rest of the sum is 0 mod q. The trick is to note that other solutions contain some elements $m_j \not\equiv 1 (\mathrm{mod}\ q)$ and cyclic permutations of these m_j.

f) Note that this result allows one to prove the QRL immediately from Lemma 3 without doing Lemma 4.

Some Extensions of the Quadratic Reciprocity Law

Next we want to give some discussion, mostly through exercises, of the extensions of the quadratic reciprocity law needed to write a program to compute the Legendre symbol.

Exercise. Show that

$$\left(\frac{2}{p} \right) = (-1)^{\frac{p^2-1}{8}}. \tag{7}$$

Hint. Let $\alpha = \exp(2\pi i/8)$, a primitive eighth root of unity, and set $y = \alpha + \alpha^{-1}$. Note that $y^2 = 2$. Then show that

$$p \equiv \pm 1 (\mathrm{mod}\ 8) \quad \text{implies} \quad y^p \equiv y(\mathrm{mod}\ p),$$
$$p \equiv \pm 3 (\mathrm{mod}\ 8) \quad \text{implies} \quad y^p \equiv -y(\mathrm{mod}\ p).$$

Then use Euler's criterion to see that

$$\left(\frac{2}{p} \right) \equiv 2^{(p-1)/2} \equiv y^{p-1} (\mathrm{mod}\ p).$$

Finally multiply by y^2.

Exercise. Use the quadratic reciprocity law to find

$$\left(\frac{103}{163} \right) \quad \text{and} \quad \left(\frac{101}{163} \right).$$

Note. You may want to wait to do this exercise as we will have an even better algorithm once we have extended the quadratic reciprocity law to the Jacobi symbol that allows any odd number on the bottom.

Exercise. Show that the number of solutions to the congruence $x^2 \equiv a \pmod{p}$, for any odd prime p, is given by $1 + \left(\frac{a}{p}\right)$.

Exercise. Use the quadratic reciprocity law to find the primes for which 7 is a quadratic residue. Do the same for 15.

If you want to write a program to compute Legendre symbols, you may want to avoid all the factorizations. To do that, it makes sense to extend the meaning of the Legendre symbol to all $\left(\frac{a}{b}\right)$ for odd positive integers b assuming a and b have no common factors. This is called the Jacobi symbol. There is a further extension called the Kronecker symbol, but we omit that one.

Definition. Suppose that a is any integer and that b is an odd positive integer with no factors in common with a. If $b = p_1 p_2 \cdots p_r$, where the p_j are (not necessarily distinct) odd primes, then the *Jacobi symbol* $\left(\frac{a}{b}\right)$ is defined to be

$$\left(\frac{a}{b}\right) = \left(\frac{a}{p_1}\right)\left(\frac{a}{p_2}\right)\cdots\left(\frac{a}{p_r}\right).$$

Warning. $\left(\frac{a}{b}\right) = 1$ does not necessarily imply that a is a square mod b, when b is not prime. For example, if b is a square, then

$$\left(\frac{a}{c^2}\right) = \left(\frac{a}{c}\right)^2 = 1$$

for a relatively prime to c. But, for example, the squares mod 9 are $a \equiv 1$, $4, 7 \pmod{9}$, while $a \equiv 2, 5, 8 \pmod{9}$ are nonsquares.

Exercise. Throughout this exercise, always assume that b is an odd positive integer and is relatively prime to the top numbers in the Jacobi symbols.

a) Show that $a \equiv a' \pmod{b}$ implies that

$$\left(\frac{a}{b}\right) = \left(\frac{a'}{b}\right).$$

b) Show that

$$\left(\frac{aa'}{b}\right) = \left(\frac{a}{b}\right)\left(\frac{a'}{b}\right).$$

c) Show that

$$\left(\frac{a}{b}\right)\left(\frac{a}{c}\right) = \left(\frac{a}{bc}\right).$$

Exercise.

a) Show that if $p_1, p_2, \ldots, p_r$ are odd primes (not necessarily distinct), then

$$\frac{p_1 - 1}{2} + \frac{p_2 - 1}{2} + \cdots + \frac{p_r - 1}{2} \equiv \frac{p_1 p_2 \cdots p_r - 1}{2} \pmod{2}.$$

b) Prove that

$$\left(\frac{-1}{b}\right) = (-1)^{(b-1)/2}, \quad \text{for any odd integer } b.$$

Exercise.

a) Show that if $p_1, p_2, \ldots, p_r$ are odd primes (not necessarily distinct), then

$$\frac{p_1^2 - 1}{8} + \frac{p_2^2 - 1}{8} + \cdots + \frac{p_r^2 - 1}{8} \equiv \frac{(p_1 p_2 \cdots p_r)^2 - 1}{8} \pmod{2}.$$

b) Prove that

$$\left(\frac{2}{b}\right) = (-1)^{(b^2-1)/8}.$$

Exercise.

a) Show that if p_i and q_j are odd primes, for $i = 1, 2, \ldots, r$ and $j = 1, 2, \ldots, s$, then

$$\sum_{i=1}^{r}\sum_{j=1}^{s} \frac{p_i - 1}{2}\frac{q_j - 1}{2} \equiv \frac{p_1 p_2 \cdots p_r - 1}{2}\frac{q_1 q_2 \cdots q_s - 1}{2} \pmod{2}.$$

b) Prove the Theorem 2 below.

Theorem 2. Quadratic Reciprocity Law for the Jacobi Symbol. Suppose that m and n are relatively prime odd integers. Then

$$\left(\frac{m}{n}\right)\left(\frac{n}{m}\right) = (-1)^{\frac{(m-1)(n-1)}{4}}.$$

Exercise.

a) Compute a table of Jacobi symbols $(\frac{a}{3533})$, for $1 \le a \le 177$. You can use Mathematica to do this (or write your own program using the QRL for the Jacobi symbol to eliminate the factorizations we had to use before).

b) Compare the speeds of various methods for computing a table of $(\frac{a}{p})$ by:
    i) squaring all the numbers from 1 to $p - 1$;
    ii) using Euler's criterion, reducing powers mod p at each stage to avoid overflowing the computers' registers;
    iii) using only the Legendre symbol and factorization;
    iv) using the Jacobi symbol with no factorization except for powers of 2.

Exercise. Compute $(\frac{-a}{p})$, for $a = 163$ and all odd primes $p \le 43$. Then try $a = 77683$ and finally try $a = 111763$. You should find that all the symbols are -1. Weird! We took these examples from Shanks [1985].

Remarks. Shanks [1972, p. 53] states that $p = 26437\,68047\,3689$ is prime and that

$$\left(\frac{p}{q}\right) = 1, \quad \text{for all primes } q = 3, 5, \ldots, 149.$$

Actually he includes $q = 2$ (the Kronecker symbol).

This example and those in the preceding exercise are of interest for algebraic number theory. Long strings of 1s or $-$1s affect the size of the Dirichlet L-functions attached to the Kronecker symbol defined by:

$$L_d(s) = \sum_{n=1}^{\infty} \left(\frac{d}{n}\right) n^{-s}, \quad \text{for Re } s > 1. \tag{8}$$

This L-function is an analogue of the Riemann zeta function. It is the sort of zeta function Dirichlet used to prove that there are infinitely many primes in arithmetic progressions:

$$a, a \pm d, a \pm 2d, \ldots, a \pm kd, \ldots,$$

provided that a and d have no common factor. See Ireland and Rosen [1982].

There are other interesting questions about quadratic residues (squares in $(\mathbb{Z}/n\mathbb{Z})$. For example, it has been conjectured by Vinogradov that if $n(p)$ denotes the least positive quadratic nonresidue for a given prime p, then $n(p) < Cp^e$, where e is an arbitrary positive number. It turns out that this question is connected with the "extended" Riemann hypothesis for Dirichlet L-functions.

These questions are still open. See Chowla [1965], Guy [1984], and LeVeque [1974] for some discussion.

Exercise. Pólya–Vinogradov Inequality. Show that for all $n, m \in \mathbb{Z}$, we have

$$\left| \sum_{j=m}^{n} \left(\frac{j}{p} \right) \right| \leq p^{1/2} \log p. \qquad (9)$$

Hint. See Davenport [1980].

This last exercise implies that quadratic residues and nonresidues are equally distributed, in the sense that if a and b are fixed and p is large, there are almost as many quadratic residues as nonresidues in the interval (ap, bp), since $p^{1/2} \log p$ is small when compared with p. It also gives an estimate for the least positive quadratic nonresidue for the prime p of $n(p) < C p^{1/2} \log p$.

Gauss Sums

We have seen Gauss sums in Chapters 2, 4, and 5, as well as in the proof of the quadratic reciprocity law. There is still much more that is known about them and their generalizations. In fact, they even have an application to the design of concert halls as we shall see. But first let's look at a slightly different formula for them than that used in the proof of the quadratic reciprocity law.

Definition.
The *Gauss sum G_p* is defined by

$$G_p(x) = \sum_{k \in \mathbb{Z}/p\mathbb{Z}} \exp\left(\frac{2\pi i k^2 x}{p} \right). \qquad (10)$$

The Gauss sum defined by formula (6) is actually a special case of that defined by formula (10) as the following exercise shows.

Exercise. Show that for an odd prime p not dividing x, we have

$$\hat{h}_p(-x) = G_p(x).$$

Hint. Use an earlier exercise that says that the number of solutions to the congruence $x^2 \equiv a \pmod{p}$ is $1 + \left(\frac{a}{p} \right)$.

Exercise. Note that the definition of $G_p(x)$ above makes sense even if p is not prime, assuming that p is an odd positive integer. Show then that

$$G_{pq}(1) = G_p(q)G_q(p), \quad \text{for all distinct odd primes } p, q.$$

Hint. Compute the left-hand side from the definition using the fact that every integer $k \bmod pq$ has a unique representation as $k = ap + bq$, where $a \in \mathbb{Z}/q\mathbb{Z}$ and $b \in \mathbb{Z}/p\mathbb{Z}$.

Exercise. Show that the Gauss sum $G_p(1)$ above is the trace of the Fourier transform on $\mathbb{Z}/p\mathbb{Z}$.

So now we know that $G_p(1) = \hat{h}_p(-1) = g =$ the constant defined in formula (6), which figured so prominently in our proof of the quadratic reciprocity law. Note that

$$\left(\frac{p}{q}\right)\left(\frac{q}{p}\right) = \frac{G_q(p)G_p(p)}{G_q(1)G_p(1)} = \frac{G_{pq}(1)}{G_q(1)G_p(1)}.$$

How do we know that the Gauss sums in the denominator do not vanish?

As we noted in Chapter 5, Gauss actually evaluated g and thus obtained another proof for the quadratic reciprocity law. Gauss worked on the evaluation of the sum g for four years (between 1801 and 1805). The answer came to him suddenly "the way lightning strikes." The result he obtained for odd integers k is

$$g = G_k(1) = \sqrt{k}(i)^{[(k-1)/2]^2} = \sqrt{k}\begin{cases} 1, & k \equiv 1 \pmod 4, \\ i, & k \equiv 3 \pmod 4. \end{cases} \tag{11}$$

Here we will only consider the case $k = p = $ prime. Since we know $g^2 = (-1)^{(p-1)/2}p$, it follows that $g = \pm i^{(p-1)/2}\sqrt{p}$. Thus, in proving (11), we must just decide which sign is correct. This problem only has meaning if we specify the primitive pth root of unity as $w = \exp(2\pi i/p)$ rather than as a root of $x^p = 1$, which generates the multiplicative group of pth roots of unity, that is, w is specified by analysis rather than by algebra.

Exercise. Show that formula (11) implies the quadratic reciprocity law.
Hint. It helps to use the congruence

$$\left(\frac{pq-1}{2}\right)^2 - \left(\frac{p-1}{2}\right)^2 - \left(\frac{q-1}{2}\right)^2 \equiv \frac{(p-1)(q-1)}{2} \pmod 4.$$

We will soon give a sketch of a proof of (11), for odd primes p, from Auslander and Tolimieri [1985]. There are other proofs of (11) in Rademacher [1964, p. 92] and Ireland and Rosen [1993, p. 75]. For a proof involving theta functions, see Dym and McKean [1972, pp. 224–6] and Terras [1985, Vol. I, p. 195]. The latter proof uses the *Landsberg – Schaar reciprocity law* for Gauss sums:

$$2p^{-1/2}G_p(q) = (2q)^{-1/2}e^{\pi i/4}G_{4q}(-p).$$

See also B. Berndt and R. Evans [1981].

In fact, the evaluation of Gauss sums can be extended to replace p by any number n.

*Theorem 3. **Evaluation of Gauss Sums.*** Suppose that the Gauss sum G_n is defined as above by replacing p by n. Then

$$G_n(1) = \begin{cases} (1+i)\sqrt{n}, & \text{for } n \equiv 0 \,(\text{mod } 4), \\ \sqrt{n}, & \text{for } n \equiv 1 \,(\text{mod } 4), \\ 0, & \text{for } n \equiv 2 \,(\text{mod } 4), \\ i\sqrt{n}, & \text{for } n \equiv 3 \,(\text{mod } 4). \end{cases}$$

See T. Nagell [1964] or E. Landau [1927] for a proof.

Carlitz [1959] used this theorem to find the characteristic polynomial $c_{F_n}(x)$ for the discrete Fourier transform matrix

$$F_n = (\omega^{rs})_{0 \le r, s \le n-1}, \quad \text{where } \omega = \exp\left(\frac{2\pi i}{n}\right).$$

The characteristic polynomial of $F_n = \det(xI - F_n) = c_{F_n}(x)$ is found to be

$c_{F_n}(x) =$
$$\begin{cases} (x - \sqrt{n})^2(x - i\sqrt{n})(x + \sqrt{n})(x^4 - n^2)^{(n-4)/4}, & \text{for } n \equiv 0 \,(\text{mod } 4), \\ (x - \sqrt{n})(x^4 - n^2)^{(n-1)/4}, & \text{for } n \equiv 1 \,(\text{mod } 4), \\ (x^2 - n)(x^4 - n^2)^{(n-2)/4}, & \text{for } n \equiv 2 \,(\text{mod } 4), \\ (x - i\sqrt{n})(x^2 - n)(x^4 - n^2)^{(n-3)/4}, & \text{for } n \equiv 3 \,(\text{mod } 4). \end{cases}$$

Exercise. Prove that Theorem 3 gives the preceding evaluation of the characteristic polynomial of the DFT.

Next we briefly examine a method which goes back to Schur in 1921 for evaluating the Gauss sum $G_p(1)$ for odd primes p. We take our discussion from Auslander and Tolimieri [1985].

Let g be a primitive root mod p, that is, g gives a generator of the multiplicative group $(\mathbb{Z}/p\mathbb{Z})^*$. Define the multiplicative character χ_a of $(\mathbb{Z}/p\mathbb{Z})^*$ by

$$\chi_a(g^b) = \exp\left(\frac{2\pi i a b}{p-1}\right), \quad \text{for } a, b = 0, 1, 2, \ldots, p-2.$$

Then define $\chi_a(0) = 0$. Note that setting $s = (p-1)^{-1/2}$ we have that

$$\mathscr{B}_1 = \{\delta_0, s\chi_0, \ldots, s\chi_{p-2}\}$$

is an orthonormal basis of $L^2(\mathbb{Z}/p\mathbb{Z})$.

Let $\mathscr{M}_p = p^{-1/2}\mathscr{F}_p$, where $\mathscr{F}_p$ is the discrete Fourier transform defined in Chapter 2. By the inversion formula for the DFT (in Theorem 1 of Chapter 2), we have

$$\mathscr{M}_p^2 f(x) = f(-x) \quad \text{for all } f \in L^2(\mathbb{Z}/p\mathbb{Z}) \text{ and } x \in \mathbb{Z}/p\mathbb{Z}.$$

Another basis of $L^2(\mathbb{Z}/p\mathbb{Z})$ is obtained by setting $r = (p-1)/2, s = (p-1)^{-1/2}$:

$$\mathscr{B}_2 = \{\delta_0, s\chi_0, s\chi_1, s\mathscr{M}_p\chi_1, \ldots, s\chi_{r-1}, s\mathscr{M}_p\chi_{r-1}, s\chi_r\}.$$

The matrix of the normalized DFT $\mathscr{M}_p$ with respect to basis $\mathscr{B}_1$ is

$$p^{-1/2} \begin{pmatrix} 1 & \sqrt{p-1} & 0 & \cdots & 0 \\ \sqrt{p-1} & -1 & 0 & \cdots & 0 \\ 0 & 0 & 0 & & G(\chi_{p-2}) \\ \vdots & \vdots & \vdots & \reflectbox{$\ddots$} & \vdots \\ 0 & 0 & G(\chi_1) & \cdots & 0 \end{pmatrix}.$$

Here $G(\chi_j)$ denotes the Gauss sum

$$G(\chi_j) = \sum_{u \in (\mathbb{Z}/p\mathbb{Z})^*} \chi_j(u) \exp\left(\frac{-2\pi i u}{p}\right) = p^{1/2}\mathscr{M}_p\chi_j(1).$$

To see that the matrix above is correct, you just need to express $\mathscr{M}_p v$, $v \in \mathscr{B}_1$, as a linear combination of basis vectors from $\mathscr{B}_1$. For example, when

$x \not\equiv 0 (\mathrm{mod}\, p)$, setting $u = xy$, we have

$$\mathcal{M}_p \chi_a(x) = p^{-1/2} \sum_{y \in (\mathbb{Z}/p\mathbb{Z})^*} \chi_a(y) \exp\left(\frac{-2\pi i x y}{p}\right)$$

$$= p^{-1/2} \sum_{y \in (\mathbb{Z}/p\mathbb{Z})^*} \chi_a(x^{-1}u) \exp\left(\frac{-2\pi i u}{p}\right)$$

$$= p^{-1/2} \chi_a(x^{-1}) \sum_{y \in (\mathbb{Z}/p\mathbb{Z})^*} \chi_a(u) \exp\left(\frac{-2\pi i u}{p}\right)$$

$$= p^{-1/2} \chi_{p-1-a}(x) G(\chi_a).$$

And when $x \equiv 0 (\mathrm{mod}\, p)$, we have

$$\mathcal{M}_p \chi_a(x) = p^{-1/2} \sum_{y \in (\mathbb{Z}/p\mathbb{Z})^*} \chi_a(y) = 0, \quad \text{if } a \not\equiv 0 \,(\mathrm{mod}\, p - 1).$$

Thus $\mathcal{M}_p \chi_a = p^{-1/2} G(\chi_a) \chi_{p-1-a}$. This explains all but the top two rows and left-hand columns. We leave that to you.

The matrix of $\mathcal{M}_p$, the normalized DFT, with respect to the basis $\mathcal{B}_2$, is

$$\begin{pmatrix} M_0 & 0 & \cdots & 0 & 0 \\ 0 & M_1 & & 0 & 0 \\ \vdots & \vdots & \ddots & \vdots & \vdots \\ 0 & 0 & & M_{r-1} & 0 \\ 0 & 0 & \cdots & 0 & G(\chi_r) \end{pmatrix},$$

where

$$M_j = \begin{pmatrix} 0 & (-1)^j \\ 1 & 0 \end{pmatrix}, \quad \text{for } j \neq 0$$

and

$$M_0 = p^{-1/2} \begin{pmatrix} 1 & \sqrt{p-1} \\ \sqrt{p-1} & -1 \end{pmatrix}.$$

To see this formula, you need to note that

$$\mathcal{M}_p^2 \chi_a(x) = \chi_a(-x) = \chi_a(-1) \chi_a(x).$$

Suppose g is our chosen generator of $(\mathbb{Z}/p\mathbb{Z})^*$ which we used to define χ_a above. Then

$$-1 = g^{(p-1)/2} \,(\mathrm{mod}\, p).$$

It follows that

$$\chi_a(-1) = \exp\left(\frac{2\pi i a(p-1)}{2(p-1)}\right) = (-1)^a.$$

So the determinants of the M_j alternate in sign. Thus they cancel out if $r = (p-1)/2$ is even and give -1 if r is odd.

It follows that

$$\det(\mathcal{M}_p) = \begin{cases} G(\chi_r)(-1)^m, & \text{if } p = 4m+1, \\ -G(\chi_r)(-1)^m, & \text{if } p = 4m+3. \end{cases}$$

To complete Schur's evaluation of the Gauss sum $G(\chi_r)$ given by (11) above, you need only note that the determinant of the matrix $p^{1/2}\mathcal{M}_p = \mathcal{F}_p$ is a Vandermonde determinant, which was evaluated in an exercise of Chapter 2 to give

$$\det(\mathcal{M}_p) = \begin{cases} (-1)^m, & \text{if } p = 4m+1, \\ -i(-1)^m, & \text{if } p = 4m+3. \end{cases}$$

This implies the desired formula for the Gauss sum $G_p(1)$. ∎

Exercise. Check the details of the preceding proof.

An Application to the Design of Concert Halls

Schroeder [1986, p. 181] notes that there is an application of residues to the design of concert hall ceilings. Such ceilings are designed with the aim of obtaining a constant power spectrum. One wants to diffuse sound waves to improve the sound quality. Schroeder proposes the ceiling design pictured in Figure I.26 based on a quadratic residue sequence for the prime 17:

$$1, 4, 9, 16, 25 - 17 = 8, \quad 36 - 2 \cdot 17 = 2, \quad 49 - 2 \cdot 17 = 15, 13,$$

and it repeats backwards.

Let

$$f(n) = \exp\left(\frac{2\pi i n^2}{p}\right).$$

The *periodic correlation sequence* of f is

$$c(k) = \sum_{n=0}^{p-1} f(n)\overline{f(n+k)} = p\delta_0(k).$$

The DFT of $f(n)$ has absolute value $p^{1/2}$, since it is essentially the Gauss sum.

Ceiling

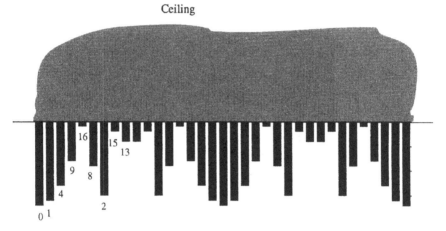

Figure I.26. Proposed concert hall ceiling design, a reflection phase-grating based on a quadratic residue sequence for the prime 17 from Schroeder [1986].

That is,

$$A(m) = \sum_{n=0}^{p-1} f(n)\exp\left(\frac{2\pi i n m}{p}\right) = \exp\left(\frac{-2\pi i m^2}{4p}\right) \sum_{n=0}^{p-1} f(n).$$

The *power spectrum* is

$$|A(m)|^2 = \text{the DFT of the correlation sequence} = p.$$

A constant power spectrum is called "flat" or "white." It is useful, for instance, a good loud speaker should have one (or a concert hall ceiling). Make a 2-dimensional grid, with rectangle at (n, m) of height $(n^2 + m^2) \pmod{p}$ to get a surface that scatters incident radiation widely with little of the total reflected energy going in any given direction. See Figures I.22 and I.23 in Chapter 5.

Last Remarks and Exercises

● *Another Application of Quadratic Residues*
Quadratic residues are of importance in the theory of error-correcting codes. See Chapter 16 of Jessie MacWilliams and Sloane [1988]. We will discuss such codes soon.

Exercise. Suppose $\mu(n)$ is the Möbius function defined in formula (8) of Chapter 1. Show that

$$\mu(n) = \sum_{\substack{1 \le k < n \\ (k,n)=1}} \exp\left(\frac{2\pi i k}{n}\right).$$

Here $(k, n) = $ g. c. d.(k, n).

Hint. Note that both sides are multiplicative functions of n as defined in Chapter 1. This reduces the proof to the case that $n = p^e$, for some prime p.

• *Gauss Sums as Finite Gamma Functions*
As noted in Chapter 5, the Gauss sum

$$\sum_{a \in (\mathbb{Z}/p\mathbb{Z})^*} \left(\frac{a}{p}\right) e^{2\pi i a x/p}$$

can be viewed as a finite analogue of the Gamma function. Many other special functions have finite analogues. We have already seen that Kloosterman sums are finite analogues of Bessel functions. See formula (12) in Chapter 5. For more examples, see R. Evans [1986,1994] and Angel, Celniker, Poulos, Terras, Trimble, and Velasquez [1994]. We will say more about such things in Chapter 19 on finite upper half planes.

Exercise. Gauss Sums over Rings of Prime Power Order. Let $q = p^r$, where $p \geq 3$ is prime and $r \geq 1$. For $v \in \mathbb{Z}_q = \mathbb{Z}/q\mathbb{Z}$, $v \neq 0$, define the *Gauss sum*

$$G_v^{(r)} = \sum_{y \in \mathbb{Z}_q} \exp(2\pi i v y^2/q).$$

Prove that $G_v^{(r)} = p G_v^{(r-2)}$. It follows that

$$G_v^{(r)} = \begin{cases} p^{r/2}, & r \text{ even}, \\ p^{(r-1)/2} G_v^{(1)}, & r \text{ odd}. \end{cases}$$

One can then evaluate $G_v^{(r)}$, using the case of Theorem 3 proved above – Gauss's formula for his Gauss sum mod p. Compare with formula (11).

These Gauss sums appear in the study of finite euclidean graphs over rings. See Medrano et al. [1998]. A different sort of Gauss sums appears in Odoni [1973].

• *Discrete Analogues of Weyl's Theorem on Uniform Distribution*
This has been a chapter devoted mostly to a topic in the theory of numbers. There are other questions coming from number theory with finite analogues. For example, we define a sequence $\{\alpha_n\}_{n \geq 1}$ of numbers $\alpha_n \in [0, 1] \cong \mathbb{R}/\mathbb{Z} \cong \mathbb{T}$ to be *uniformly distributed* if for any interval $(a, b) \subset [0, 1]$, we have

$$\lim_{n \to \infty} \frac{|\{\alpha_k \in (a, b) \mid k = 1, \ldots, n\}|}{n} = b - a.$$

H. Weyl [1968] showed in 1916 that it is equivalent to prove that for every step function f of period 1 we have

$$\lim_{n \to \infty} \frac{\sum_{k=1}^{n} f(\alpha_k)}{n} = \int_a^b f(x)\, dx.$$

Then Weyl shows it is equivalent to check the limit for every bounded Riemann integrable function f, or for every trigonometric polynomial f, or for every character $\exp(2\pi i b x)$, for $b \in \mathbb{Z}$. Thus the sequence $\{\alpha_n\}_{n \geq 1}$ is uniformly distributed iff

$$\lim_{n \to \infty} \frac{\sum_{k=1}^{n} \exp(2\pi i b \alpha_k)}{n} = \int_a^b \exp(2\pi i b x)\, dx = \begin{cases} 1, & b = 0, \\ 0, & b \neq 0. \end{cases}$$

For example, one can use these criteria to see that if β is an irrational number and $\alpha_n = n\beta$ (modulo 1), then the sequence $\{\alpha_n\}_{n \geq 1}$ is uniformly distributed in [0,1). Ajtai et al. [1990] find a discrete analogue of Weyl's theorem and then construct a set with a small DFT. This has applications to computer science and graph theory.

Exercise. Write a report on the investigations of Ajtai et al. [1990] mentioned above on finite analogues of Weyl's equidistribution theorem.

Chapter 9
The Fast Fourier Transform or FFT

It [the FFT] also dominates the search for offshore oil. Water layers produce "ringing" that masks the signals from oil or gas; those primary reflections could not be seen before 1960. Now virtually all exploration data is digital and more than 10^9 systems of equations are solved every year – partly by the FFT and partly by Levinson's algorithm for constant-diagonal (Toeplitz) matrices. The key problem is deconvolution...

Strang [1986, p. 467].

Despite a fear of oil slicks on the beach in Encinitas, we want to sketch the theory of the fast Fourier transform or FFT. The simplest version of this transform allows one to compute the DFT on a finite cyclic group of order $n = 2^k$ elements in $n \log n$ operations rather than n^2 operations. Other algorithms exist if n is not a power of 2 but we will not discuss them here. The FFT has revolutionized digital signal processing.

We are mostly following the discussion in G. Strang [1986]. The FFT is usually attributed to Cooley and Tukey [1965]. However, Heidemann et al. [1984] note that the Cooley–Tukey algorithm was essentially known to Gauss in 1805, which is two years before Fourier and 160 years before Cooley and Tukey.

Brigham [1974] says that R. L. Garwin asked Tukey to give him a rapid way to compute the Fourier transform during a meeting of the President's Scientific Advisory Committee. Then Garwin went to the computing center at IBM Research in Yorktown Heights where J. W. Cooley programmed the Fourier transform, because he had nothing better to do. After receiving many requests for the program, Cooley and Tukey published their paper in 1965.

It turned out that many other people besides Gauss were using similar techniques (e.g., Danielson and Lanczos [1942], Good [1958], Runge [1903], Runge and König [1964], Stumpff [1939], and Thomas [1963]).

Other references for the FFT are Blahut [1991], Elliott and Rao [1982], Nussbaumer [1982], Tolimieri et al. [1989], and Walker [1991].

The discrete Fourier transform on $\mathbb{Z}/n\mathbb{Z}$ can be viewed as a matrix. Recall that the DFT is defined by

$$\hat{f}(x) = \sum_{y=0}^{n-1} f(y) \exp(-2\pi i x y/n).$$

Set $w = w_n = \exp(2\pi i/n)$ and

$$f = {}^t(f(0), f(1), \ldots, f(n-1)), \quad g = {}^t(\hat{f}(0), \hat{f}(1), \ldots, \hat{f}(n-1)).$$

Then

$$g = F_n f,$$

where

$$F_n = \begin{pmatrix} 1 & 1 & 1 & \cdots & 1 \\ 1 & w & w^2 & \cdots & w^{n-1} \\ \vdots & \vdots & \vdots & & \vdots \\ 1 & w^{n-2} & w^{2(n-2)} & \cdots & w^{(n-2)(n-1)} \\ 1 & w^{n-1} & w^{2(n-1)} & \cdots & w^{(n-1)(n-1)} \end{pmatrix} = (w^{ij})_{0 \le i, j \le n-1}.$$

Computing $\hat{f}(x)$ requires n^2 multiplications. We will ignore the additions, as they are cheaper. If $n = 2m$, you can factor the matrix F_n and save time and money. To do this, note that

1) $n = 2m$ implies $w_n^2 = w_m$, since $\exp\left(\frac{2\pi i \cdot 2}{2m}\right) = \exp\left(\frac{2\pi}{m}\right)$.
2) We can split the vector f of values of the function $f(x)$ into even and odd components:

$$f = (f', f''), \quad \text{where } f' = {}^t(f(0), f(2), \ldots, f(n-2)) \quad \text{and}$$
$$f'' = {}^t(f(1), f(3), \ldots, f(n-1)).$$

Then, for $0 \le j \le n - 1$, we have

$$g_j = (F_n f)_j = \sum_{k=0}^{n-1} w_n^{kj} f(k)$$
$$= \sum_{r=0}^{m-1} w_n^{2rj} f(2r) + \sum_{r=0}^{m-1} w_n^{(2r+1)j} f(2r+1)$$

$$= \sum_{r=0}^{m-1} w_m^{rj} f(2r) + w_n^j \sum_{r=0}^{m-1} w_m^{rj} f(2r+1)$$

$$= (F_m f')_j + w_n^j (F_m f'')_j$$

$$= g_j' + w_n^j g_j'', \quad \text{if } g' = F_m f' \quad \text{and} \quad g'' = F_m f''.$$

3) If $0 \le j \le m-1$ and we look at g_{j+m}, we obtain

$$g_{j+m} = \sum_{k=0}^{m-1} w_m^{k(j+m)} f_k' + w_n^{j+m} \sum_{k=0}^{m-1} w_m^{k(j+m)} f_k''.$$

Now $w_m^{k(j+m)} = w_m^{kj}$ and $w_n^{j+m} = w_n^j w_{2m}^m = -w_n^j$. Thus

$$g_{j+m} = (F_m f')_j - w_n^j (F_m f'')_j.$$

Summary of the Beginning of the FFT Algorithm

For $n = 2m$ and $0 \le j \le m-1$, we have

$$f' = (f(0), f(2), \ldots, f(n-2)), \quad f'' = (f(1), f(3), \ldots, f(n-1)),$$

$$g' = F_m f', \quad g'' = F_m f'',$$

$$g_j = g_j' + w_n^j g_j'', \quad \text{and} \quad g_{j+m} = g_j' - w_n^j g_j''. \tag{1}$$

Theorem 1. What is the Number of Multiplications Needed to Compute F_n?
After using the Cooley–Tukey procedure r times, we get $\frac{n}{2}(r+2)$ multiplications, which is less than $n \log n$ and much less than n^2.

Proof. Without the FFT, we had $(2m) \cdot (2m) = 4m^2$ multiplications. After one step given by formula (1), we have $2m^2 + m$ multiplications. Next suppose that $n = 2^r$ and repeat the procedure r times. The number of multiplications needed without the FFT would be $n^2 = 2^{2r} = n2^r$.

To prove that the multiplication count is correct, use mathematical induction on r. If $r = 1$, we have $m = 1$ and Cooley–Tukey gives three operations rather than four, as we said above.

The induction step goes from $m = 2^r$ to $n = 2^{r+1}$. Let $M(n)$ be the number of multiplications needed to compute F_n. We have

$$M(n) = 2M(m) + m = \frac{n}{2}(r+2) + 2^r = 2^r(r+2) + 2^r = 2^r(r+3),$$

which is the correct formula. ∎

Example. When $m = 2$ and $n = 4 = 2m$, we have

$$F_2 = \begin{pmatrix} 1 & 1 \\ 1 & -1 \end{pmatrix}.$$

Then

$$\begin{pmatrix} f(0) \\ f(1) \\ f(2) \\ f(3) \end{pmatrix} \xrightarrow{\alpha} \begin{pmatrix} f(0) \\ f(2) \\ f(1) \\ f(3) \end{pmatrix} = \begin{pmatrix} f' \\ f'' \end{pmatrix} \xrightarrow{\gamma} \begin{pmatrix} F_2 f' \\ F_2 f'' \end{pmatrix} \xrightarrow{\delta} g.$$

And formula (1) gives the factorization of F_4 as a composition $\delta \circ \gamma \circ \alpha$:

$$F_4 = \begin{pmatrix} 1 & 1 & 1 & 1 \\ 1 & i & i^2 & i^3 \\ 1 & i^2 & i^4 & i^6 \\ 1 & i^3 & i^6 & i^9 \end{pmatrix}$$

$$= \begin{pmatrix} 1 & 0 & 1 & 0 \\ 0 & 1 & 0 & i \\ 1 & 0 & -1 & 0 \\ 0 & 1 & 0 & -i \end{pmatrix} \begin{pmatrix} 1 & 1 & 0 & 0 \\ 1 & -1 & 0 & 0 \\ 0 & 0 & 1 & 1 \\ 0 & 0 & 1 & -1 \end{pmatrix} \begin{pmatrix} 1 & 0 & 0 & 0 \\ 0 & 0 & 1 & 0 \\ 0 & 1 & 0 & 0 \\ 0 & 0 & 0 & 1 \end{pmatrix}.$$

Exercise.

a) Check the preceding factorization of F_4.

b) Obtain the analogous factorization of F_6.

Rader [1968] showed how to speed up the DFT on $\mathbb{Z}/p\mathbb{Z}$ when p is a prime. This makes use of the fact that the multiplicative group

$$U_p = \{a \ (\mathrm{mod} \ p) \mid p \text{ does not divide } a\}$$

consists of all powers mod p of a primitive element g mod p. See Theorem 4 in Chapter 1. Tolimieri et al. [1989] has more information on Rader's approach.

The FFT is used to approximate the usual infinite Fourier series as well as to transform the problem of multiplication of two polynomials into a more rapidly computable problem. The asymptotically fastest known algorithm for multiplication of integers is obtained using the FFT. See Aho, Hopcraft, and Ullman [1974], Knuth [1981], and Press et al. [1986].

One can extend the idea in the FFT algorithm from the modulus $n = 2^r$ to a modulus of the form $n = rs$ and then to products of arbitrary numbers of factors. Here one uses the fact that any integer k modulo rs has the unique representation:

$$k = k_2 r + k_1 \quad \text{with } k_1 \in \mathbb{Z}/r\mathbb{Z} \quad \text{and} \quad k_2 \in \mathbb{Z}/s\mathbb{Z}. \qquad (2)$$

Thus the DFT on $\mathbb{Z}/rs\mathbb{Z}$ has the following expression, setting $w = \exp(2\pi i/rs)$, $m = m_1 s + m_2$, with $m_1 \in \mathbb{Z}/r\mathbb{Z}$ and $m_2 \in \mathbb{Z}/s\mathbb{Z}$:

$$F_{rs} f(k) = \sum_{m_2 \in \mathbb{Z}/s\mathbb{Z}} \sum_{m_1 \in \mathbb{Z}/r\mathbb{Z}} f(m) w^{skm_1} w^{km_2}.$$

We then have

$$skm_1 = k_2 m_1 rs + sk_1 m_1, \quad \text{while } km_2 = k_2 m_2 r + k_1 m_2.$$

So

$$F_{rs} f(k) = \sum_{m_2 \in \mathbb{Z}/s\mathbb{Z}} \sum_{m_1 \in \mathbb{Z}/r\mathbb{Z}} f(m) \, w^{sk_1 m_1} w^{(k_2 r + k_1)m_2}$$

$$= \sum_{m_2 \in \mathbb{Z}/s\mathbb{Z}} \left(\sum_{m_1 \in \mathbb{Z}/r\mathbb{Z}} f(m) \, w^{sk_1 m_1} \right) w^{(k_2 r + k_1)m_2}.$$

The inner sum is a DFT on $\mathbb{Z}/r\mathbb{Z}$ and the outer one is a DFT on $\mathbb{Z}/s\mathbb{Z}$ of the inner one times $w^{k_1 m_2}$. You can view the matrix of the whole mess as a tensor product. See Tolimieri et al. [1989, Chapter 3] or Brigham [1974, Chapter 12] for more discussion. Use can also be made of the Chinese remainder Theorem (see Elliott and Rao [1982, Chapter 5]). Generalizations of the FFT to noncommutative finite groups have been found by Diaconis and Rockmore [1988].

An Application of the FFT – Digital Spectral Analysis

Newland [1993, p. 165] writes: "Since 1965, when the FFT was brought onto the scene, it has profoundly changed our approach to digital spectral analysis."

The history of digital spectral analysis is a long one. Let us sketch a bit of it. See Marple [1987, Chapter 1] for more details. People have always needed to know when the rainy season would come, when the tide would turn, when the eclipse would occur. After Fourier in the 1800s, it was possible to build machines to do harmonic analysis. For example, in 1872, Sir William Thomson had built a machine to predict the tides. In 1898, Schuster created a method to find hidden periodicities in time series such as sunspot numbers. This work is the basis for the well-known theory that there is an 11-year sunspot cycle. Many books on time series and digital spectral analysis consider the sunspot example in great detail. See Bloomfield [1976], Brillinger [1981], and Marple [1987]. Matlab has a nice demo of the use of the FFT to find the period in this example. The basic idea of the method, in modern language is: Given $x(t), t = 1, 2, \ldots, n$, look at a graph of $|\hat{x}(s)|^2$ (called the power spectral density or periodogram). Find the values s_0 of s making this power spectral density a maximum. These should be the dominant periods. The Matlab demo does come up with period $\cong 11$ years for the sunspot data.

However, the early researchers often found periodograms confusing, even if one increased the data length. This led to unhappiness with the method. For one needs to use smoothing methods from Fourier analysis as well as statistical tests of significance in order to find hidden periods with much hope of success.

Bloomfield [1976, p. 6] relates, for example, the case of the investigator in 1897 who claimed to discover periods related to the lunar cycle in some Japanese earthquakes. Schuster showed in 1897 that the results were not statistically significant.

Economic time series analysis seems even harder. Granger and Newbold [1986, p. 58] report that for economic data it was found in the early 1960s that "most of the estimated spectra had identical shapes, being very high at very low frequencies, falling steeply as frequency increased, becoming flat at high frequencies (ω near π) and with only the occasional peak at a seasonal frequency to break up the inherent smoothness. . . . The basic reason for the typical spectral shape arising is the very high correlation observed between adjacent values of the levels of most economic series when observed frequently . . ."

To get some feeling for the subject, let's look at an example of interest to someone living in a semidesert with a yard full of rain forest plants such as avocado trees.

Example of Spectral Analysis of L.A. Rainfall Data Using Matlab and the FFT

Look at $w(t) =$ inches of rainfall at the Los Angeles Civic Center from January to December in year t. See Table I.13. Note that since the rainy season is the winter, this measuring period is rather unnatural. The graph in Figure I.27 shows the data from Table I.13.

Table I.13. *Los Angeles rainfall (measured in inches at L.A. Civic Center from January through December). The first number (top left) is from the year 1992 and the last (lower right) from 1878 (from the Los Angeles Times, Jan. 7, 1993, p. B2)*

22.65	15.07	6.49	4.56	11.57	9.11	18	8.92	8.9	34.04
14.41	10.92	26.33	17	30.57	14.97	11.01	10.7	16.69	17.45
6.54	9.26	16.54	26.32	7.58	23.66	12.91	26.81	7.98	12.31
15.37	5.83	9.57	6.23	17.49	13.24	13.62	11.89	13.69	4.08
24.95	14.33	7.38	10.63	7.59	4.12	16.22	12.78	17.45	22.57
7.4	31.28	20.26	12.06	27.16	17.97	18.24	14.49	14.67	18.76
10.72	18.93	13.02	8.32	8.69	18.63	18.56	8.94	8.11	6.25
15.27	19.85	11.18	8.82	17.49	8.45	23.29	16.67	23.21	17.17
9.78	17.85	4.89	23.92	13.74	15.3	21.46	19.19	11.88	14.77
13.12	11.96	11.3	8.69	4.83	14.28	11.8	12.55	7.51	21.96
18.72	12.84	12.69	33.26	20.82	16.02	16.72	10.53	40.29	14.14
10.74	5.53	18.65	17.41	20.86					

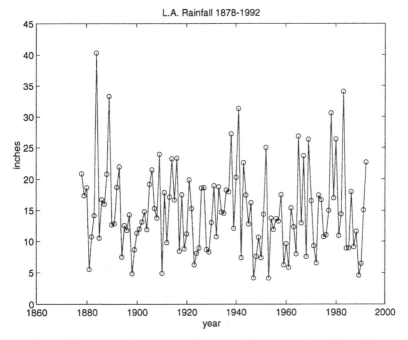

Figure I.27. A plot of the vector of L.A. rainfall data $w(t)$ from Table I.13.

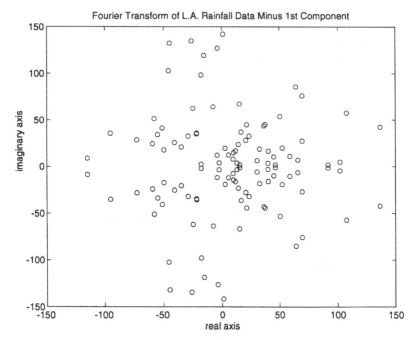

Figure I.28. Plot of the values $\hat{w}(t)$ of the Fourier transform of the L.A. rainfall data with the first component deleted.

If we imitate the Matlab FFT demo, which analyzes sunspot data, we need to take the FFT of the vector w from Table I.13. The vector $z = w$ has complex values. The first value $\hat{w}(0)$ is the sum of the values of the vector $w(t)$. We leave it out in the graph given in Figure I.28. The Matlab command used to obtain Figure I.28 is

$$plot(z, \text{'co'}).$$

The Student Edition of Matlab, Version 4 User's Guide (see Math Works [1995, p. 399]) has the following to say about the FFT algorithm that they use:

When the sequence length is a power of two, a high-speed radix-2 fast Fourier transform algorithm is employed .... When the sequence length is not an exact power of two, an alternate algorithm finds the prime factors of the sequence length and computes the mixed-radix discrete Fourier transform of the shorter sequences.... For example, on one machine a 4096-point real FFT takes 2.1 seconds and a complex FFT of the same length takes 3.7 seconds. The FFTs of neighboring sequences of length 4095 and 4097, however, take 7 seconds and 58 seconds, respectively.

You probably cannot tell much from Figure I.28. So look instead at Figure I.29, which is the periodogram or plot of $|\hat{w}(t)|^2$. There is a maximum at 12.67

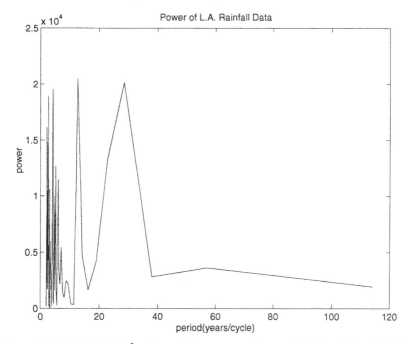

Figure I.29. Plot of $|\hat{w}(t)|^2$ – the periodogram or power spectral density for the L.A. rainfall data. Maximum occurs at 12.67 years.

years. Does this mean anything? To answer this question you would need to read Marple [1987] or some of the other references mentioned earlier.

The references on digital spectral analysis and time series analysis tell us that we should use some sort of smoothing or filtering methods to improve the accuracy of our period numbers. Matlab gives a way to take advantage of Welch's method of averaging periodograms of overlapping sections. See The Matlab Version 4 Student Edition User's Guide (Math Works [1995, pp. 776–9]) and Welch [1967]. We used the Matlab Signal Processing Toolbox command psd (included in the Version 4 Student Edition but not the latest Professional Edition without the Signal Processing Toolkit) to compute the power spectral density in Figure I.30. This leads to a maximum at the period 28.75 years. Does this have more meaning than the previous number 12.67 years? We leave it as an exercise for the reader to comment on this question.

Figure I.30 was generated using the Version 4 Student Matlab commands

$$[p, \ f] = \text{psd}\,(w);$$
$$f\,(0) = [\,]; \ p\,(0) = [\,];$$
$$\text{per} = 1./f;$$

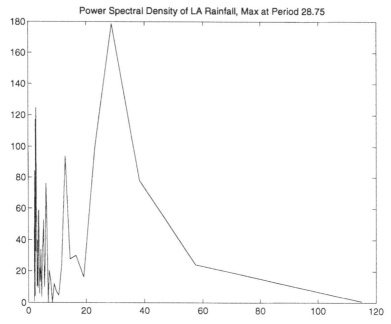

Figure I.30. Power spectral density of the L.A. rainfall data. Maximum occurs at period of 28.75 years.

$$\text{plot (per, } p)$$
$$\text{index} = \text{find} \, (p = \max \, (p))$$
$$p \, (\text{index})$$

Note that this does not work in Professional Matlab unless you have the Signal Processing Toolkit.

Chapter 10

The DFT on Finite Abelian Groups – Finite Tori

There are, in fact, many higher dimensions. One of these higher
dimensions is time, another higher dimension is the direction in
which space is curved, and still another higher dimension may
lead toward some utterly different universes existing parallel to
our own.

At the deepest level, our world can be regarded as a pattern
in infinite-dimensional space, a space in which we and our
minds move like fish in water.

R. Rucker [1984, p. 3]

In this chapter we step into higher dimensional, yet finite universes, such as
the n-dimensional vector space $\mathbb{F}_p^n$ over the finite field $\mathbb{F}_p$. Or to put it a little
differently, we move from finite cyclic groups to general finite abelian groups
(which are, by Theorem 1 below, direct products of cyclic groups). There are
many applications, for example, in digital image processing and in the theory
of error-correcting codes. The theory of error-correcting codes is needed to
transmit pictures from Mars to Earth, as well as for communications networks
on Earth. In our search for Ramanujan graphs, we have already needed Fourier
analysis on the group $\mathbb{F}_p^n$ (see Chapter 5).

So we will study products of finite circles such as

$$G = (\mathbb{Z}/m_1\mathbb{Z}) \oplus \cdots \oplus (\mathbb{Z}/m_r\mathbb{Z}),$$

which consists of vectors $v = {}^t(v_1, \ldots, v_n)$, with $v_j \in \mathbb{Z}/m_j\mathbb{Z}$. We can call this
a "finite torus" as at the end of Chapter 1; we called $\mathbb{Z}/3\mathbb{Z} \oplus \mathbb{Z}/5\mathbb{Z}$ a finite torus
even though it was also isomorphic to $\mathbb{Z}/15\mathbb{Z}$, which we think of as a finite
circle. Anyway, here we will view any finite abelian group as a finite torus and
will discuss the Fourier transform on an arbitrary finite abelian group G. But
first we need to know that an arbitrary finite abelian group is really a finite torus.

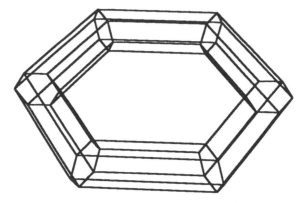

Figure I.31. The finite torus $(\mathbb{Z}/6\mathbb{Z})^2$.

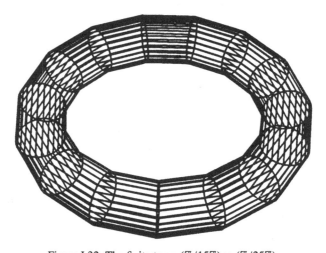

Figure I.32. The finite torus $(\mathbb{Z}/15\mathbb{Z}) \times (\mathbb{Z}/25\mathbb{Z})$.

Figures I.31 and I.32 show $(\mathbb{Z}/6\mathbb{Z})^2$ and $\mathbb{Z}/15\mathbb{Z} \times \mathbb{Z}/25\mathbb{Z}$, respectively. Figure I.32 was produced with the Mathematica commands:

$$\ll \text{Graphics'Shapes'}$$

Show [WireFrame [Graphics3D [Torus [5,1,15,25]]], Boxed →False, View-Point → {2, 1, 1}].

Figure I.37 below shows $(\mathbb{Z}/2\mathbb{Z})^4$ as the 4-dimensional cube or tesseract. It was also drawn using Mathematica commands.

Fundamental Theorem of Abelian Groups

Some references for this topic are: Dym and McKean [1972], Herstein [1964], Hungerford [1974], Lang [1984], Ono [1990], and Schreier and Sperner [1959].

Theorem 1. **Fundamental Theorem of Abelian Groups.** Every finite abelian group G is isomorphic to a direct product of cyclic groups, that is,

$$G \cong (\mathbb{Z}/m_1\mathbb{Z}) \oplus \cdots \oplus (\mathbb{Z}/m_r\mathbb{Z}),$$

where the group on the right consists of all vectors $(a_1, \ldots, a_r)$ where $a_j \in \mathbb{Z}/m_j\mathbb{Z}$. And the addition of two vectors is componentwise, that is,

$$(a_1, \ldots, a_r) + (b_1, \ldots, b_r) = (a_1 + b_1, \ldots, a_r + b_r).$$

Moreover, we can take the m_j such that m_j divides m_{j+1}, for all j. Then the list of m_j is uniquely determined by G.

Proof.

● *Step 1. Preliminaries Involving Free Abelian Groups.*

Let $g_1, \ldots, g_n$ be a system of generators of G. Thus any element x of G has the form

$$x = \sum_{i=1}^{n} a_i g_i, \quad \text{for } a_i \in \mathbb{Z}.$$

Let F be a *free abelian group of rank n* with $\mathbb{Z}$-basis $u_1, \ldots, u_n$, that is, F is generated by $u_1, \ldots, u_n$ and if we have a relation

$$\sum_{i=1}^{n} a_i u_i = 0, \quad \text{for } a_i \in \mathbb{Z},$$

it follows that all $a_i = 0$. Any element of F has the unique expression

$$\sum_{i=1}^{n} a_i u_i, \quad a_i \in \mathbb{Z}.$$

The group F is isomorphic with $\mathbb{Z}^n$. The rank n of a free abelian group F is uniquely associated to F. See Hungerford [1974].

Define a homomorphism $T : F \to G$ by

$$T(a_1 u_1 + \cdots + a_n u_n) = a_1 g_1 + \cdots + a_n g_n, \quad \text{for } a_i \in \mathbb{Z}.$$

Set $K = \ker T$. Then $G \cong F/K$. Since K is a subgroup of the free abelian group F of rank n, it follows that K is itself a free abelian group of rank $m \leq n$. See Hungerford [1974, p. 73]. ∎

Exercise. Give a detailed discussion of the preceding statements.

Note that K consists of all $\sum_{i=1}^{n} a_i u_i$ so that $\sum_{i=1}^{n} a_i g_i = 0$. That is, K is the set of relations for G. Since K is also a free abelian group, we have a $\mathbb{Z}$-basis $v_1, \ldots, v_m$ of K. Let A be an $n \times n$ matrix of integers determined by the relation

$$(v_1, \ldots, v_m, 0, \ldots, 0) = (u_1, \ldots, u_n)A. \tag{1}$$

• *Step 2. Elementary Divisor Theory.*

We can perform *elementary row and column operations* on the matrix A without losing the property that it is the matrix relating a $\mathbb{Z}$-basis of K with a $\mathbb{Z}$-basis of F.

Elementary Row (Column) Operations

1) Permute row (or column) i and row (or column) j.
2) Multiply row (or column) i by ± 1.
3) Replace row i by [(row i)+k*(row j)], for some integer k or the same for columns i and j.

These operations correspond to replacing A by UAV, where U and V are in $GL(n, \mathbb{Z})$, the general linear group of all $n \times n$ integer matrices of determinant ± 1. These operations also correspond to changing bases of the $\mathbb{Z}$-modules involved, for each operation is invertible by an elementary row or column operation.

Claim

By use of the elementary row and column operations on A, we can bring our matrix A into the following form:

$$UAV = \begin{pmatrix} e_1 & 0 & \cdots & 0 & 0 & \cdots & 0 \\ 0 & e_2 & \cdots & 0 & 0 & \cdots & 0 \\ \vdots & \vdots & \cdots & \vdots & \vdots & \cdots & \vdots \\ 0 & 0 & \cdots & e_r & 0 & \cdots & 0 \\ 0 & 0 & \cdots & 0 & 0 & \cdots & 0 \\ \vdots & \vdots & \cdots & \vdots & \vdots & & \vdots \\ 0 & 0 & \cdots & 0 & 0 & \cdots & 0 \end{pmatrix}, \tag{2}$$

where $0 < e_i$ divides e_{i+1}.

To see this, let e_1 be the smallest positive integer that occurs in any matrix obtainable from our original matrix by a finite number of elementary row and column operations. We can permute e_1 into the upper left-hand corner. We can then use the division algorithm and the third type of elementary row and column operation to replace everything in the first row and column by its remainder upon division by e_1. Since e_1 is the smallest positive number obtainable by such operations, it follows that all the remainders are 0. That explains the first row and column of the matrix (2). We can also show that e_1 divides all remaining elements of our matrix, for we can add the first row to the ith row and then add any integer multiple of the first column to the jth column to replace the i, j element by its remainder upon division by e_1. So the i, j element must be divisible by e_1.

Thus after this first step we have a matrix

$$\begin{pmatrix} e_1 & 0 \\ 0 & C \end{pmatrix}, \quad \text{where all entries of } C \text{ are divisible by } e_1.$$

If not all of the elements of the matrix C are 0, we continue the process as before. Let e_2 be the smallest positive number which occurs as a result of a finite number of elementary row and column operations on the matrix C. As before, we move e_2 to the top left place in C and argue that we can make the remaining elements of the first row and column of the new matrix vanish. All elements of the new matrix must be divisible by e_2. Use the third type of operation to replace all nondiagonal elements of the second row and column by 0. After this second step we have a matrix

$$\begin{pmatrix} e_1 & 0 & 0 \\ 0 & e_2 & 0 \\ 0 & 0 & D \end{pmatrix}, \quad \text{where all entries of } D \text{ are divisible by } e_2.$$

Continue until the matrix has the desired form (2).

- *Step 3. Final Claim.*

Suppose $e_1 = \cdots = e_k = 1$, then

$$G \cong (\mathbb{Z}/e_{k+1}\mathbb{Z}) \oplus \cdots \oplus (\mathbb{Z}/e_n\mathbb{Z}).$$

Also $r = n$ in formula (2), that is, the matrix A has no zeros on the diagonal.

Proof. According to our diagonalization of the matrix A in (2), we can choose new sets of generators of K and F, respectively:

$$(w_1, \ldots, w_n) = (u_1, \ldots, u_n)U^{-1} \quad \text{and}$$
$$(z_1, \ldots, z_n) = (v_1, \ldots, v_m, 0, \ldots, 0)V.$$

Then, using (1) and (2), we have

$$(z_1, \ldots, z_n) = (w_1, \ldots, w_n) \begin{pmatrix} e_1 & 0 & \cdots & 0 & 0 & \cdots & 0 \\ 0 & e_2 & \cdots & 0 & 0 & \cdots & 0 \\ \vdots & \vdots & \cdots & \vdots & \vdots & & \vdots \\ 0 & 0 & \cdots & e_r & 0 & \cdots & 0 \\ 0 & 0 & \cdots & 0 & 0 & \cdots & 0 \\ \vdots & \vdots & \cdots & \vdots & \vdots & & \vdots \\ 0 & 0 & \cdots & 0 & 0 & \cdots & 0 \end{pmatrix}.$$

This means that

$$z_j = e_j w_j, \quad j = 1, \ldots, r, \quad z_j = 0, \quad \text{for } j = r+1, \ldots, n.$$

We know that $G \cong F/K$, by the fundamental homomorphism theorem, and that

$$F = (\mathbb{Z}w_1) \oplus \cdots \oplus (\mathbb{Z}w_n), \qquad K = (\mathbb{Z}e_1 w_1) \oplus \cdots \oplus (\mathbb{Z}e_r w_r).$$

It follows that

$$G \cong F/K \cong (\mathbb{Z}/e_1\mathbb{Z}) \oplus \cdots \oplus (\mathbb{Z}/e_r\mathbb{Z}) \oplus \mathbb{Z} \oplus \cdots \oplus \mathbb{Z}.$$

Since G is finite, r must equal n. If $e_1 = \cdots = e_k = 1$, we see that we can omit the first k terms in the sum, since they are the 0 group. This completes the proof of the theorem except to note that if e_j divides e_{j+1} for all j, the representation of G is unique. We leave that as an exercise. ■

Remarks. The method of the preceding proof is often called *elementary divisor theory*. It works if we replace $\mathbb{Z}$ by any euclidean domain, that is, an integral domain having a division algorithm. An example of such a domain other than $\mathbb{Z}$ itself is the ring $F[x]$ of polynomials over a field F. Elementary divisor theory applied to a matrix A with entries in $F[x]$ gives a diagonal matrix called the Smith normal form of A. This also gives a constructive way to find the Jordan normal form of a matrix M starting with the Smith normal form of $M - xI$. See Schreier and Sperner [1959] for a discussion of the Smith normal form. Maple has a command to compute it.

Elementary divisor theory is also very useful in the study of ideals in algebraic number fields.

Theorem 1 also holds for any finitely generated abelian group if you add on a number of copies of $\mathbb{Z}$. The proof works too.

We do not have to assume that, in $G \cong (\mathbb{Z}/m_1\mathbb{Z}) \oplus \cdots \oplus (\mathbb{Z}/m_r\mathbb{Z})$, the m_j divide m_{j+1}. We could also assume that the ms are powers of primes, by the Chinese remainder theorem.

Discrete Fourier Transform on Finite Abelian Groups

By Theorem 1, we may suppose that our finite abelian group is

$$G = (\mathbb{Z}/m_1\mathbb{Z}) \oplus \cdots \oplus (\mathbb{Z}/m_r\mathbb{Z}), \tag{3}$$

where the addition is componentwise. As usual, $L^2(G)$ denotes the vector space of all $\mathbb{C}$-valued functions on G with the *inner product*

$$\langle f, g \rangle = \sum_{x \in G} f(x)\overline{g(x)}.$$

Our aim is to find a Fourier transform on G and for that we need the characters of G.

Definition. A *character* χ of a finite abelian group G is a group homomorphism from G into the multiplicative group $\mathbb{T}$ of complex numbers of norm 1.

For a group G of the form (3), the following products of exponentials give characters of G:

$$\chi(x) = e_a(x) = e_{a_1}(x_1) \cdots e_{a_r}(x_r), \quad \text{where } e_{a_i}(x_i) = \exp(2\pi i a_i x_i/m_i), \tag{4}$$

for $a, x \in G$. It should be clear, by the standard properties of exponentials, that e_a is indeed a *group character* of G for each $a \in G$.

Define the *dual group* $\hat{G}$ to be the set of all characters of G. The group operation in $\hat{G}$ is pointwise multiplication of functions. This means that

$$\hat{G} = \{\chi \mid \chi \text{ is a character of } G\}, \tag{5}$$

with $\chi\psi(x) = \chi(x)\psi(x)$, for all $x \in G$.

If G is given by (3) above, we can identify G with a subgroup of $\hat{G}$ by identifying a with e_a in (4). Here we use the fact that $e_a e_b = e_{a+b}$. To see that every character of G given by (3) must have the form (4), you need to know Theorem 2 below, which says that $L^2(G) \cong L^2(\hat{G})$ so that $|G| = |\hat{G}|$.

Thus, for finite abelian groups, the dual group $\hat{G}$ is isomorphic to the original group G. This is the same thing that happened for the cyclic group $\mathbb{Z}/n\mathbb{Z}$. In this situation, we say that G is *self-dual*. Another such self-dual group is the additive group of real numbers or euclidean n-space $\mathbb{R}^n$.

Definition. The *discrete Fourier transform* or DFT on $f \in L^2(G)$ is defined by

$$\mathscr{F}f(\chi) = \hat{f}(\chi) = \sum_{a \in G} f(a)\overline{\chi(a)} = \langle f, \chi \rangle, \quad \text{for } \chi \in \hat{G}. \tag{6}$$

This is the same definition as in formula (5) of Chapter 2. And the basic properties are just the same as in Theorem 1 of that chapter.

Theorem 2. **Basic Properties of the DFT on a Finite Abelian Group.**

1) $\mathscr{F}: L^2(G) \to L^2(\hat{G})$ is a 1-1, onto linear map.
2) **Convolution.** Defining convolution by

$$(f * g)(x) = \sum_{y \in G} f(y)g(x - y), \quad \text{for } x \in G,$$

we have

$$\mathscr{F}(f * g)(\chi) = \mathscr{F}f(\chi) \cdot \mathscr{F}g(\chi), \quad \text{for all } \chi \in \hat{G}.$$

3) **Inversion.**

$$f(x) = \frac{1}{|G|} \sum_{\chi \in \hat{G}} \mathscr{F}f(\chi)\chi(x) = \frac{1}{|G|} \sum_{a \in G} \langle f, \chi \rangle \chi(x).$$

4) **Plancherel Theorem or Parseval Equality.**

$$\langle f, f \rangle = \frac{1}{|G|} \langle \mathscr{F}f, \mathscr{F}f \rangle.$$

Here the inner product on $F, H \in L^2(\hat{G})$ is

$$\langle F, H \rangle = \sum_{\chi \in \hat{G}} F(\chi)\overline{H(\chi)}.$$

5) **Translation.** Define for $x \in G$, $f^s(x) = f(s + x)$. Then

$$\mathscr{F}(f^s)(\chi) = \chi(s)\mathscr{F}f(x).$$

Proofs. The proofs are left as exercises.

You just imitate the proofs in Chapter 2 usually word for word. You will need the lemma. ■

Lemma. **Orthogonality Relations for the Characters.** Using the definition above of characters χ, ψ of the finite abelian group G, we have

$$\langle \chi, \psi \rangle = \begin{cases} |G|, & \chi = \psi, \\ 0, & \text{otherwise.} \end{cases}$$

Proof. The proof of Chapter 2 works again. You can reduce to the case $\psi = 1$, by noting

$$\langle \chi, \psi \rangle = \langle \chi \psi^{-1}, 1 \rangle,$$

where 1 denotes the *trivial character*, which is the constant function taking everything in G to 1.

Let

$$S = \langle \chi, 1 \rangle = \sum_{a \in G} \chi(a).$$

If χ is not the trivial character $\chi(b) \neq 1$ for some $c \in G$. Multiply this equation by $\chi(b)$ and obtain

$$\chi(b)S = \chi(b) \sum_{a \in G} \chi(a) = \sum_{a \in G} \chi(b)\chi(a)$$

$$= \sum_{a \in G} \chi(b+a) = \sum_{c \in G} \chi(c) = S.$$

Here, for the last equality, we set $c = b + a$ and note that the variable c runs over G as fast as the variable a does, since G is a group. Thus we have

$$\chi(b)S = S, \quad \text{with } \chi(b) \neq 0.$$

It follows that $S = 0$. ∎

Example. Let us consider a two-dimensional analogue of the example in Figure I.9 of Chapter 2. We want to graph the Fourier transform of a function $g(x, y)$ such that

$$g(-x, y) = g(x, -y) = g(x, y).$$

This will ensure that the Fourier transform of g is real-valued, which makes it much easier to produce a graph. We take our space to be $V = (\mathbb{Z}/32\mathbb{Z})^2$. Define $g : V \to \mathbb{R}$ by $g = \delta_S$, where $S = \{(0, 0), (\pm 1, 0), (0, \pm 1), (\pm 1, \pm 1)\}$. Identifying $\mathbb{Z}/32\mathbb{Z}$ with the set $\{0, 1, \ldots, 31\}$, we can also represent g by a 32×32 matrix m of values. The matrix m has entries zero except for

$$m(1, 1) = m(1, 2) = m(1, 32) = 1,$$
$$m(2, 1) = m(2, 2) = m(2, 32) = 1,$$
$$m(32, 1) = m(32, 2) = m(32, 32) = 1.$$

Here we must use indices starting at 1 so that the index 1 really represents 0 while 2 corresponds to 1 and so on. If you are not careful about indices here, you will not produce the same DFT at all.

We plot the values of m as a surface over the $(x, y) \in (\mathbb{Z}/32\mathbb{Z})^2$ plane using the Matlab command mesh(m). See Figure I.33 for the result. You can also use the Matlab command surf(m) to get a brightly colored surface.

We use the Matlab command fft $2(m)$ to compute the two-dimensional fast Fourier transform of g. The result is real but Matlab does not seem to know it. However, we can plot the real matrix n corresponding to the values of $\hat{g}(x, y)$ for $x, y \in \mathbb{Z}/32\mathbb{Z}$ as seen in Figure I.34.

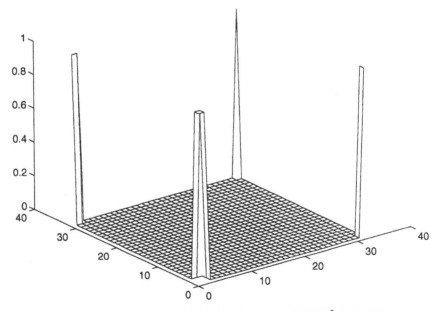

Figure I.33. Matlab mesh graph of the function g on $(\mathbb{Z}/32\mathbb{Z})^2$ defined by a matrix m of values where all values are 0 except for $m(1, 1) = m(1, 2) = m(1, 32) = 1$; $m(2, 1) = m(2, 2) = m(2, 32) = 1$; $m(32, 1) = m(32, 2) = m(32, 32) = 1$.

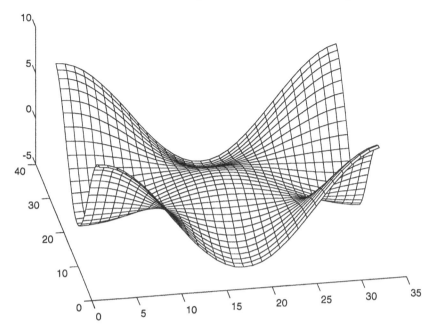

Figure I.34. Matlab mesh plot of the two-dimensional fast Fourier transform of the function represented in Figure I.33.

The graph in Figure I.34 is exactly the same as that of its continuous analogue – the function

$$h(x, y) = \frac{1}{9}(1 + 2\cos(2\pi x))(1 + 2\cos(2\pi y)), \quad (x, y) \in [0, 1]^2.$$

This is the two-dimensional analogue of the result of Figure I.9 in which the Fourier transform can be seen to be close to the graph of $(1+2\cos(2\pi x))*(1/3)$. It should come as no surprise that the computer cannot distinguish the discrete from the continuous.

Exercise. Suppose that G and H are two finite abelian groups under addition. Consider the *direct product* of these groups

$$G \times H = \{(g, h) \mid g \in G, h \in H\},$$

with componentwise sum, that is, $(a, b) + (c, d) = (a + c, b + d)$. Show that, if the dual group $\hat{G}$ of characters of G is defined by Equation (5) above, then

$$(G \times H)\hat{} \cong \hat{G} \times \hat{H}.$$

Here $\cong$ denotes group isomorphism.

Exercise. Give a detailed proof that the dual group $\hat{G}$ is isomorphic to G, if G is a finite abelian group.

The Case $(\mathbb{Z}/2\mathbb{Z})^n$

Example. $G = (\mathbb{Z}/2\mathbb{Z})^n = \mathbb{F}_2^n$.

For this group, the dual group consists of exponentials of the form

$$e_a(x) = \prod_{j=1}^{n} \exp(2\pi i a_j x_j / 2) = \prod_{j=1}^{n} (-1)^{a_j x_j} = (-1)^{{}^t a \cdot x},$$

where

$$^t a \cdot x = \sum_{j=1}^{n} a_j x_j.$$

Thus the DFT is

$$\mathscr{F}f(a) = \sum_{x \in \mathbb{F}_2^n} (-1)^{{}^t a \cdot x} f(x).$$

This means that the matrix of the DFT is the following matrix (if we list the vectors in the usual way, identifying the vectors in $\mathbb{F}_2^n$ with numbers written in binary):

$$H_{2^n} = ((-1)^{{}^t a \cdot x})_{a, x \in \mathbb{F}_2^n}. \tag{7}$$

This is called a *Hadamard matrix*. It is of interest for the theory of error-correcting codes. A reference is Ball and Coxeter [1987].

Note that

$$H_2 = \begin{pmatrix} 1 & 1 \\ 1 & -1 \end{pmatrix}, \qquad H_{2^{n+1}} = \begin{pmatrix} H_{2^n} & H_{2^n} \\ H_{2^n} & -H_{2^n} \end{pmatrix} = H_{2^n} \otimes H_2. \tag{8}$$

The Hadamard matrix H_m is an $m \times m$ matrix of ± 1s such that $H_m {}^t H_m = m I$. The matrix also produces some interesting graphics. See Figure I.35 drawn using Mathematica's ListDensityPlot command on H_{2^n}, for some small values of n. For H_m to exist, 4 must divide m or $m = 1, 2$. Furthermore, it is possible to use quadratic residues to find H_m for $m = p^r + 1$, where $p = $ prime, $4 \mid (p^r + 1)$. See MacWilliams and Sloane [1988, pp. 44–8]. Such matrices lead to useful codes. The smallest m without a construction of H_m (according to MacWilliams and Sloane) is $m = 268$.

Why did Hadamard study these matrices? He wanted a matrix such that $|\det H_m|$ is maximal (i.e., $m^{m/2}$).

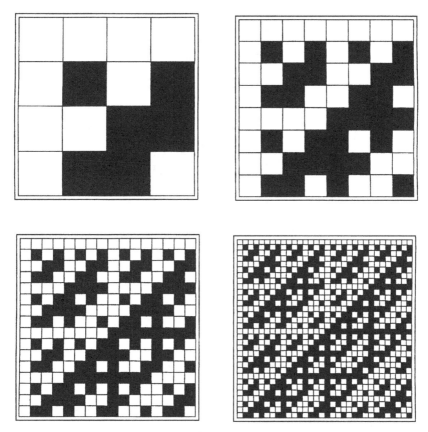

Figure I.35. Mathematica ListDensityPlots of the Hadamard matrices H_4, H_8, H_{16}, H_{32} as defined by formula (7).

Hadamard matrices can be used to construct codes such as that which was used in the [1969] *Mariner* spacecraft, which took pictures of Mars. The picture associated to $\left(\begin{smallmatrix} H_{32} \\ -H_{32} \end{smallmatrix}\right)$ was given in Figure I.1. Actually the lower half of the matrix in Figure I.1 was multiplied by -1 and rotated. Figure I.1 is a redrawing by Mathematica of a rather blurry picture in Posner's article in Mann's book on error-correcting codes [1968, p. 23]. We shall say more about error-correcting codes in the next chapter.

Coxeter [1933] shows that the construction of a Hadamard matrix H_m is equivalent to choosing m of the 2^{m-1} vertices of the cube in $(m-1)$-dimensional space so as to form a simplex.

Harmuth [1972] notes that the Hadamard matrices correspond to a system of orthogonal functions on the interval $[-1/2, 1/2]$ called *Walsh functions*. He

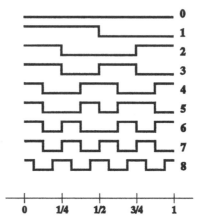

Figure I.36. The first nine Walsh functions. Of these, the first eight come from reordering the rows of the Hadamard matrix H_8 in Figure I.29.

reorders the rows of H_n and gets the functions pictured in Figure I.36. The Walsh functions have values ± 1 and form an orthogonal family on $[-1/2, 1/2]$ with respect to the usual inner product obtained by integrating

$$\langle f, g \rangle = \int_{-1/2}^{1/2} f(x)g(x)dx.$$

The Walsh functions go back to J. L. Walsh in [1923]. See also Elliott and Rao [1982, Ch. 8].

Harmuth [1972, pp. 2–3] notes: "Walsh functions are presently the most important example of non-sinusoidal functions in communications. They have been used for the transposition of conductors' open wires for more than 60 years. Rademacher functions, which are a subset of the Walsh functions, were used for this purpose toward the end of the 19th century." Recently with the advent of wavelets, Harmuth's statement no longer seems valid. However, Walsh functions are connected with Haar wavelets. See Coifman and Wickerhauser [1993, p. 138].

Beauchamp [1975, p. vii] says that the advantage of these functions is "the reduction in speed of computation and storage when compared with sines and cosines."

Harmuth explains how the Walsh functions can be used to do the information processing work of sines and cosines. He applies this to various practical engineering problems (e.g., multiplexing). In amplitude modulation, m signals are passed through frequency low-pass filters and amplitude modulated onto m carriers, which are added and transmitted. Then the receiver must separate out the signals.

Harmuth [1972, p. 9] notes that the Walsh functions have been known to coding theorists under the name of Reed–Muller codes, which are among the few codes that are actually used.

Mathematically, Walsh functions are actually characters on the dyadic group, which is a local field and is not finite. Thus we will not consider it here. Of course, most of the engineering references avoid speaking about the topology of the field. Beauchamp [1975, p. 141] notes: "Practical applications of Walsh function theory in communications have shown that, in most cases, it is sufficient to use a finite set of discrete Walsh functions."

Harmuth [1, pp. 85–9] does consider the idea of dyadic time and provides a picture of it. He says: "There may never be a use for it [dyadic time], but one cannot be sure since there are many animals in physics that have no need and no means to distinguish between past and future."

See Fine [1949, 1950] or Vilenkin [1968] for a more mathematical discussion of Walsh functions.

Exercise. Write a report on the Walsh functions. What local group do they live on?

Applications to Hamming Graphs

Example 1. Hypercubes in $\mathbb{F}_2^n$

Let's consider $G = \mathbb{F}_2^n$ and the Cayley graph $X_n = X(\mathbb{F}_2^n, \{u_1, \ldots, u_n\})$, where u_j is the vector with 1 in the jth entry and 0 everywhere else. We call X_n a *Hamming graph*. So for $n = 3$ we get a cube and for $n = 4$ we get a hypercube or tesseract. See Figure I.37.

Banchoff [1990, p. 74] notes: "Moving a cube perpendicular to itself creates a hypercube." You should be able to see this happening in Figure I.37. Banchoff [1990, pp. 11, 67–8, 74–5, 115, 119, 122] has many more pictures of four-dimensional cubes. There is also a movie. See Coxeter [1991] and Rucker [1984] as well.

Now we want to ask some of the standard questions that were asked in Chapters 4 and 5 about these hypercube graphs. It will turn out that these graphs are of use in many places (e.g., the theory of error-correcting codes, the

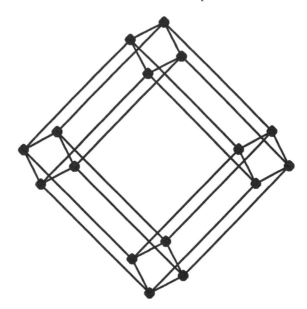

Figure I.37. $X(\mathbb{F}_2^n, \{u_j \mid j = 1, 2, 3, 4\})$. The hypercube or tesseract.

theory of switching circuits, and in modeling heat diffusion). See the next three chapters.

The following definitions were made by Hamming in connection with the theory of error-correcting codes.

Definition. The *Hamming weight* of a vector in $\mathbb{F}_2^n$ is given by

$$|v| = \text{the number of entries in } v \in \mathbb{F}_2^n \text{ that are different from } 0. \qquad (9)$$

Then we can define a *distance*

$$d(v, w) = |v - w|, \quad \text{for } v, w \in \mathbb{F}_2^n.$$

The unit basis vectors u_j are the vectors with $|u_j| = 1$.

Exercise. Show that the Hamming distance is a metric, that is, for all $v, w, z \in \mathbb{F}_2^n$, we have the following three properties:

a) $d(v, w) \geq 0$ and $d(v, w) = 0$ iff $v = w$;
b) $d(v, w) = d(w, v)$
c) $d(v, w) \leq d(v, z) + d(z, w)$.

Exercise. Show that if p is a prime such that $p > n$, then defining the Hamming weight on $x \in \mathbb{F}_p^n$ by $|x| = \#\{j \mid x_j \neq 0\}$, we have

$$|x| = x_1^{p-1} + \cdots + x_n^{p-1}.$$

So the generating set for our graph $X_n = X(\mathbb{F}_2^n, \{u_1, \ldots, u_n\})$ is

$$S = S(1) = \left\{ s \in \mathbb{F}_2^n \mid |s| = 1 \right\} = \{u_1, \ldots, u_n\},$$

where $|s|$ is the Hamming weight defined by (9) and u_j are the unit basis vectors in $\mathbb{F}_2^n$. Thus the *adjacency operator* of the Cayley graph $X_n = X(\mathbb{F}_2^n, \{u_1, \ldots, u_n\})$ is given for $x \in \mathbb{F}_2^n$ and $f \in L^2(\mathbb{F}_2^n)$ by

$$
\begin{aligned}
Af(x) &= \sum_{s \in S} f(x+s) = \sum_{|x-y|=1} f(y) = (f * \delta_S)(x) \\
&= f(x+u_1) + f(x+u_2) + \cdots + f(x+u_n),
\end{aligned}
$$

where u_j is the jth unit basis vector in $\mathbb{F}_2^n$.

Next we investigate the spectrum of A.

Spectrum of A for X $(\mathbb{F}_2^n, \{u_1, \ldots, u_n\})$

For $a \in \mathbb{F}_2^n$, let $e_a(x) = (-1)^{'a \cdot x}$ and suppose that, as usual, the u_j are the unit basis vectors of $\mathbb{F}_2^n$. Then the eigenvalues λ_a of the adjacency operator A for the Cayley graph $X(\mathbb{F}_2^n, \{u_1, \ldots, u_n\})$ are given by

$$\lambda_a = \sum_{j=1}^{n} e_a(u_j) = \sum_{j=1}^{n} (-1)^{a_j} = n - 2|a|. \tag{10}$$

To see the last equality, note that in the second sum you get $+1$ a total of $(n - |a|)$ times and -1 a total of $|a|$ times.

So we have showed that

$$\text{spec } A = \{n, n-2, n-4, \ldots, -n+2, -n\}.$$

It follows that:

- The graph is bipartite.
- The graph can only be Ramanujan for small n; e.g., $n = 2, \ldots, 6$.
- $0 \in \text{spec } A$ iff n is even.

It is also possible to consider Cayley graphs built on shells.

Example 2. Cayley Graph for Shell of Hamming Distance r

Let $|v|$ denote the Hamming weight of $v \in \mathbb{F}_2^n$ and define the shell $S(r)$ by

$$S(r) = \left\{ v \in \mathbb{F}_2^n \,\middle|\, |v| = r \right\}.$$

So the case $r = 1$ is the generating set used in Example 1.

The Cayley graph to be considered now is $X(\mathbb{F}_2^n, S(r))$. It has an adjacency operator with eigenvalues λ_a, $a \in \mathbb{F}_2^n$:

$$\lambda_a = \sum_{\substack{v \in \mathbb{F}_2^n \\ |v| = r}} (-1)^{'a \cdot v} = K_r(|a|; n), \tag{11}$$

where $K_r(x; n)$ denotes the Krawtchouk polynomial defined below.

Definition. The *Krawtchouk polynomial* $K_r(x; n)$ is defined by

$$K_r(x; n) = \sum_{j=0}^{r} (-1)^j \binom{x}{j} \binom{n - x}{r - j}, \tag{12}$$

where

$$\binom{x}{j} = \frac{x(x - j) \cdots (x - j + 1)}{j!}, \quad \text{for } x \in \mathbb{R}, \ j \in \mathbb{Z}^+,$$

$$\binom{x}{0} = 1, \quad \binom{x}{j} = 0, \quad \text{for } j \notin \{0, 1, 2, 3, \ldots\}.$$

The first few Krawtchouk polynomials are

$$K_0(x; n) = 1; \quad K_1(x; n) = n - 2x; \quad K_2(x; n) = \binom{n}{2} - 2nx + 2x^2.$$

The Krawtchouk polynomial is a special case of the *Gauss hypergeometric function*:

$$K_r(x; n) = {}_2F_1\left(\begin{matrix} -r, -x \\ -n \end{matrix} \,\middle|\, 2 \right).$$

See Lebedev [1972] for more information on the Gauss hypergeometric function.

Before proving formula (11), we list a few important properties of the Krawtchouk polynomials. Most important for us is the *three-term recurrence relation*:

$$(k + 1)K_{k+1}(x; n) = (n - 2x)K_k(x; n) - (n - k + 1)K_{k-1}(x; n). \tag{13}$$

The Krawtchouks form an *orthogonal family* on $\{0, 1, \ldots, n\}$:

$$\sum_{i=0}^{n} \binom{n}{i} K_a(i; n) K_b(i; n) = 2^n \binom{n}{a} \delta_a(b). \tag{14}$$

There is a *generating function*, $0 \le i \le n, i \in \mathbb{Z}$:

$$(1+z)^{n-i}(1-z)^i = \sum_{k=0}^{n} K_k(i; n) z^k. \tag{15}$$

Finally there is the formula

$$\binom{n}{i} K_k(i; n) = \binom{n}{k} K_i(k; n). \tag{16}$$

Exercise. Prove as many formulas for Krawtchouk polynomials as you can stand to prove (e.g., (13)–(16) above).

Hint. One reference for the Krawtchouks is MacWilliams and Sloane [1988, Chapter 5].

Proof of (11). We prove the formula by a straightforward counting argument. Without loss of generality, we may assume that

$$a_i = \begin{cases} 1, & \text{for } 1 \le i \le k, \\ 0, & \text{otherwise.} \end{cases}$$

Then

$$\lambda_a = \sum_{\substack{v \in \mathbb{F}_2^n \\ |v| = r}} (-1)^{'a \cdot v}$$

Look at Table I.14.
  This tells you that

$$\lambda_a = \sum_{i=0}^{k} \binom{k}{i} \binom{n-k}{r-i} (-1)^i = K_r(k; n).$$

∎

Table I.14. *Computation of* λ_a

	$\overbrace{}^{k}$	
a	$11111\cdots 1111$	$0000\cdots 0000000$
v	$\underbrace{1\cdots 11}_{i}\,00\cdots 0$	$\underbrace{111\cdots 111}_{r-i}\,00\cdots 00$

The proof of (11) does not really answer the question: Why use the Krawt-chouk polynomials? If you are not a devotee of special functions, you may want a more group theoretical approach.

To do this, we need to enlarge the group from the abelian group $X = \mathbb{F}_2^n$ to the nonabelian group

$$G = \mathbb{F}_2^n \propto \mathscr{S}_n, \qquad (17)$$

where $\mathscr{S}_n$ denotes the symmetric group of permutations of $\{1, 2, \ldots, n\}$ .The symbol $\propto$ denotes the *semidirect product*, meaning that the group operation is not quite componentwise but instead for $x, y \in \mathbb{F}_2^n$ and $\sigma, \tau, \in \mathscr{S}_n$, we define

$$(x, \sigma) \cdot (y, \tau) = (x + \sigma y, \sigma \tau) \quad \text{with } (\sigma y)_i = y_{\sigma(i)}.$$

This works like multiplication of matrices

$$\begin{pmatrix} \sigma & x \\ 0 & 1 \end{pmatrix} \begin{pmatrix} \tau & y \\ 0 & 1 \end{pmatrix}.$$

Because the group defined by (17) is not abelian, it does not really fit into Part 1. So we will be rather sketchy in what follows. See Part 2 for more details on symmetric spaces G/K and spherical functions. The basic examples of symmetric spaces and spherical functions are continuous – the sphere and Laplace spherical harmonics. In general, a *symmetric space* is a quotient G/K of a group G and a subgroup K such that we get a commutative algebra from the space $L^2(K \backslash G/K)$, consisting of K-bi-invariant functions on G under convolution product. In the continuous case, a (zonal) *spherical function* is a function on G/K which is a K-bi-invariant eigenfunction of the G-invariant differential (or integral) operators on G/K with value 1 at the coset of the identity of G. When $G = O(3)$ and $K = O(2)$, G/K is the sphere in $\mathbb{R}^3$. The G- invariant differential operators are polynomials in the Laplacian on the sphere. And the spherical functions are the classical Laplace spherical harmonics, which are Legendre polynomials in $\cos \theta$ (using the usual spherical coordinates). See Terras [1985, Chapter 2].

The analogous theory has been worked out for a large number of real symmetric spaces G/K. Many people have been involved, (e.g., Cartan, Weyl, Gelfand, Godement, Harish-Chandra, Helgason, and Tamagawa). See Terras [1985].

Finite symmetric spaces have already been introduced in Chapter 5. They have been investigated by a wide variety of authors (e.g., Bannai, Stanton, Soto-Andrade). We will have much more to say about them when we consider nonabelian groups in Part 2.

When G/K is a symmetric space, (i.e., when $L^1(K \backslash G/K)$ is a commutative algebra under convolution on G), we shall say that (G, K) is a *Gelfand pair*.

Definition. Suppose that G/K is a symmetric space or (G, K) is a Gelfand pair. A *spherical function* is a function $f : G \rightarrow \mathbb{C}$ satisfying

1) *K-invariance:* $f(kg) = f(g)$, for all $k \in K$, $g \in G$;
2) *eigenfunction for G-invariant operators:* $Lf = \lambda f$, for all G-invariant operators L on G/K, where $\lambda \in \mathbb{C}$;
3) *normalized to be 1 at the identity:* $f(e) = 1$, where e is the identity of G.

In the case under consideration at present, we have G defined by formula (17) and $K = S_n$, so that $G/K \cong \mathbb{F}_2^n$. Then a *spherical function* $f : \mathbb{F}_2^n \rightarrow \mathbb{C}$ is a permutation – invariant eigenfunction of the adjacency operators A_i of the Cayley graph $X(\mathbb{F}_2^n, S(i))$, where

$$S(r) = \left\{ v \in \mathbb{F}_2^n \, \big| \, |v| = r \right\}.$$

Thus $f(x) = f^*(|x|)$; that is, f depends only on the Hamming distance to the origin. We will also assume that $f(0) = 1$.

The theory of group representations can help to analyze spherical functions (see Chapter 20 or Diaconis [1988], p. 54). In the cases that we studied in Terras [1985], we could avoid group representations in our discussion (mainly because we had some simple explicit formulas for the spherical functions which were special cases of a result of Harish-Chandra). But that is not the case for finite nonabelian groups.

Highly Regular Graphs and Collapsed Adjacency Matrices

Some references for this subject are Bollobás [1979, p. 158] and Godsil [1993, p. 75].

Instead of making use of the representations of the nonabelian group $G = \mathbb{F}_2^n \propto \mathscr{S}_n$, let's consider a more elementary discussion.

Definition. A connected graph X is *highly regular with collapsed adjacency matrix* C iff for every vertex $x \in X$ there is a partition of X into blocks $V_1 = \{x\}, V_2, \ldots, V_r$ such that each vertex $y \in V_j$ is adjacent to exactly c_{ij} vertices in V_i. The collapsed adjacency matrix is obtained from the ordinary adjacency matrix A of X as follows: Write $X = \{v_1, \ldots, v_n\}$ and $A = (a_{ij})_{1 \le i, j \le n}$. Then $C = (c_{ij})_{1 \le i, j \le r}$ with

$$c_{ij} = \sum_{v_s \in V_i} a_{st}, \quad \text{if } v_t \in V_j. \tag{18}$$

The above sum is independent of the choice of v_t in V_j, by definition.

*Theorem 3. **Collapsed Adjacency Matrices.*** Suppose X is a highly regular graph with adjacency matrix A and collapsed adjacency matrix C. Then both A and C have the same minimal polynomial. Thus λ is an eigenvalue of A iff λ is a root of the characteristic polynomial of C.

Proof. Define

$$P = (p_{ij})_{1 \le i \le r, 1 \le j \le n}, \quad \text{where } p_{ij} = \begin{cases} 1, & \text{if } v_j \in V_i, \\ 0, & \text{otherwise.} \end{cases}$$

Note that

$$PA = CP.$$

So if x is an eigenvector of A corresponding to the eigenvalue λ, we have

$$\lambda P x = P \lambda x = P A x = C P x.$$

Thus $Px \ne 0$ implies that Px is an eigenvector of C corresponding to the eigenvalue λ.

To see that $Px \ne 0$, note that $x \ne 0$ implies $x_j \ne 0$ for some j. Then take $V_1 = \{v_j\}$. We are assuming that this is possible. So

$$(Px)_1 = \sum_k p_{1k} x_k = x_j \ne 0.$$

■

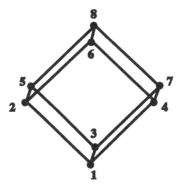

Figure I.38. The cube $\mathbb{F}_2^3$.

Example. The Cube $\mathbb{F}_2^3$ is Highly Regular. Label the vertices of the cube as indicated in Figure I.38. Then the adjacency matrix is

$$A = \begin{pmatrix} 0 & 1 & 1 & 1 & 0 & 0 & 0 & 0 \\ 1 & 0 & 0 & 0 & 1 & 1 & 0 & 0 \\ 1 & 0 & 0 & 0 & 1 & 0 & 1 & 0 \\ 1 & 0 & 0 & 0 & 0 & 1 & 1 & 0 \\ 0 & 1 & 1 & 0 & 0 & 0 & 0 & 1 \\ 0 & 1 & 0 & 1 & 0 & 0 & 0 & 1 \\ 0 & 0 & 1 & 1 & 0 & 0 & 0 & 1 \\ 0 & 0 & 0 & 0 & 1 & 1 & 1 & 0 \end{pmatrix}.$$

The block $S(i)$ consists of all vertices at a distance i from the origin either in the sense of the Hamming distance or in the sense of the graph-theoretic distance in the cube. So our blocks which partition the vertices are

$$S(0) = \{1\}, \quad S(1) = \{2, 3, 4\}, \quad S(2) = \{5, 6, 7\}, \quad S(3) = \{8\}.$$

Then the collapsed adjacency matrix is

$$C = \begin{pmatrix} 0 & 3 & 0 & 0 \\ 1 & 0 & 2 & 0 \\ 0 & 2 & 0 & 1 \\ 0 & 0 & 3 & 0 \end{pmatrix}.$$

The eigenfunctions of C satisfy a *three-term recurrence relation*:

$$(3 - r)h(r + 1) + rh(r - 1) = \lambda h(r), \quad \text{for } r = 0, 1, 2, 3.$$

Compare this with the three-term recurrence for the Krawtchouk polynomials in Equation (13) above.

The solutions are, for $k = 0, 1, 2, 3$:

$$\lambda_k = 3 - 2k = K_1(k; 3),$$

$$h_k(r) = K_k(r; 3) = \sum_{j=0}^{k} (-1)^j \binom{r}{j} \binom{3-r}{k-j}.$$

Note that both the eigenvalue λ_k and the eigenfunction h_k are Krawtchouk polynomials. This gives another proof of formula (11) for $n = 3$.

Example. $\mathbb{F}_2^n$. It is possible to generalize this result to $X(\mathbb{F}_2^n, S(1))$. Once more we have a highly regular graph whose vertices are partitioned by the shells

$$S(i) = \{x \in \mathbb{F}_2^n \,|\, |x| = i\}, \quad i = 0, 1, 2, \ldots, n.$$

Clearly any spherical function f is constant on a block $S(i)$. Since f is an eigenfunction of the adjacency operator $Af = \delta_{S(1)} * f$, we have

$$Af(x) = \sum_{|x-y|=1} f(y) = \lambda f(x).$$

For $x \in S(i)$, we need to know the *collapsed adjacency matrix*

$$C = (c_{ij})_{1 \leq i, j \leq n}, \qquad c_{ij} = |\{y \in S(j) \,|\, |x - y| = 1\}|.$$

The number c_{ij} is independent of the choice of $x \in S(i)$. Since $f(x) = f^*(|x|) = f^*(i)$, it follows that $Af = \lambda f$ can be rewritten as

$$\sum_{j=0}^{n} c_{ij} f^*(j) = \lambda f^*(i), \quad \text{for } 0 \leq i \leq n.$$

Thus we have replaced the original $(2^n \times 2^n)$-matrix eigenvalue problem with an $(n+1) \times (n+1)$-matrix eigenvalue problem. We could also do the analogous problem with A_1 replaced with A_i.

The adjacency operator A_1 can be collapsed using this partition into the tridiagonal matrix:

$$
C = \begin{pmatrix}
0 & n & 0 & 0 & \cdots & \cdots & \cdots & \cdots & \cdots & 0 & 0 \\
1 & 0 & n-1 & 0 & \cdots & \cdots & \cdots & \cdots & \cdots & 0 & 0 \\
0 & 2 & 0 & n-2 & \cdots & \cdots & \cdots & \cdots & \cdots & 0 & 0 \\
\vdots & \vdots & \vdots & \vdots & \cdots & \vdots & \vdots & \vdots & \cdots & \vdots & \vdots \\
0 & 0 & 0 & 0 & \cdots & i & 0 & n-i & \cdots & 0 & 0 \\
\vdots & \vdots & \vdots & \vdots & \cdots & \vdots & \vdots & \vdots & \cdots & \vdots & \vdots \\
0 & 0 & 0 & 0 & \cdots & \cdots & \cdots & \cdots & \cdots & 0 & 1 \\
0 & 0 & 0 & 0 & \cdots & \cdots & \cdots & \cdots & \cdots & n & 0
\end{pmatrix}.
$$

The eigenvalues of these matrices are $n - 2k = K_1(k; n)$ with eigenfunctions

$$
f_k(x) = K_k(|x|; n), \quad \text{for } k = 0, 1, \ldots, n.
$$

There is still much more to be said about the structure of the adjacency operators

$$
A_i f = f * \delta_{S(i)}.
$$

Note that if $\mathscr{S}_n$ denotes the symmetric group of permutations of n elements then for some fixed $y \in S(i)$:

$$
A_i f(x) = \sum_{\substack{y \in \mathbb{F}_2^n \\ |x-y|=i}} f(y) = \frac{1}{i!(n-i)!} \sum_{\sigma \in \mathscr{S}_n} f(x + \sigma y_0).
$$

That is $A_i/n!$ is a *mean-value operator* obtained by averaging over $K = \mathscr{S}_n$.

*Theorem 4. **Eigenvalues are Eigenfunctions.*** Suppose that $f : \mathrm{K \backslash G / K} \to \mathbb{C}$ is a K bi-invariant eigenfunction of the mean-value operator

$$
M_y f(x) = \frac{1}{|K|} \sum_{\sigma \in K} f(x + \sigma y), \quad \text{for fixed } y.
$$

That is, suppose $M_y f = \lambda_y f$ for $\lambda_y \in \mathbb{C}$. Then $f(0) \neq 0$ implies that $\lambda_y = f(y)/f(0)$.

Proof. Set $x = 0$ in

$$
\frac{1}{|K|} \sum_{\sigma \in K} f(x + \sigma y) = \lambda_y f(x).
$$

Since $f(\sigma y) = f(y)$ for all $\sigma \in K = \mathscr{S}_n$, this means that each term in the sum is the same and so the sum is $f(y) = \lambda_y f(0)$. ∎

Remarks. The adjacency operators $A_i f = \delta_{S(i)} * f, i = 0, 1, \dots, n$, commute and thus generate a commutative algebra of operators. The spherical functions are actually simultaneous eigenfunctions of the entire algebra, as we know

$$A_i f_j(x) = \lambda_i f_j(x), \quad \lambda_i(j) = K_i(j; n), \quad f_j(x) = K_j(|x|; n).$$

The three-term recursion for the Krawtchouk polynomials comes from a similar result for the A_is. See the following references for a more general study of such algebras: E. Bannai [1990], E. Bannai and T. Ito [1984], and D. Stanton [1981, 1984, 1990].

This is the theory of *association schemes* begun by Delsarte in [1973]. The graphs that we are studying are called *distance regular graphs*. See Brouwer, Cohen, and Neumaier [1989] for more information on such graphs. Another reference is Godsil [1993].

After we have discussed error-correcting codes, we will see that the Krawtchouk polynomials are of great importance in the subject. See the next chapter. We will have more to say about spherical functions in Chapter 20.

Chapter 11

Error-Correcting Codes

Even when that chance is one in a billion, a computer running at 25 megaHertz – the speed of last year's laptops, ponderously slow by supercomputing standards – is going to screw up 90 times an hour. What's to be done?

The answer, researchers found, lay in what are known as error-correcting codes. . . . Error-correcting code technology is nowadays as common as compact disks: it's what allows your favorite Mozart or Madonna CD to play perfectly even though your cat's been clawing the disk.

B. Cipra [1994, Vol. 2, p. 37]

We want to consider an application of the previous section – the sort of codes one uses to make communication as error free as possible (the opposite of secret codes). Some references are: Dornhoff and Hohn [1978], Jessie MacWilliams and Neil Sloane [1988], Vera Pless [1989], N. Sloane [1975], H. B. Mann [1968], J. H. van Lint [1982], R. E. Blahut [1983, 1992], A. Poli and L. Huguet [1992], and O. Pretzel [1992].

Suppose that one must send a message of 0s and 1s over long distances, for example, from my computer on Earth to Mr. Spock's computer on Vulcan (see Figure I.39). No doubt there will be errors and some random 0 in the original signal $s \in \mathbb{F}_2^n$ will be mischievously changed to a 1. In order for Mr. Spock to interpret my message correctly, there must be some redundancy built in. The error-correcting code builds in that redundancy by replacing the signal s with $x \in \mathbb{F}_2^{n+r}$. If errors are added in transmission, Mr. Spock uses a decoder to find the most likely original signal $s' \in \mathbb{F}_2^n$. See the article by Posner in the reference by Mann above for a nice discussion of the code used in the 1969 *Mariner* Mars spacecraft. The code came from the Hadamard transform – the DFT on $\mathbb{F}_2^n$. We will say more about this later. But first we need some definitions.

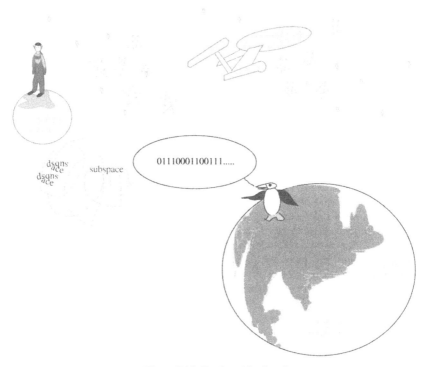

Figure I.39. Earth to Mr. Spock.

Definition. A *binary linear code* C is a vector subspace C of $\mathbb{F}_2^n$. If $\dim_{\mathbb{F}_2^n} C = k$, we call C an $[n, k]$-*code*.

As in the previous chapter, we define the *Hamming weight* of $x \in \mathbb{F}_2^n$ by

$$|x| = \#\{j \mid x_j \neq 0\}. \tag{1}$$

And the *distance* between x and y in $\mathbb{F}_2^n$ is $d(x, y) = |x - y|$. This is the same as the distance in the Cayley graph for the n-cube $G = \mathbb{F}_2^n$ with generating set

$$S = \{x \in G \mid |x| = 1\}.$$

The decoder will decode the received message as the closest codeword using the Hamming distance.

If C is an $[n, k]$-code such that the minimum distance of a code word from 0 is d, we say C is an $[n, k, d]$-*code*. If $d = 2e + 1$, such a code corrects e errors. To see this, suppose that we have a code C such that any $x, y \in \mathbb{F}_2^n$ have distance $d(x - y) \geq 3$. Then, if the received word r has at most 1 error, it cannot be in the Hamming ball of radius 1 about both x and y since $|x - y| \geq 3$ (by the triangle inequality). So the code can correct 1 error.

If $\mathbb{F}_2^n$ is the disjoint union of Hamming balls of fixed radius e centered at the elements of a code C, we call C a *perfect* code. Thus, given a received word r, there is a *unique* code word $c \in C$ within Hamming distance e of r. So we decode r as c. Perfect codes are the most efficient. On $\mathbb{F}_2^n$, they have been classified. We will say more about this later.

For rapid transmission of messages, one wants $[n, k, d]$ such that n is small. And one wants k large for efficiency and d large to correct lots of errors. The coding theorist's first question is: Given n and d, how large can k be? See the references for more information on this question.

We could replace $\mathbb{F}_2$ by an arbitrary finite field $\mathbb{F}_q$, but we won't. Other metrics have been used also, for example, the Lee metric (see the article of Golumb in Mann [1968]) and the arithmetic distance of Lenstra [1977–78]. Or see D. Gordon [1987]. Another reference is van Lint [1982].

For a nice historical survey, see the MacWilliams article in Mann [1968]. We find the following story there, for example:

For some time. . . computers were designed to use a parity check bit. The first computer of my acquaintance was so designed; if the word it was digesting contained an odd number of ones it stopped. This computer was not transistorized, it was not even vacuum-tubed, it operated with genuine relays. It had to be left running all night to get the roots of a 5th degree equation, and in the morning one would find it had stopped with a parity check failure. It is my belief that Hamming codes were invented or rediscovered by Dick Hamming because he got so aggravated with this computer.

So far in our discussion, Fourier analysis hasn't told us much about codes. But, be patient. In a later chapter, we will find that the finite analogue of the Poisson summation formula leads to Jessie MacWilliams theorem on weight enumerators of codes. It is also true that the DFT can be useful for decoding.

Next let's consider some examples of codes. Our examples will be *cyclic codes*. This means that if the word $c = {}^t(c_0, c_1, c_2, \ldots, c_{n-1})$ is in the code $C \subset \mathbb{F}_2^n$, then so is the *cyclic shift*

$$c' = {}^t(c_{n-1}, c_0, c_1, c_2, \ldots, c_{n-2}) \in C.$$

Let the *polynomial corresponding to the codeword c* be

$$p_c(x) = \sum_{j=0}^{n-1} c_j x^j, \quad \text{for } c_j \in \mathbb{F}_2. \tag{2}$$

Then the polynomial corresponding to the shifted codeword c' is

$$p_{c'}(x) = x p_c(x) \bmod (x^n - 1).$$

That is, we compute among polynomials in the quotient ring

$$R_n = \mathbb{F}_2[x]/(x^n - 1). \tag{3}$$

This ring R_n is analogous in many ways to $\mathbb{Z}/m\mathbb{Z}$.

A set of elements $I_C \subset R_n$ corresponds to the cyclic code C iff I_C is an ideal in the ring R_n, that is, I_C is a subgroup closed under multiplication by R_n. See the references on the list at the beginning of this chapter for more information on this correspondence between cyclic codes and ideals in quotient rings of the ring of polynomials $\mathbb{F}_2[x]$. It turns out that any ideal I in R_n is principal, that is

$$I = g(x) R_n = \{g(x) h(x) \mid h(x) \in R_n\}.$$

Here we are writing $h(x)$ instead of the class of $h(x)$ in $R_n = \mathbb{F}_2[x]/(x^n - 1)$. To see that $I = g(x) R_n$, take $g(x)$ to be the polynomial of least degree in I. To get a nontrivial ideal, $g(x)$ should be a factor of $x^n - 1$. Thus we define a cyclic code by taking some irreducible factor $g(x)$ of $x^n - 1$ in $\mathbb{F}_2[x]$. We say that $g(x)$ is the *generator polynomial for the code*. We could also just prescribe the codewords somehow, for example, as the image of a linear transformation or the kernel (i.e., via generator matrices or parity-check matrices). See Dornhoff and Hohn [1978, Chapters 5, 9].

Example 1. The Hamming [7, 4, 3]-Code. Here we are following Sloane [1975]. First note that we have the factorization

$$x^7 - 1 = (x - 1)(1 + x + x^3)(1 + x^2 + x^3) \quad \text{in } \mathbb{F}_2[x].$$

Let's take $g(x) = 1 + x + x^3$ to generate our ideal in R_n corresponding to the

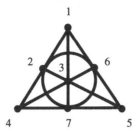

Figure I.40. Projective plane of order 2: 7 points and 7 lines.

code as described above. The codewords in $\mathbb{F}_2^7$ are:

$$
\begin{array}{ll}
0\,0\,0\,0\,0\,0\,0 & 1\,1\,1\,1\,1\,1\,1 \\
1\,1\,0\,1\,0\,0\,0 & 0\,0\,1\,0\,1\,1\,1 \\
0\,1\,1\,0\,1\,0\,0 & 1\,0\,0\,1\,0\,1\,1 \\
0\,0\,1\,1\,0\,1\,0 & 1\,1\,0\,0\,1\,0\,1 \\
0\,0\,0\,1\,1\,0\,1 & 1\,1\,1\,0\,0\,1\,0 \\
1\,0\,0\,0\,1\,1\,0 & 0\,1\,1\,1\,0\,0\,1 \\
0\,1\,0\,0\,0\,1\,1 & 1\,0\,1\,1\,1\,0\,0 \\
1\,0\,1\,0\,0\,0\,1 & 0\,1\,0\,1\,1\,1\,0
\end{array}
$$

The Hamming [7, 4, 3] code is actually perfect. Codewords 2 through 8 form the incidence matrix of the *projective plane of order* 2 pictured in Figure I.40. This means if $P_j = j$th point and $L_i = i$th line,

$$
a_{ij} = \begin{cases} 1, & \text{if } P_j \in L_i, \\ 0, & \text{otherwise.} \end{cases}
$$

Example 2. Quadratic Residue Codes. Let ω be some primitive pth root of unity in some field containing $\mathbb{F}_2$, where 2 is a quadratic residue mod p. This means that p must be congruent to $\pm 1 \pmod 8$. Let $\square$ denote the set of squares in $\mathbb{F}_p^*$ and $\square'$ be the set of nonsquares. Define the polynomials

$$
q(x) = \prod_{r \in \square} (x - \omega^r) \quad n(x) = \prod_{r \in \square'} (x - \omega^r). \tag{4}
$$

Exercise. Show that the polynomials $q(x)$ and $n(x)$ have coefficients in $\mathbb{F}_2$ and that

$$
x^p - 1 = (x - 1)q(x)n(x).
$$

Let R be the ring $\mathbb{F}_2[x]/(x^p - 1)$. The cyclic codes with generators

$$q(x), (x - 1)q(x), n(x), \quad \text{or} \quad (x - 1)n(x), \quad \text{respectively}$$

are called the *QR codes*

$$Q, \overline{Q}, N, \overline{N}, \quad \text{respectively.}$$

In the special case $p = 7$, Q has the generator polynomial

$$(x + \omega)(x + \omega^2)(x + \omega^4) = x^3 + x + 1.$$

So Q is the [7, 4, 3] Hamming code. If we take $p = 23$, then Q is the [23, 12, 7] Golay code. This is also a perfect code.

One can extend the QR code Q by adding a parity check. That is, we form a new code $\hat{Q}$ by adding a 0 at the end of every codeword in Q with even weight and adding a one at the end of every codeword of odd weight. Gleason and Prange have proved (see MacWilliams and Sloane [1988, Ch. 16, Section 5]) that the extended quadratic residue code is fixed by the projective special linear group $PSL(2, \mathbb{F}_p) = SL(2, \mathbb{F}_p)/ \pm I$, where $SL(2, \mathbb{F}_p)$ consists of all 2×2 matrices with entries in $\mathbb{F}_p$ and determinant 1.

In a cyclic code corresponding to an ideal $g(x)R_n$, $R_n = \mathbb{F}_2[x]/(x^n - 1)$, the code words correspond to polynomials of the form $g(x)h(x), h(x) \in R_n$. If you write this out in terms of the coefficients

$$g(x) = \sum_{i=0}^{n-1} c_i x^i, \qquad h(x) = \sum_{i=0}^{n-1} d_i x^i, \quad \text{with } c_i, d_i \text{ in } \mathbb{F}_2,$$

you see that

$$g(x)h(x) = \sum_{i=0}^{n-1} u_i x^i,$$

where $u_i = \sum_{k=0}^{n-1} c_k d_{i-k} = c * d(i) = $ convolution of c and d as functions on $\mathbb{Z}/n\mathbb{Z}$. Thus the discrete Fourier transform on $\mathbb{Z}/n\mathbb{Z}$ turns $c * d$ into $\hat{c} \cdot \hat{d}$, which is useful for decoding (see Blahut [1983]).

Example 3. Codes from the DFT or Hadamard Transform. We take this discussion from Posner's article in Mann [1]. The article considers the code used in the 1969 *Mariner* probe to Mars. Look at the DFT on $\mathbb{F}_2^5$, that is, the Hadamard matrix

$$H_5 = [(-1)^{{}^t u \cdot v}]_{u,v \in \mathbb{F}_2^5},$$

with u, v ordered as for the corresponding numbers in binary. Form the new matrix

$$G = \Phi \begin{bmatrix} H_5 \\ -H_5 \end{bmatrix},$$

where Φ replaces 1s with 0s and -1s with 1s. Thus ϕ is built up using the group homomorphism below applied to each matrix element:

$$\Phi : \{-1, 1\} \longrightarrow \mathbb{F}_2 \text{ defined by } \phi(-1) = 1, \phi(1) = 0.$$

Here the group operation for $\{-1, 1\}$ is multiplication while that for $\mathbb{F}_2$ is addition.

The rows of G are the codewords of the $[32, 6, 7]$ code used in the 1969 *Mariner* Mars probe. See Figure I.1 for a picture of the codewords. Posner's article describes how the decoding was done using the FFT.

Exercise.
a) Why is the $[32, 6, 16]$ code described above actually six dimensional?
b) How many errors does it correct?

Hints.

a) Set

$$K_1 = \begin{bmatrix} 0 \\ 1 \end{bmatrix} \quad \text{and} \quad K_{n+1} = \begin{bmatrix} 0 & K_n \\ 1 & K_n \end{bmatrix}.$$

Show that $\Phi(H_n) = K_n\,{}^tK_n$ and $\Phi(-H_n) = J - \phi(H_n)$, where J is an $n \times n$ matrix of 1s.
b) The minimum distance between codewords is 16.

Lloyd's Theorem on the Existence of Perfect *e*-Error Correcting Codes

This theorem says that if a binary perfect *e*-error correcting code C of length n exists, then the Krawtchouk polynomial

$$K_e(x - 1; n - 1) = \psi_e(x) \tag{5}$$

has e distinct zeros in $\{1, 2, \ldots, n\}$. It was first proved by Lloyd in 1957.

A sketch of the proof goes as follows. See van Lint [1982, p. 93] for more details. Assume that there exists a perfect $[n, k, e]$ code C. Let A be the adjacency matrix of the graph of $G = \mathbb{F}_2^d$, with generating set the unit vectors

$$S = \{v \in G \,||v| = 1\}.$$

The eigenvalues of A are

$$\{2j - n \,|\, 0 \leq j \leq n\}.$$

Reorder the rows and columns of A as follows. First take those with codewords in $C = C_0$. Then take those in

$$C_i = \{v \in G \,|\, d(v, C) = i\}, \quad 0 \leq i \leq e.$$

Since C is perfect, the sets C_i partition G. So the adjacency matrix A is partitioned into blocks B of the form

$$\begin{bmatrix} 0 & n & 0 & 0 & 0 & \cdots & 0 & 0 & 0 & 0 \\ 1 & 0 & n-1 & 0 & 0 & \cdots & 0 & 0 & 0 & 0 \\ 0 & 2 & 0 & n-2 & 0 & \cdots & 0 & 0 & 0 & 0 \\ \vdots & \vdots & \vdots & \vdots & \vdots & \cdots & \vdots & \vdots & \vdots & \vdots \\ 0 & 0 & 0 & 0 & 0 & \cdots & 0 & e-1 & 0 & n-e+1 \\ 0 & 0 & 0 & 0 & 0 & \cdots & 0 & 0 & e & n-e \end{bmatrix}.$$

Substitute $x = n - 2y$ in $\det(B - xI_{e+1})$ and find $2y(-1)^e e! \psi_e(y)$.

One can use this result to classify the binary perfect e-error correcting codes. See MacWilliams and Sloane [1988, Chapter 6]. There are two necessary conditions for the existence of a perfect e-error correcting code. One comes from Lloyd's theorem. The other is called the *sphere packing condition* and says that for a perfect $[n, k, d]$-code to correct e errors,

$$\sum_{i=0}^{e} \begin{bmatrix} n \\ i \end{bmatrix} = 2^{n-k}. \tag{6}$$

See Dornhoff and Hohn [1978, p. 232].

A reference on the zeros of Krawtchouk polynomials is Laura Chihara and Dennis Stanton [1990].

The following binary perfect $[n, k, d]$ codes are known: $d = 2e + 1$. In 1973 Tietavainen showed using work of van Lint that this list is complete. In fact, he worked with codes over general finite fields. And he did not assume that the codes were linear subspaces. He did assume that the alphabet had prime power order. For $e > 2$, $e \neq 6, 8$, M. R. Best showed that this last restriction can be dropped using an idea of Bannai. The cases $e = 6, 8$ were settled by Hong in 1984 and 1986. The question is still open for $e = 2$. Bannai [1990] says: it is "very unlikely that any new example exists" for $e = 2$.

Complete List of Binary Perfect [n, k, d] Codes

- *Trivial*: $k = 0$, $e \geq n$.
- *Binary $(2e + 1)$-times repetition $[2e + 1,1]$ codes*: This code repeats each k-digit block $2e + 1$ times. It corrects e errors.
- *Hamming codes*: $[2^r - 1, 2^r - 1 - r, 3]$ and other nonlinear codes with these parameters. If $r = 3$, this is the Hamming $[7, 4, 3]$ code mentioned earlier.
- *Golay* $[23, 12, 7]$ code discussed earlier.

Exercise. Give a complete proof of Lloyd's theorem.
Hint. You can find a discussion in many of the references listed above.

We have not discussed the basic problems of encoding and decoding signals. See the references given at the beginning for solutions to these problems. One method for decoding cyclic and quadratic residue codes is to make use of the group of permutations of coordinates leaving the code invariant. See Pless [1989, p. 100]. More in keeping with our theme, one can make use of the DFT. See Blahut [1983, 1992].

Other interesting topics are Goppa and multivariable codes. See Poli and Huguet [1992] and Pretzel [1992]. These topics require some knowledge of algebraic geometry.

Two separate groups of researchers have shown that codes over finite rings like $\mathbb{Z}/4\mathbb{Z}$ may lead to new advances in the theory of error-correcting codes, for codes that are nonlinear viewed as codes over $\mathbb{F}_2$ become linear when viewed over $\mathbb{Z}_4$. The researchers are R. Hammons, V. Kumar, R. Calderbank, N. Sloane, and P. Solé. See Calderbank et al. [1993] and B. Cipra [1994, Vol. II, pp. 37–40].

Cipra [1994, Vol. 11, p. 38] notes:

Coding theorists have known for decades that nonlinear codes of a given length can have more code words than their linear counterparts. Among linear codes of length 16, for example, the best double-error correcting code has 128 code words. But in 1967, A. W. Nordstrom and John Robinson constructed a nonlinear code containing 256 words (Nordstrom was a high-school student at the time, Robinson an electrical engineer at the University of Iowa). In effect, only 7 digits in each code word of the linear code carry information (the other 9 do the error correcting), whereas in the Nordstrom-Robinson code, 8 digits carry information, an improvement of approximately 14%.

The map from the linear code over $\mathbb{Z}/4\mathbb{Z}$ to $(\mathbb{Z}/2\mathbb{Z})^2$ takes 0 to $(0, 0)$, 1 to $(0, 1)$, 2 to $(1, 1)$, and 3 to $(1, 0)$. Cipra [1994, Vol. II, p. 39]: says: "The surprise is that this trick accounts for essentially all of the nonlinear codes that theorists have studied so far, including the Nordstrom–Robinson, Kerdock, and Preparata codes."

In the next chapter on the Poisson summation formula we will have a few more things to say about error-correcting codes.

Chapter 12

The Poisson Sum Formula on a Finite Abelian Group

> We come now to reality. The truth is that the digital computer
> has totally defeated the analog computer. The input is a
> sequence of numbers and not a continuous function. The output
> is another sequence of numbers, whether it comes from a digital
> filter or a finite element stress analysis or an image processor.
>
> Strang [1986, p. 290]

Now we seek finite analogues of a famous formula of continuous Fourier analysis called the Poisson summation formula (see formula (3) below). This is the formula which connects Fourier analysis on the circle $\mathbb{R}/\mathbb{Z} \cong \mathbb{T}$ with Fourier analysis on the real line $\mathbb{R}$ (i.e., it connects Fourier series and Fourier integrals). One sums the function on $\mathbb{R}$ over translates by elements of $\mathbb{Z}$. This formula has a large number of applications. See Terras [1985, Ch. 1].

The Poisson summation formula may also be viewed as a close relative of the method of images in partial differential equations. This method allows one to write down a fundamental solution of the Laplace equation on the sphere from a fundamental solution of the Laplace equation in 3-space, for example. One sums the latter over the symmetry group. And in this context, one sees that the Poisson sum formula is a special case of Selberg's trace formula.

In the context of finite abelian groups G, the Poisson summation formula relates a sum over a subgroup H of G and a sum over $H^{\#}$ = the dual to H in G (which can be identified with $(G/H)\hat{\ }$, i.e., the characters χ on G, which are trivial on H). The formula has many applications. We will consider some of the applications in coding theory, the theory of switching functions, and work on the Ising model in statistical mechanics. There are also more recent applications in wavelet analysis (see Holschneider [1995b]).

We need to recall a definition.

Definition. Suppose G is a finite abelian group. The *dual group* $\hat{G}$ is

$$\hat{G} = \{\chi : G \to \mathbb{T} \mid \chi(ab) = \chi(a)\chi(b) \text{ for all } a, b \in G\}. \tag{1}$$

Here $\mathbb{T}$ denotes the multiplicative group of complex numbers z with $|z| = 1$. The dual group is the group of characters of G under the operation of pointwise multiplication, that is,

$$\chi\mu(a) = \chi(a)\mu(a), \quad \text{for } \chi, \mu \text{ in } \hat{G}, a \in G.$$

Recall from Chapter 10 that the dual group $\hat{G}$ of a finite abelian group is isomorphic to G.

Lemma 1. If H is a subgroup of the finite abelian group G, then the quotient G/H is also a finite abelian group and its *dual in G* is

$$H^{\#} = \{\chi \in \hat{G} \mid \chi(h) = 1 \text{ for all } h \in H\} \cong (G/H)\hat{\ }. \tag{2}$$

Note that $H^{\#}$ is not the same as $\hat{H}$.

Proof. Suppose that χ is a character of G which is identically 1 on H. Then $\chi(gh) = \chi(g)$ for all $h \in H$. Thus χ defines a character on G/H.

Conversely, suppose χ is a character of G/H. Define a character ψ on G via $\psi(g) = \chi(gH)$. Then ψ is identically 1 on H, which implies that $\psi \in H^{\#}$. ∎

Note $\{I\}^{\#} = \hat{G}$.

The next theorem provides an analogue of the Poisson summation formula. The simplest version of the classical result says that for a smooth rapidly decreasing function $f : \mathbb{R} \to \mathbb{C}$ we have

$$\sum_{n \in \mathbb{Z}} f(n) = \sum_{n \in \mathbb{Z}} \hat{f}(n), \tag{3}$$

where

$$\hat{f}(a) = \int_{-\infty}^{\infty} f(x) \exp(-2\pi i a x) \, dx = \text{the Fourier transform of } f. \tag{4}$$

Poisson summation can be viewed as relating Fourier analysis on $\mathbb{R}$ and that on $\mathbb{R}/\mathbb{Z}$. It can be used to speed convergence of series and to discover unexpected symmetries such as the functional equation for Riemann's zeta function. See Terras [1985, Vol. I] for more information. In the language of the theory of distributions the Poisson summation formula says that the Fourier transform of a delta comb is again a delta comb. Here a delta comb means a line of deltas located at the integer points on the real line. It is also often called the picket fence or shah functional. A favorite function $f(x)$ on $\mathbb{R}$ to plug into Poisson summation is the heat kernel $f(x) = e^{-\pi t x^2}$, for $t > 0$. This leads to Jacobi's transformation formula for the theta function.

If you view Poisson summation as a formula connecting Fourier analysis on $\mathbb{R}$ and $\mathbb{R}/\mathbb{Z}$, you have to replace $f(x)$ by $f(x + y)$ and obtain

$$\sum_{n \in \mathbb{Z}} f(n + y) = \sum_{n \in \mathbb{Z}} \hat{f}(n) \exp(2\pi i n y). \tag{5}$$

The function on the left is a period 1 function of y and the series on the right is its Fourier series. Thus we have connected Fourier analysis on $\mathbb{R}$ with that on $\mathbb{R}/\mathbb{Z}$ (i.e., Fourier integrals with Fourier series).

You can view the right-hand side of Poisson summation as a sum over the dual to $\mathbb{R}/\mathbb{Z}$, which can be identified with $\mathbb{Z}$ itself. In what follows we replace $\mathbb{R}$ with a finite abelian group G and $\mathbb{Z}$ with a subgroup H of G. Then we will obtain equality between a sum over H and a sum over

$$H^{\#} \cong (G/H)\hat{},$$

the dual of H in G (not to be confused with $\hat{H}$). We will see that if H is a code, $H^{\#}$ is the dual code. In fact, $H^{\#}$ arises in many other notions of duality that have arisen in various contexts.

If the group G has order $|G| = n$ and the subgroup H has order $|H| = m$, then we are saying that a sum over m elements equals another sum over n/m elements. This may be useful if m is large and n/m is small.

Theorem 1. **Poisson Summation Formula on a Finite Abelian Group.** Suppose H is a subgroup of the finite abelian group G and $f : G \to \mathbb{C}$. Then

$$\frac{1}{|H|} \sum_{h \in H} f(gh) = \frac{1}{|G|} \sum_{\chi \in H^{\#}} \hat{f}(\chi) \chi(g),$$

where $H^{\#} = \{\chi \in \hat{G} \mid \chi(h) = 1 \text{ for all } h \in H\}$ and

$$\hat{f}(\chi) = \sum_{g \in G} f(g) \overline{\chi(g)} = \text{the DFT of } f \text{ on } G.$$

Proof. Start by defining $f^{\#}$ to be $|H|$ times the left-hand side of our formula, giving a function on G/H. This function $f^{\#}$ has a Fourier expansion on G/H

given for $g \in G$ by

$$f^{\#}(gH) = \sum_{h \in H} f(gh) = \frac{|H|}{|G|} \sum_{\chi \in (G/H)^{\hat{}}} \langle f, ^{\#}\chi \rangle_{G/H} \chi(g).$$

Here the Fourier coefficient is

$$\langle f^{\#}, \chi \rangle_{G/H} = \sum_{y \in G/H} f^{\#}(y)\, \overline{\chi(y)} = \sum_{y \in G/H} \sum_{h \in H} f(yh)\, \overline{\chi(y)}$$

$$= \sum_{x \in G} f(x)\overline{\chi(x)} = \hat{f}(\chi),$$

where $\hat{f}$ denotes the discrete Fourier transform on G. Here we have used the fact that G is a disjoint union of cosets of H. And we need to know that $\chi(h) = 1$ for all $h \in H$. Substitute

$$\langle f^{\#}, \chi \rangle_{G/H} = \hat{f}(\chi)$$

into the Fourier expansion of $f^{\#}$, on G/H to complete the proof. ∎

Set $g = I$ in the theorem to obtain the following corollary.

Corollary. Under the same hypotheses as the theorem, we have

$$\frac{1}{|H|} \sum_{h \in H} f(h) = \frac{1}{|G|} \sum_{\chi \in H^{\#}} \hat{f}(\chi).$$

Exercise. Take $G = \mathbb{F}_p \times \mathbb{F}_p$ and $H = \mathbb{F}_p \times \{0\}$. Plug the function

$$f_r(x) = \exp\left(\frac{2\pi i r}{p} (x_1^2 + x_2^2) \right)$$

into the corollary of Poisson's summation formula. Deduce that if G_1 is the Gauss sum defined by formula (10) of Chapter 8, then

$$G_1^2 = \left(\frac{-1}{p} \right) p,$$

where $\left(\frac{a}{p} \right)$ is the Legendre symbol of formula (1) in Chapter 8.

Poisson Summation as a Trace Formula

We saw in Chapter 2 that the inversion formula for the DFT on $\mathbb{Z}/n\mathbb{Z}$ is a trace formula. There is a similar result for Poisson summation. Let's look at a finite abelian group G with a subgroup H.

Then let $k : G \to \mathbb{C}$ act by $*$ on functions $f : G/H \to \mathbb{C}$. We have the operator

$$L_k f(x) = (k * f)(x) = \sum_{y \in G} k(x - y) f(y)$$

$$= \sum_{y \in G/H} \sum_{h \in H} k(x - h - y) f(y). \tag{6}$$

It follows that the trace of L_k restricted to $V = L^2(G/H)$ is

$$\mathrm{Tr}\, L_k \big|_V = \sum_{y \in G/H} \sum_{h \in H} k(y - h - y) = \frac{|G|}{|H|} \sum_{h \in H} k(h), \tag{7}$$

and this must also equal the sum of the eigenvalues of the operator L_k restricted to V.

As in Chapter 2, the characters of G that are trivial on H (i.e., $\chi \in H^{\#} \cong (G/H)^{\widehat{}}$) form a complete orthonormal set of eigenfunctions of the operator L_k restricted to V. This is seen by the orthogonality relations and the lemma below, which we should have proved earlier.

*Lemma 2. **Characters are Eigenfunctions of Convolution Operators.*** Suppose $k \in L^2(G)$ and $\chi \in \hat{G}$. Then

$$(k * \chi)(a) = \langle k, \chi \rangle \chi(a), \quad \text{for all } a \in G.$$

Here

$$\langle k, \chi \rangle = \hat{k}(\chi) = \text{DFT of } k \text{ at } \chi = \sum_{y \in G} k(y) \overline{\chi(y)},$$

as in formula (6) of Chapter 10. It follows that χ is an eigenfunction of the convolution operator with eigenvalue $\hat{k}(\chi)$.

Proof. The lemma is proved by computing the convolution:

$$(k * \chi)(a) = \sum_{y \in G} k(y) \chi(a - y) = \chi(a) \sum_{y \in G} k(y) \chi(-y) = \chi(a) \langle k, \chi \rangle. \quad \blacksquare$$

Now we return to our discussion of the Poisson summation formula as a trace formula. It follows from (7) and the lemma that the trace of L_k restricted to $V = L^2(G/H)$ is

$$\mathrm{Tr}\, L_k \big|_V = \frac{|G|}{|H|} \sum_{h \in H} k(h) = \sum_{\chi \in H^{\#}} \hat{k}(\chi), \tag{8}$$

where $\hat{k}(\chi)$ is as in the lemma above. This is Poisson summation at the identity element of G. In Chapters 22, 23, 24 we will consider analogues of this for non-commutative groups. To obtain the general result, replace $k(y)$ by $k_g(y) = k(g + y)$.

Exercise. Check this last remark. ■

Application to Coding Theory: The MacWilliams Identities

To apply the Poisson summation formula, we need to make it more explicit. Suppose $G = \mathbb{F}_2^n$ and $H = C$ is a linear code (i.e., a vector subspace). Then

$$C^\# = H^\# = \{\chi : G \to \mathbb{T} \mid \chi(h) = 1 \text{ for all } h \in C\}. \tag{9}$$

Now $\chi = \chi_a$ for some $a \in G$, where $\chi_a(x) = (-1)^{'a \cdot x}$, as in the preceding chapter. So we have

$$C^\# = \{\chi_a \mid a \in \mathbb{F}_2^n \text{ and } {}'au = 0 \text{ for all } u \in C\}. \tag{10}$$

Example. Cyclic Codes. Recall the definition of cyclic codes in the last chapter. Suppose that C is a cyclic code with generator polynomial $g(x)$ and

$$x^n - 1 = g(x)h(x).$$

Then if m is the degree of h, the polynomial $x^m h(x^{-1})$ is the generator polynomial for the code $C^\#$.

Exercise. Prove the last statement.
Hint. See Pless [1989, p. 75].

Thus the Poisson sum formula becomes a formula for a code C in $\mathbb{F}_2^n$ and its dual code $C^\#$:

$$\frac{1}{|C|} \sum_{a \in C} f(a) = 2^{-n} \sum_{u \in C^\#} \hat{f}(\chi_u), \tag{11}$$

where

$$\hat{f}(\chi_u) = \sum_{x \in \mathbb{F}_2^n} (-1)^{'a \cdot x} f(x).$$

To put (11) into the form used by MacWilliams and Sloane [1988], replace C by $C^\#$. Then we have

$$\frac{1}{|C^\#|} \sum_{a \in C^\#} f(a) = 2^{-n} \sum_{u \in C} \hat{f}(\chi_u). \tag{12}$$

This implies, since $|C^\#| = |G|/|C| = 2^n/|C|$,

$$\sum_{a \in C^\#} f(a) = |C|^{-1} \sum_{u \in C} \hat{f}(\chi_u). \tag{13}$$

Next we need some definitions.

Definitions. If C is an $[n, k]$ binary code (i.e., a k-dimensional subspace of $\mathbb{F}_2^n$) define the *weight distribution function* of C by

$A_i = $ the number of codewords of Hamming weight i in C.

Define the *weight enumerator polynomial* by

$$w_C(x, y) = \sum_{i=0}^{n} A_i x^{n-i} y^i.$$

The weight distributions are important for decoding (see Pless [1988, Ch. 8]).

Finally we can prove the MacWilliams theorem. van Lint [1982, p. 39] calls this theorem "one of the most fundamental results in coding theory." MacWilliams proved the result in 1963 as part of her Ph.D. thesis at Harvard. See also MacWilliams and Sloane [1988].

Theorem 2. MacWilliams Identity for Weight Enumerators of Binary Codes. The following identity relates the weight enumerator polynomial defined above for the binary linear code C in $\mathbb{F}_2^n$ and that of its dual code $C^\#$ defined above:

$$w_{C^\#}(x, y) = |C|^{-1} w_C(x + y, x - y).$$

Proof. Apply formula (13) to the function

$$f(u) = x^{n-|u|} y^{|u|}.$$

Here we are thinking that x and y are fixed complex numbers. To complete the proof, it suffices to show that

$$\hat{f}(u) = (x + y)^{n-|u|} (x - y)^{|u|}.$$

To see this, note that

$$\hat{f}(u) = \prod_{i=1}^{n} \left(\sum_{v_i=0}^{1} (-1)^{u_i v_i} x^{1-v_i} y^{v_i} \right).$$

If $u_i = 0$, the inner sum is $x + y$. If $u_i = 1$, the inner sum is $x - y$. ∎

Exercise. Consider the [7, 4, 3] Hamming code C defined above. Show that the weight enumerators of this code and its dual are

$$w_C(x, y) = x^7 + 7x^4 y^3 + 7x^3 y^4 + y^7$$

and

$$w_{C^*}(x, y) = x^7 + 7x^3 y^4.$$

Note that the symmetry given by the MacWilliams identities is not obvious.

Remarks on Applications to Statistical Mechanics. The Ising Model

B. Cipra [1987, p. 937] writes:

In spite of their familiarity, phase transitions are not well understood. One purpose of the Ising model is to explain how short-range interactions between, say, molecules in a crystal give rise to long-range, correlative behavior, and to predict in some sense the potential for a phase transition. The Ising model has also been applied to problems in chemistry, molecular biology, and other areas where "cooperative" behavior of large systems is studied. These applications are possible because the Ising model can be formulated as a *mathematical* problem.

References for applications of the Poisson sum formula in mathematical physics are the papers of Ginibre [1970] and H. P. McKean [1964], as well as the collection of papers edited by Pool [1972]. A nice reference for the Ising model is B. Cipra [1987]. The physicists want to understand why substances change from liquid to solid at some temperature, for example. The model of Ising ferromagnets – first studied by Ising in 1925 – was created to help.

In fact, Ising studied "spontaneous magnetization" in his doctoral thesis. One holds a lattice (such as $\mathbb{Z}$) of magnetic material at constant temperature in a magnetic field. The external magnetic field is slowly turned off. At temperatures below some critical temperature the lattice keeps some magnetism. One wants to understand this critical temperature. Ising studied a one-dimensional model and found no phase transition. He also had some arguments that the same thing would happen in three dimensions and became discouraged. But about ten years later it was shown by R. Peierls that a phase transition would occur in two dimensions. The three-dimensional case is still open.

The finite Ising model is a finite set L in a lattice such as $\mathbb{Z}^d$ with spins $\sigma_x = \pm 1$ attached to each site, $x \in L$, and bonds joining neighboring pairs of sites. McKean [1964] discusses the high–low temperature duality of Kramers–Wannier for the two-dimensional Ising model.

We take our discussion mostly from McKean [1964] and also Cipra [1987], Hurt [1983], and Kasteleyn's article in Luck et al. [1990, pp. 195–202 and pp. 264–75]. The *energy* or *Hamiltonian* of the set L is

$$H(\sigma) = -E \sum_{\text{bonds}} \sigma_x \sigma_y - J \sum_{\text{sites}} \sigma_x. \tag{14}$$

Here "bonds" means neighboring pairs $x, y \in L$ and "sites" means elements $x \in L$. The constant E is a positive coupling constant for a ferromagnet while J is an external magnetic field, which we will assume is 0.

The thermodynamic properties of the system can be derived from the *partition function*, which is defined to be

$$Z = Z(\beta, E, J, N) = \sum_{\sigma} \exp\{-\beta H(\sigma)\}. \tag{15}$$

The sum is over all spins, which are functions $\sigma : L \to \{\pm 1\}$. Here N is the number of lattice sites. Let $C = E\beta$, where $\beta = 1/kT$, where k is Boltzmann's constant and T is the Kelvin temperature. Then

$$Z(C) = \sum_{\text{spins } \sigma} \exp\left\{C \sum_{\text{bonds}} \sigma_x \sigma_y\right\}.$$

Exercise. Consider the special case of three lattice sites in a line. Assume that the Hamiltonian is

$$H = -E(\sigma_1 \sigma_2 + \sigma_2 \sigma_3) - J(\sigma_1 + \sigma_2 + \sigma_3).$$

Compute the partition function in the case that $J = 0$. Then generalize to N lattice sites.

Answer for Three Lattice Sites.

$$Z = 8 \cosh^2(\beta E).$$

(See Cipra [1987, p. 942].)

The partition function tells one the important physical constants of the system. For example one finds (see Cipra [1987, p. 942]) that the internal energy is

$$U = \frac{\partial \log Z}{\partial \beta}.$$

Thus it is natural to look at the log of Z. The *Ising problem* is to find an analytic formula for the (thermodynamic) limit

$$F = \lim_{|L| \to \infty} \frac{1}{N} \log Z(C, N).$$

The hope is that the phase transitions will be the discontinuities of F or a derivative of F. Lee and Yang gave the first proof that the limit exists in 1952. Few explicit solutions seem to be known. Ising did a 1-dimensional problem in 1925. Onsager did a 2-dimensional one in 1944.

The *Kramers–Wannier duality* result says that for an infinite lattice:

$$Z(C) = Z(C^*) \left(\frac{\sinh 2C^*}{2} \right)^{-B/2}. \tag{16}$$

Here B is the number of bonds of L and C^* is the positive root of

$$\sinh 2C \sinh 2C^* = 1.$$

This relation is only approximate for a finite lattice.

McKean obtains the result for $\mathbb{Z}^2$ by looking at finite lattices L and Poisson summation for G = the group of paths on L with coefficients from $\mathbb{Z}/2\mathbb{Z}$. This is the group $(\mathbb{Z}/2\mathbb{Z})^B$. And H is the subgroup of closed paths or cycles. Use is made of Van der Waerden's formula for the partition function, which says

$$Z = \left(\frac{\sinh 2C}{2} \right)^{B/2} \sum_{\text{spins}} \prod_{\text{bonds}} (e^{C_*} + \sigma_x \sigma_y e^{-C_*}).$$

What is C_*? It comes from the equation

$$e^{\pm C} = \left(\frac{\sinh 2C}{2} \right)^{1/2} (e^{C_*} \pm e^{-C_*}).$$

This is all we will have to say about the Ising model. We leave it to the interested reader to get more information from the references mentioned above.

Switching Functions

> Currently computer chips can have a density of some 10
> million devices per square centimeter.
>
> Likharev and Claeson [1992, p. 85]

> I am writing this article on a computer that contains some 10
> million transistors.
>
> Keyes [1993, p. 70]

> The most basic digital devices are *gates*. . . . the three most
> important kinds of gates [are and, or, not]. . . . Any digital
> function can be realized using just these three kinds of
> gates. . . . A collection of one more more gates fabricated on a
> single silicon chip is called an *integrated circuit* (IC). Large ICs
> with millions of transistors or more may be half an inch or more
> on a side. . . .Any IC with over 500,000 transistors is definitely
> VLSI [very large-scale integration]. . . . Prices of VLSI devices
> can range from well under $10 (for 1Mbit memories) to
> hundreds of dollars (for the 3-million transistor Intel Pentium
> microprocessor). Perhaps the next level of integration will be
> called RLSI – ridiculously large-scale integration.
>
> Wakerly [1994, pp. 6–11]

Here we wish to say a bit about logic circuits. It is a bit frightening how quickly the circuits we use have become so complicated. The references of the 1970s spoke of thousands rather than millions of transistors. References for this application include those mentioned above plus Dewdney [1989], Dornhoff and Hohn [1978], Gilbert [1976], Harrison [1965], and Lechner's article in Mukhopadhyay [1971]. Logic or switching circuits were once combinations of transistors, resistors, etc. connected by wires. Now one semiconductor chip may contain millions of these elements. And, no doubt, by the time you read this, the chips will have even more elements.

Keyes [1993, p. 78] asks: "What . . . would one want with a technology that can etch a million transistors into a grain of sand or put a supercomputer in a shirt pocket?" He answers this question by looking to a time when we can all carry a copy of the Library of Congress. The chips we have now are managing not just our methods of communicating information but also our automobiles' carburetors, traffic lights, CD players, household appliances, . . .

We do not want to talk about the very mutable engineering or physics of these chips, but of the logic that rules them. This goes back to George Boole who invented Boolean algebra in 1854. Claude E. Shannon showed in 1938 how to use Boolean algebra to analyze circuits built from relays. This switching algebra involves variables x that can only have values 0 or 1 – representing relay contact open or closed. Today this may represent voltage high or low, or some other physical attributes.

If the circuit has n inputs and 1 output, it can be viewed as a *switching or Boolean function*

$$f : \mathbb{F}_2^n \rightarrow \{0, 1\}.$$

We write the range as $\{0, 1\}$ since the usual addition of Boolean functions is neither that of $\mathbb{C}$ nor that of $\mathbb{F}_2 = \mathbb{Z}/2\mathbb{Z}$. For if f and g are two switching functions, then $(f + g)(x) = 1$ means that either $f(x) = 1$ or $g(x) = 1$ (or both). Thus $1 + 1 = 1$ and not 0 or 2. We frequently use the symbol $+$ in this section to denote Boolean addition. If we want to use the addition mod 2, we use the symbol $\oplus : 1 \oplus 1 = 0$.

Let F_n denote the set of *all switching functions* $f : \mathbb{F}_2^n \rightarrow \{0, 1\}$. Clearly $|F_n| = 2^{2^n}$, which gets large quite rapidly. We turn F_n into a *Boolean algebra* by defining the following operations for $f, g \in F_n$:

$$(f + g)(x) = \begin{cases} 1, & \text{if } f(x) \text{ or } g(x) \text{ or both} = 1, \\ 0, & \text{otherwise}, \end{cases}$$

$$(f \cdot g)(x) = \begin{cases} 1, & \text{if both } f(x) \text{ and } g(x) = 1, \\ 0, & \text{otherwise}, \end{cases}$$

$$(f \oplus g)(x) = \begin{cases} 1, & \text{if } f(x) \text{ or } g(x) \text{ but not both} = 1, \\ 0, & \text{otherwise}, \end{cases}$$

Look at F_2, for example. Table I.15 gives truth tables for $x + y$, $x \oplus y$, and xy.

The Boolean algebra F_n is a *poset* or partially ordered set under the relation $f \leq g$ iff $f \cdot g = f$. Note that this means that $f(x) = 1$ implies $g(x) = 1$, that is, f implies g.

Table I.15. *Truth tables for*
$x + y$, $x \oplus y$, *and* xy

x	y	$x+y$	$x \oplus y$	xy
0	0	0	0	0
0	1	1	1	0
1	0	1	1	0
1	1	1	0	1

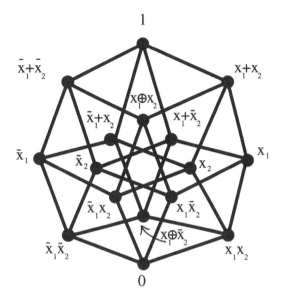

1

$\bar{x}_1+\bar{x}_2$

x_1+x_2

$x_1\oplus x_2$

$\bar{x}_1+x_2$

$x_1+\bar{x}_2$

$\bar{x}_1$

$\bar{x}_2$

x_1

x_2

$\bar{x}_1x_2$

$x_1\bar{x}_2$

$x_1\oplus\bar{x}_2$

$\bar{x}_1\bar{x}_2$

x_1x_2

0

Figure I.41. Poset diagram for F_2, the switching functions of two variables.

The *poset diagram* of F_n is a graph with vertices corresponding to elements of F_n. You draw a rising line between two vertices f and g iff $f < g$ and there is no h with $f < h < g$. In Figure I.41 you see the poset F_2. It is another drawing of a four-dimensional cube or tesseract. Compare with Figure 37.

Since $|F_n| = 2^{2^n}$ blows up so rapidly with n, one has problems trying to simplify the switching circuits (e.g., to speed up the operations). Much effort has been devoted to counting elements of F_n inequivalent under permutation of the inputs, for example. See the references mentioned above. One also wants to write a given switching function in some "simplest" form. This leads to a search for the *prime implicants* p of $f \in F_n$. This is a product of letters (or literals)

$$P = abc \cdots r$$

such that $P \leq f$ and such that there is no other product of letters Q with $P < Q < f$.

First note that $P \leq f$ means that $P^{-1}(1) \subset f^{-1}(1)$. Secondly note that P is a product of letters iff $P^{-1}(1)$ is a subcube of $\mathbb{F}_2^n$.

Example.

$$P = xy \in \mathbb{F}_3$$

means

$$P^{-1}(1) = \{(1, 1, z) \mid z \in \mathbb{F}_2\} = (1, 1, 0) \oplus H,$$

where H is a one-dimensional subspace of $\mathbb{F}_2^3$ given by

$$H = \{(0, 0, z) \mid z \in \mathbb{F}_2\}.$$

Therefore the search for prime implicants of $f \in F_n$ is a search for sets $c \oplus H$, where H is some k-dimensional subspace of $\mathbb{F}_2^n$ such that

$$c \oplus H \subset f^{-1}(1).$$

Lechner (in Mukhopadhyay [1971]) gives the following theorem, which tells one how to write a program to do the search for the prime implicants. There are many other methods which can be used (as described in the references above).

Theorem 3. **Lechner's Theorem on the Extraction of Prime Implicants of Switching Functions.** Suppose that F_n is the Boolean algebra of n-input switching functions

$$f : \mathbb{F}_2^n \to \{0, 1\}$$

and H is any subspace of $\mathbb{F}_2^n$. Let $c \in \mathbb{F}_2^n$. Then $c \oplus H = P^{-1}(1)$ for some prime implicant P of f means

$$c \oplus H \subset f^{-1}(1).$$

This is equivalent to the following equality:

$$\sum_{a \in H^\#} \hat{f}(a)(-1)^{'a \cdot c} = 2^n.$$

Here $H^\# = \{w \mid {}^t w x = 0, \text{ for all } x \in H\}$, and $\hat{f}$ denotes the discrete Fourier transform of f on $\mathbb{F}_2^n$.

Proof. This result is a straightforward consequence of Theorem 1. If we let $G = \mathbb{F}_2^n$ and $H = $ some k-dimensional subspace of G, then

$$1 = 2^{-k} \sum_{h \in H} f(c + h) = 2^{-n} \sum_{a \in H^\#} \hat{f}(a)(-1)^{'a \cdot c}.$$

Here we use the fact that $(G/H)\hat{\,}$ is the set of the $e_a(x) = (-1)^{'a \cdot c}$ that are identically 1 on H. This means $a \in \mathbb{F}_2^n$ such that ${}^t a x = 0$ for all $x \in H$ (i.e., $a \in H^\#$).

We use the fact that the values of f are only 0 and 1 to see the converse. ∎

From Theorem 3, Lechner obtains an algorithm to extract prime implicants of f, starting with $\hat{f}$. There are 2^n Fourier coefficients of f. Check the subspaces H of $\mathbb{F}_2^n$ successively in order of descending dimensions, which is ascending in order of number of variables in the corresponding prime implicant P. The prime implicants with one letter will be extracted first, those with two letters second, etc. Lechner also makes use of a fast Fourier transform algorithm depending on a recursive definition of the DFT on $\mathbb{F}_2^n$.

Chapter 13

Some Applications in Chemistry and Physics

> In 1882 Schuster formulated the extraordinarily prescient
> suggestion that the main function of the study of spectra would
> be to obtain information about the structure of atoms and
> molecules and the nature of the forces that bind them together,
> and coined the name "spectroscopy" for this new science.
>
> Sternberg [1994, p. 382]

> The vibrational spectra of about a quarter of a million chemical
> compounds have been investigated, and tens of thousands are
> tabulated in a form accessible to an electronic computer . . . .
>
> Sternberg [loc. cit., p. 406]

Our goal in this chapter is to find some very concrete applications of our results
in chemistry and physics. We will look at Ehrenfest's urn model in statistical
mechanics, systems of vibrating masses with cyclic groups of symmetries, and,
finally, Hückel's theory of the aromatic compounds like benzene.

In the following examples we make use of the fact that the DFT diagonalizes
the adjacency operator for the Cayley graph of a finite abelian group. This is the
generalization of Theorem 2 in Chapter 3. Perhaps we should restate it for the
more general situation.

Theorem 1. Suppose that G is a finite abelian group under $+$ with dual group

$$\hat{G} = \{\chi : G \to \mathbb{T} \mid \chi(a+b) = \chi(a)\chi(b), \text{ for all } a, b \in G\},$$

where $\mathbb{T}$ is the multiplicative group of complex numbers z with $|z| = 1$. Then
if S is a symmetric set of generators of G (meaning that $s \in S$ implies $-s \in S$)

the Cayley graph $X(G, S)$ has adjacency operator A with spectrum

$$\text{spec}(A) = \{\mathscr{F}\delta_s(\chi) \mid \chi \in \hat{G}\}.$$

Here $\delta_s(x) = 1$ for $x \in S$ and 0 otherwise. And $\mathscr{F}\delta_s(\chi)$ is the discrete Fourier transform of δ_s.

Proof. This is left as an exercise.

Hint. Imitate the proof of Theorem 2 of Chapter 3 in the case that G is cyclic or just note that the characters of G form a complete orthogonal set of eigenfunctions of A.

Ehrenfest's Urn Model in Statistical Mechanics

Random walks on $\mathbb{F}_2^d$ are of interest because of a connection with a problem in statistical mechanics. Ehrenfest's urn model involves two urns and d balls. A ball is chosen and moved to the other urn. This process is repeated k times. One wants to know the limiting distribution of balls as $k \to \infty$. This can be viewed as a model for heat diffusion. See M. Kac [1947].

At time t, a state of the system corresponds to a vector $v \in \mathbb{F}_2^d$, where

$$v_i = \begin{cases} 1, & \text{if the } i\text{th ball is in the right-hand urn,} \\ 0, & \text{otherwise.} \end{cases}$$

The transition from time t to $t + 1$ can be viewed as giving a random walk on $\mathbb{F}_2^d$. The state changes by picking a coordinate v_i and changing it to $v_i + 1$. This means that the transition matrix of the Markov chain is $T = \frac{1}{d}A$, where A is the adjacency operator of the Cayley graph $X(\mathbb{F}_2^d, S(1))$ and

$$S(1) = \{x \in \mathbb{F}_2^d \mid |x| = 1\},$$

if $|x|$ is the Hamming weight from formula (9) in Chapter 10. We know from formula (10) of Chapter 10 that the eigenvalues of T are

$$\left\{1, 1 - \frac{2}{d}, 1 - \frac{4}{d}, \dots, -1 + \frac{2}{d}, -1\right\}.$$

Question. What does this say about $\lim_{k \to \infty} T^k p$, for a probability vector p?

Answer. It does not exist.

So we must modify the random walk slightly. Compare with Diaconis [1988]. We allow the ball the possibility of staying in one place. That is, we consider the Cayley graph $X(\mathbb{F}_2^d, S(1) \cup \{0\})$. This means we have a Markov transition matrix $T^0 = \frac{1}{d+1}(I + A)$, where A is the adjacency operator of $X(\mathbb{F}_2^d, S(1))$. So the spectrum of T^0 is

$$\left\{ 1, 1 - \frac{2}{d+1}, 1 - \frac{4}{d+1}, \ldots, -1 + \frac{2}{d+1} \right\}.$$

Clearly now we have

$$\lim_{k \to \infty} T^k p = u = 2^{-d} = \text{uniform},$$

for a probability vector p as in Theorem 1 of Chapter 6. And we can obtain a more precise result as in the Theorem 2 of Chapter 6. We find

$$\|T^m p - u\|_1 \le 2^{d/2} \beta^m, \quad \text{where } \beta = 1 - \frac{2}{d+1}. \tag{1}$$

Exercise. How big must m be to make

$$2^{d/2} \left(1 - \frac{2}{d+1} \right)^m$$

small?

Hint.

$$1 - \frac{2}{d+1} < e^{-1/(d+1)}.$$

So

$$2^{d/2} \left(1 - \frac{2}{d+1} \right)^m < e^{-c} \quad \text{if } m > (d+1)^2 + (d+1)c.$$

This should be compared with the result of Diaconis [1988, p. 72] (which is better).

Vibrating Springs and Things

Some references for this section are: Fan Chung and S. Sternberg [1992, 1993], Coulson, O' Leary and Mallion [1978], Courant and Hilbert [1953], Cvetković, Doob, and Sachs [1980, Chapter 8], Fässler and Stiefel [1992], Keener [1988,

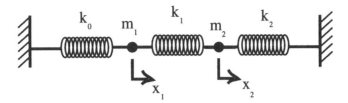

Figure I.42. Vibrating system of two masses.

p. 24], Lomont [1993, pp. 99 ff], Starzak [1989], Sternberg [1994], and Strang [1976, pp. 284–6, 343 ff].

Consider a vibrating system governed by a Hooke's law potential; for example, think of springs connecting two masses as in Figure I.42.

According to Newton's law of motion $F = ma$ (neglecting gravity), the system of ordinary differential equations governing the vibrating masses in Figure I.42 is

$$m_1 \frac{d^2 x_1}{dt^2} = k_1(x_2 - x_1) - k_0 x_1,$$

$$m_2 \frac{d^2 x_2}{dt^2} = -k_1(x_2 - x_1) - k_2 x_2.$$

One way to obtain this system of ordinary differential equations uses the calculus of variations. See Courant and Hilbert [1953]. Form the Lagrangian $L = T - V$, where

$$2T = m_1 \dot{x}_1^2 + m_2 \dot{x}_2^2, \quad \text{if } \dot{x}_i = \frac{dx_i}{dt},$$

$$2V = k_0 x_1^2 + k_2 x_2^2 + k_1(x_1 - x_2)^2 = k_0' x_1^2 - 2k_1' x_1 x_2 + k_2' x_2^2,$$

$$\text{where } k_0' = k_0 + k_2, \ k_2' = k_2 + k_1, \ k_1' = k_1.$$

Then Hamilton's *Principle of Least Action* says that the masses will move so as to minimize the action functional:

$$A[x] = \int_{t_1}^{t_2} (T - V) \, dt.$$

The Euler–Lagrange equations in the calculus of variations is then the system of ordinary differential equations:

$$\frac{d}{dt}\left(\frac{\partial L}{\partial \dot{x}_i}\right) - \frac{\partial L}{\partial x_i} = 0, \quad i = 1, 2,$$

where

$$2L = 2T - 2V = m_1 \dot{x}_1^2 + m_2 \dot{x}_2^2 - \left(k_0' x_1^2 - 2k_1' x_1 x_2 + k_2' x_2^2\right).$$

It follows that we can write the Euler–Lagrange equation as a matrix equation for the vector $x = {}^t(x_1, x_2)$ of displacements:

$$Mx'' = Kx, \quad \text{where } K = \begin{pmatrix} -k_0' & k_1' \\ k_1' & -k_2' \end{pmatrix}, \qquad M = \begin{pmatrix} m_1 & 0 \\ 0 & m_2 \end{pmatrix},$$

$$x'' = \frac{d^2 x}{dt^2} = \begin{pmatrix} x_1''(t) \\ x_2''(t) \end{pmatrix}. \tag{2}$$

Now set $x = x_0 e^{i\omega t}$ and find that

$$-\omega^2 M x_0 - K x_0 = 0.$$

This means that we have a generalized eigenvalue problem and we want

$$\det(K + \omega^2 M) = 0.$$

In fact, since M is diagonal with positive diagonal entries, we can write $M = R^2$, where

$$R = \begin{pmatrix} \sqrt{m_1} & 0 \\ 0 & \sqrt{m_2} \end{pmatrix}.$$

Then we seek to find the eigenvalues $-\omega^2$ of

$$R^{-1} K R^{-1} = \begin{pmatrix} -k_0' m_1^{-1} & k_1' (m_1 m_2)^{-1/2} \\ k_1' (m_1 m_2)^{-1/2} & -k_2' m_2^{-1} \end{pmatrix}.$$

We went to all this trouble to obtain a symmetric matrix so that we would know that we can apply the spectral theorem. Note that Keener [1988, p. 24] didn't do this.

Let's take $k_j = m_j = 1$ for simplicity. Then

$$B = R^{-1} K R^{-1} = \begin{pmatrix} -2 & 1 \\ 1 & -2 \end{pmatrix} = -2I + A,$$

$$A = \begin{pmatrix} 0 & 1 \\ 1 & 0 \end{pmatrix} = \text{adjacency matrix of graph formed by two masses}.$$

So a solution x to equation (2) is

$$x = x_0 e^{i\omega t},$$

where

$$Bx_0 = -\omega^2 x_0.$$

That is, x_0 is an eigenvector of B corresponding to the eigenvalue $-\omega^2$. In our case the eigenvalues of B are $\lambda = -1$ and -3. The corresponding eigenvectors are

$$\begin{pmatrix} 1 \\ 1 \end{pmatrix} \quad \text{and} \quad \begin{pmatrix} 1 \\ -1 \end{pmatrix}.$$

Writing $\omega^2 = -\lambda$, we find $\omega = \pm 1, \pm\sqrt{3}$. These are the natural frequencies of vibration.

If the system starts from rest and the first mass is given a unit displacement while the second mass is held in place, we have

$$x_0 = \begin{pmatrix} 1 \\ 0 \end{pmatrix}, \quad x_0' = \begin{pmatrix} 0 \\ 0 \end{pmatrix}.$$

Then we find that the solution is

$$x(t) = \frac{1}{2}\cos t \begin{pmatrix} 1 \\ 1 \end{pmatrix} + \frac{1}{2}\cos(\sqrt{3}t) \begin{pmatrix} 1 \\ -1 \end{pmatrix}.$$

The motion is the average of two pure oscillations corresponding to the two eigenvectors of M. In the first mode the masses move exactly in unison and the spring in the middle is never stretched. In the second mode, the masses move in opposite directions. The general solution is a linear combination of the two normal modes. Our solution for the special case is almost periodic but never returns exactly to the initial conditions because $\sqrt{3}$ is irrational. The masses do come arbitrarily close to the initial condition.

Many problems have this same sort of form (e.g., in circuit theory and in chemistry). If the masses or atoms or circuit elements form a Cayley graph of a group, then we can find the fundamental frequencies of vibration as well as the normal modes using the Fourier transform on the group.

Chemistry of Aromatic Compounds such as Benzene

In 1825 a young laboratory director at the Royal Institution in London was asked to analyze the liquid residue that formed during the production of lamp gas. Considered one of the best chemists of the time, the 33-year old researcher, Michael Faraday, made a discovery that would mark the beginning of a

> new branch of chemistry. He isolated from the residue a new
> hydrocarbon – a molecule consisting of only hydrogen and
> carbon atoms – and called it bicarburet of hydrogen. We now
> call the substance benzene, a compound that consists of six
> hydrogen atoms attached to six carbon atoms. It is the
> prototype of a class of molecules known as the aromatic
> compounds.
>
> Aihara [1992, p. 62]

> In 1865, staring drowsily one evening into the fire, he [Kekulé]
> saw in a dream the cyclic structure for benzene, that
> fundamental unit of all aromatic molecules, and of graphite and
> the fullerenes. In his reverie, he imagined the atoms gamboling
> before his eyes.. "one of the snakes had seized hold of its own
> tail, and the form whirled mockingly before my eyes."
>
> D. E. H. Jones (alias Daedalus)
> in Kroto and Walton [1993, p. 9]

References for this application include those just quoted plus Atkins [1987], Cvetković, Doob, and Sachs [1980, Chapter 8], and Sternberg [1994].

In the chemistry of molecules built up out of carbon and hydrogen, one can often make use of our knowledge of eigenvalues of graphs to obtain approximate information. Carbon forms bonds with four of its electrons. Three of these electrons are called σ-electrons and the remaining electron is called a π-electron. According to a theory developed by Hückel in 1931, one should look at the graph corresponding to the σ-electron carbon–carbon bonds. The vertices are carbon atoms. The edges represent the bonds. If the graph is bipartite, the chemists call the molecule an alternant hydrocarbon. If the graph is $X(\mathbb{Z}/N\mathbb{Z}, \{\pm 1\})$ (i.e., an N-cycle), then the chemists call it [N]-annulene.

Example (Benzene C_6H_6). J. Elkington et al. [1990, p. 21] write: "Among the vapors emitted during filling [of your gas tank] is benzene, a chemical that causes leukemia and blood disorders." and [loc. cit., p. 28]: "Plastics contain polyvinyl propylene, phenol, ethylene, polystyrene, and benzene, all of which are among the most hazardous air pollutants around."

As noted above, the molecule benzene must be both admired and feared. It corresponds to the finite circle graph for $\mathbb{Z}/6\mathbb{Z}$, with carbon atoms at the six vertices. See Figure I.43. Benzene is very stable. This makes it very useful (e.g., as an octane enhancer in gasoline). It is an aromatic compound, which means it has a ringlike structure and not that it has a strong odor (see Aihara [1992]).

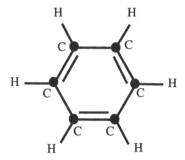

Figure I.43. Benzene.

Aihara believes that "more than just a source of stability, the benzene ring seems to be a shape nature prefers to form." He goes on to compare benzene with cyclobutadiene C_4H_4. The latter is a 4-carbon ring and is "notoriously reactive and difficult to prepare." Why should the 6-vertex ring be more stable than the 4-vertex ring? Hückel theory attempts to explain this.

One expects to get a similar problem to that of the vibrating masses linked by springs. Thus one expects a generalized eigenvalue problem for the matrix

$$H = ES, \quad S = I + \sigma A,$$
$$H = \alpha I + \beta A,$$

where A is the adjacency operator of the graph and α, β, σ are constants (assumed known).

We can relate an eigenvalue λ of A to the energy E by

$$\lambda = \frac{E - \alpha}{\beta - \sigma E}.$$

Hückel argues that the stability of the molecule is measured by the quantity (rest mass energy)

$$E = \frac{2}{n} \sum_{\substack{\text{top } n/2 \text{ eigenvalues } \lambda \text{ of } A}} \lambda.$$

We know the eigenvalues of the adjacency operator for $X(\mathbb{Z}/6\mathbb{Z}, \{\pm 1\})$ are $\pm 2, \pm 1, \pm 1$. So the value of E for benzene is $(2 + 1 + 1)/3 \cong 1.33$.

The value of E for a four-vertex ring like cyclobutadiene is one. So it is certainly less stable than benzene.

Example. Buckminsterfullerene C_{60}. Kroto and Walton [1993, inside cover] write:

In 1965, Buckminsterfullerene (fullerene -60) C_{60} [the buckyball] was discovered seren-dipitously during graphite laser vaporization experiments designed to simulate the chemistry in a red giant carbon star. The molecule was isolated for the first time in macroscopic amounts in 1990, a breakthrough which triggered an explosion of research into its chemical and physical properties. . . . It seems almost impossible to comprehend how the existence of the third well-characterized allotrope of carbon could have evaded discovery until virtually the end of the twentieth century.

References on the buckyball C_{60} include the collection of Kroto and Walton [1993] plus Aihara [1992], Baggott [1996], Chung [1996], Chung and Sternberg [1992, 1993], Curl and Smalley [1991], Kostant [1995], Pennisi [1991], and Sternberg [1994]. The A.A.A.S. called the buckyball the molecule of the year in 1991. See the Dec. 20, 1991 issue of *Science* (Vol. 254). In the editorial on p. 1705 of that issue of *Science* D. E. Koshland, Jr. said:

This molecule, and the family of fullerenes derived from it, were named after the architect whose geodesic dome provided a prophetic vision of its atomic counterpart and who was a powerful evangelist for the relation of structure to function. C_{60} burst into physics and chemistry only a few years ago and has captured the enthusiasm of experimentalists and theoreticians. It has incredible symmetry for such a large molecule, in which 60 carbon atoms are joined with a mixture of single and double bonds arranged in 20 hexagons and 12 pentagons. Its chemical versatility is astonishing, reacting with alkali metals such as potassium and rubidium, halogens such as fluorine, free radicals, and Grignard reagents. The molecule itself and many of its derivatives are readily soluble in organic solvents, but recently amino adducts have been added which make it soluble in water.

In *Science,* Vol. 253, Sept. 27, 1991, there is an article of *G.* Taubes (pp. 1467–79) titled "The disputed birth of buckyballs." The discovery was the result of a collaboration between Harry Kroto and Richard Smalley. Taubes says on p. 253 of that article: "And yet the collaboration is in tatters, offering a cautionary tale in how a partnership can go wrong." Smalley has emphasized the night he sat up trying to build the buckyball. Kroto emphasizes his inspiration by the work of Buckminster Fuller. All this is rather surprising since the buckyball is a very well-known figure (see Figure I.44): the truncated dodecahedron to a mathematician, the soccerball to an American, and the football to the rest of the world. In fact many others seem to have conjectured that the buckyball was a possibility. See the first two articles in Kroton and Walton [1993]. Perhaps the reason is that as Osawa says in his article in Kroto and Walton [1993, p. 1]: "In the 1960s and 1970s, non-benzoid aromatics were favourite targets for organic chemists. There was a prevailing dogma that aromaticity, due to the

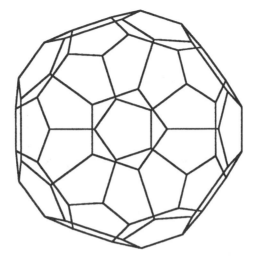

Figure I.44. The buckyball or soccerball or truncated icosahedron.

delocalization of π-electrons, is best realized in planar molecules . . . . for this reason aromaticity tacitly remained a two-dimensional concept."

One can also compute the value of the energy E for the buckyball. The answer is that $E = 1.5$ for the buckyball (see Chung and Sternberg [1992, 1993]). Thus the buckyball should be more stable than benzene. The discussion of the buckyball requires a knowledge of the representations of the nonabelian group A_5, the alternating group of even permutations of five elements. See Chapter 21. Aihara [1992, p. 64] notes that Hückel theory applies only to molecules with one ring. He develops a theory of topological resonance energy that allows one to analyze the stability of other aromatic compounds such as the buckyball.

Topics for Further Study

1. Study vibrating membranes by approximating the membrane by a graph. See Cvetković, Doob, and Sachs [1980, p. 252 ff]. This is the wave equation in two space variables. Consider the problem of whether you can hear the shape of a membrane. We will say more about this in the next chapter.
2. Study finite analogues of or approximations to the various standard partial differential equations of mathematical physics and consider what happens if there is enough symmetry to find a Cayley graph involved. We did this for the heat equation in Chapter 7. Does the finite version approximate the continuous one in some sense? We found that this was the case for the heat equation. One reference is Strickwerda [1989].

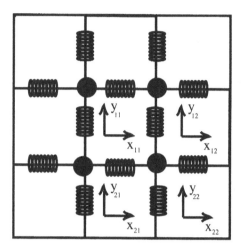

Figure I.45. A system of vibrating masses.

3. Study the Aihara [1992] theory of topological resonance energy. Use it to analyze the stability of benzene and the buckyball. He obtains the topological resonance energies per pi electron as: .0454 for benzene (very stable), .0274 for the buckyball (stable), −0.3066 for cyclobutadiene (extremely unstable).

Exercise. (From Starzak, 1989, p. 262.) Consider a system of four vibrating masses in the plane connected as indicated in Figure I.45. You may assume that the x coordinates can be considered separately. Solve the resulting problem as we did for the system of Figure I.42 using the DFT for the square.

Chapter 14
The Uncertainty Principle

To get some feeling for the size of the limit given by
Heisenberg's relation, suppose that the position of an electron
is measured to the accuracy of one part in a nanometer
(10^{-9} m); then the momentum would become so uncertain that
one could not expect that, one second later, the electron would
be closer than 100 kilometers away!

R. Penrose [1989, p. 248]

The uncertainty principle is widely known for its
"philosophical" applications: in quantum mechanics, of
course, it shows that a particle's position and momentum
cannot be determined simultaneously (Heisenberg [1930]); in
signal processing it establishes limits on the extent to which the
"instantaneous frequency" of a signal can be measured (Gabor
[1946]). However, it also has technical applications, for
example in the theory of partial differential equations
(Fefferman [1983]).

D. Donoho and P. Stark [1989, p. 906]

The classical uncertainty principle says that a function and its Fourier transform cannot both be highly localized or concentrated. In quantum mechanics, this becomes the statement that it is impossible to find a particle's position and momentum simultaneously, as Penrose makes quite explicit in the quote above. But, as Donoho and Stark remark, the uncertainty principle shows its (ugly/beautiful) face in many other areas such as signal processing and medical imaging. References for this chapter are Donoho and Stark [1989], Dym and McKean [1972], Grünbaum [1990], Smith [1990], Terras [1985, Vol. I], Elinor Velasquez [1998], and Wolf [1994].

The finite abelian analogue of the uncertainty principle can be stated quite simply. We will ultimately give three versions. See Theorems 1, 2, and 5 below.

The main goal of the paper of Donoho and Stark [1989] is to generalize the uncertainty principle so that it is not necessary to require that f and $\hat{f}$ be exactly 0 outside intervals. Instead one can ask them to be practically zero outside sets $T, B \subset G$. Then one finds that $|T||B| \geq |G|(1-\delta)^2$, where δ is "a small number bound up in the definition of the phrase 'practically zero'." We give a precise definition of this phrase before Theorem 5 below.

Donoho and Stark [1989, p. 907] note that: "The discrete-time principle proves that a wideband signal can be reconstructed from narrow-band data – provided the wide-band signal to be recovered is sparse or 'impulsive.' The classical uncertainty principle does not apply in these examples."

Definition. If $f : G \to \mathbb{C}$, define the *support of f* to be

$$\operatorname{supp} f = \{x \in G \mid f(x) \neq 0\}.$$

In this chapter, as usual, $\hat{f}$ denotes the discrete Fourier transform of f from Equation (6) in Chapter 10; that is,

$$\hat{f}(\chi) = \sum_{x \in G} f(x)\overline{\chi(x)}, \quad \text{for } \chi \in \hat{G}.$$

Theorem 1. **Uncertainty Principle – First Version.** Suppose that G is a finite abelian group with dual group $\hat{G}$. If $f : G \to \mathbb{C}$ and f is not identically 0, then

$$|\operatorname{supp} f||\operatorname{supp} \hat{f}| \geq |G|.$$

Proof. (Donoho and Stark [1989, p. 927].) Define the L^2-norm of f by

$$\|f\|_2^2 = \sum_{x \in G} |f(x)|^2.$$

Then we have

$$\|f\|_2^2 \leq \|f\|_\infty^2 |\operatorname{supp} f|, \tag{1}$$

where

$$\|f\|_\infty^2 = \max_{x \in G} |f(x)|.$$

Now we can use the Fourier inversion formula from Theorem 2 of Chapter 10, which says that

$$f(x) = \frac{1}{|G|} \sum_{\chi \in \hat{G}} \hat{f}(\chi)\chi(x).$$

Since $|\chi(x)| \leq 1$, it follows that

$$\|f\|_\infty \leq \frac{1}{G} \sum_{\chi \in \hat{G}} |\hat{f}(\chi)|.$$

Thus, by the Cauchy–Schwarz inequality, we have

$$\|f\|_\infty^2 \leq \frac{1}{|G|^2} \sum_{\chi \in \hat{G}} |\hat{f}(\chi)|^2 \sum_{\chi \in \text{supp}(\hat{f})} 1^2.$$

Now, use the Plancherel theorem, which says

$$\|f\|_2^2 = \frac{1}{|G|} \|\hat{f}\|_2^2$$

to see that

$$\|f\|_\infty^2 \leq \frac{1}{|G|} \|f\|_2^2 \, |\text{supp } \hat{f}|. \tag{2}$$

Then combine (1) and (2) to obtain

$$\|f\|_2^2 \leq \frac{1}{|G|} \|f\|_2^2 \, |\text{supp } f||\text{supp } \hat{f}|.$$

Since we are assuming that $\|f\|_2^2 \neq 0$, we can divide by it and finish the proof. ∎

It is easy to give an example for which equality occurs in Theorem 1. Let $f = \delta_e$, where e is the identity of G. Then $\hat{f} = 1$. So supp $f = \{e\}$ and supp $\hat{f} = G$. Thus we have equality in Theorem 1.

To get a slightly less trivial example, consider a subgroup H of G and let

$$f(x) = \delta_H(x) = \begin{cases} 1, & x \in H, \\ 0, & x \notin H. \end{cases}$$

So $|\text{supp } f| = |H|$. What can we say about the support of the Fourier transform of f? Let's compute it:

$$\hat{f}(\chi) = \sum_{y \in G} f(y)\overline{\chi(y)} = \sum_{y \in H} \overline{\chi(y)} = \begin{cases} |H|, & \text{if } \chi \in H^\#, \\ 0, & \text{otherwise.} \end{cases}$$

Here $H^{\#}$ is the dual to H, as defined in Chapter 12; that is,

$$H^{\#} = \{\chi \in \hat{G} \mid \chi(h) = 1, \quad \text{for all } h \in H\}.$$

Thus we have proved the following lemma.

Lemma 1. Suppose that H is a subgroup of G. Then

$$\hat{\delta}_H = |H|\delta_{H^{\#}}$$

and

$$|\text{supp } \delta_H||\text{supp } \hat{\delta}_H| = |H||G/H| = |G|.$$

This gives an example of equality for the inequality in Theorem 1.

Example. $G = \mathbb{Z}/ab\mathbb{Z}$, with a and $b > 1$.

$$H = \{nb(\text{mod } ab) \mid n \in \mathbb{Z}\} \cong \mathbb{Z}/a\mathbb{Z}.$$

That is, H is a cyclic subgroup of the additive group G generated by $b(\text{mod } ab)$. Clearly $|H| = a$.

By Lemma 1, we have

$$\hat{\delta}_H = |H|\delta_{H^{\#}} = a\delta_{H^{\#}}.$$

We can identify $\hat{G}$ with G via $\chi_c(x) = \exp(2\pi icx/n)$, $n = ab$. Thus we can identify $H^{\#}$ with the cyclic group generated by $a(\text{mod } ab)$.

Donoho and Stark [1989] view

$$\delta_H = \delta_0 + \delta_b + \delta_{2b} + \cdots + \delta_{(a-1)b} = \underset{}{\sqcup\!\sqcup}^a$$

as an analogue of the Shah functional or Dirac comb (see Terras [1985, Vol. I, p. 37]). It is nonzero at a equally spaced elements of $G = \mathbb{Z}/n\mathbb{Z}$. And the Fourier transform is a times the Shah functional for $H^{\#}$; that is, identifying everything as a subgroup of G, we have

$$\hat{\delta}_H = a\delta_{H^{\#}} = a\underset{}{\sqcup\!\sqcup}^b,$$

where $H^{\#}$ is the cyclic group of order b generated by $a(\text{mod } ab)$. This Fourier transform is nonzero at b equally spaced elements of G (identifying G and $\hat{G}$).

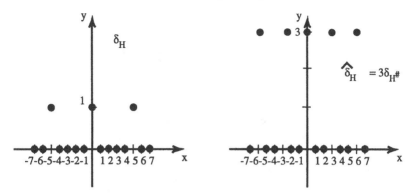

Figure I.46. Graphs of δ_H and its Fourier transform for $G = \mathbb{Z}/15\mathbb{Z}$, $H \cong \mathbb{Z}/3\mathbb{Z}$. Here $H = \{-5, 0, 5(\mathrm{mod}\ 15)\}$ and $H^{\#} \cong \{-6, -3, 0, 3, 6(\mathrm{mod}\ 15)\}$. $\hat{\delta}_H = 3\delta_{H^{\#}}$.

To illustrate what is happening take $a = 3$ and $b = 5$ and draw pictures like those in Chapter 2 (see Figure I.46).

It is rather hard to see H and its dual $H^{\#}$ inside of G; that is, it is hard to see where $f = \delta_H$ and its Fourier transform are supported in the figure. So let us use the Chinese remainder theorem (Theorem 2 of Chapter 1) to do this more easily.

If a and b are relatively prime, that is, have no common factors, the Chinese remainder theorem says that we have a ring isomorphism:

$$\mathbb{Z}/ab\mathbb{Z} \to \mathbb{Z}/a\mathbb{Z} \times \mathbb{Z}/b\mathbb{Z}$$
$$x(\mathrm{mod}\ ab) \mapsto (x(\mathrm{mod}\ a), x(\mathrm{mod}\ b)).$$

So now let us view the subgroup H as $\mathbb{Z}/a\mathbb{Z} \times \{0\}$. Then, making use of the exercise below, we see that we can identify the dual of H as the group $F = \{0\} \times \mathbb{Z}/b\mathbb{Z}$. With this identification we get three-dimensional graphs pictured in Figure I.47 for $a = 3$ and $b = 5$. In this view, both $f = \delta_H$ and $\hat{f}$ appear localized; that is, both are supported on one-dimensional subsets. Does this contradict uncertainty?

Exercise. Suppose that a and b are relatively prime integers greater than one. Let $G = \mathbb{Z}/ab\mathbb{Z}$ as above.
a) Suppose that H is the subgroup $\{nb(\mathrm{mod}\ ab) \mid n \in \mathbb{Z}\}$ of G. Show that under the isomorphism of the Chinese remainder theorem (Theorem 2 of Chapter 1), we map H to the subgroup $\mathbb{Z}/a\mathbb{Z} \times \{0\}$.
b) Recall the exercise after Figure I.34 in Chapter 10, which shows that $\hat{G}$ can be identified with the direct product of the dual groups of $\mathbb{Z}/a\mathbb{Z}$ and $\mathbb{Z}/b\mathbb{Z}$. Use this identification to show that we can identify $H^{\#}$ as $\mathbb{Z}/b\mathbb{Z}$.

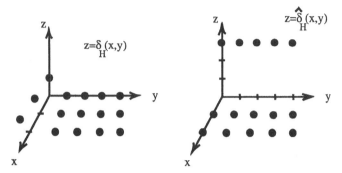

Figure I.47. A Chinese remainder theorem view of the graphs of δ_H and its Fourier transform for $G = \mathbb{Z}/15\mathbb{Z} \cong \mathbb{Z}/3\mathbb{Z} \oplus \mathbb{Z}/5\mathbb{Z}$, $H = \mathbb{Z}/3\mathbb{Z}$.

There are many other ways to view the uncertainty principle. For example, we can take a view closer to that of Slepian and Pollack (see Terras [1985, Vol. I, p. 51] and Donoho and Stark [1989]). This means that we must study the time and band-limiting operators defined as follows.

Definition. Suppose that $f \in L^2(G)$ and T is any subset of G and B any subset of $\hat{G}$. The *time-limiting operator* P_T is defined by

$$P_T f = f \cdot \delta_T. \tag{3}$$

And the *band-limiting operator* $R_B f$ is defined by

$$R_B f(x) = \frac{1}{|G|} \sum_{\chi \in B} \hat{f}(\chi)\chi(x), \tag{4}$$

where

$$\hat{f}(\chi) = \sum_{y \in G} f(y)\overline{\chi(y)}.$$

In order to study these linear operators $Q : L^2(G) \to L^2(G)$, we need to think a little about norms on the space of such operators, that is, norms on the space of $n \times n$ complex matrices, $n = |G|$. A good reference for such matrix norms is Horn and Johnson [1991, Chapter 5].

Definition. A *matrix norm* $\|A\|$, for an $n \times n$ complex matrix A must have the following properties for arbitrary $n \times n$ matrices A and B:

1) $\|A\| \geq 0$;
2) $\|A\| = 0 \quad \Leftrightarrow \quad A = 0$;
3) $\|cA\| = |c| \|A\|$,   for $c \in \mathbb{C}$;
4) $\|A + B\| \leq \|A\| + \|B\|$;
5) $\|AB\| \leq \|A\| \|B\|$.

We will need to consider two such norms.

Definition. Suppose that $Q : L^2(G) \to L^2(G)$ is a linear operator. The *operator norm* $\|Q\|$ of Q is defined as follows:

$$\|Q\| = \max \left\{ \frac{\|Qf\|_2}{\|f\|_2} \;\middle|\; f \in L^2(G), \, f \neq 0 \right\}.$$

Exercise. Show that if Q is a linear operator on $L^2(G)$, $\|Q\|$ is the operator norm defined above, and λ is the largest eigenvalue of the Hermitian positive semidefinite matrix Q^*Q, then $\|Q\| = \lambda^{1/2}$. Note that since Q^*Q is a Hermitian matrix, the spectral theorem says it has a complete orthonormal set of eigenvectors. Since Q^*Q is positive semidefinite, all of its eigenvalues λ are nonnegative.

Definition. The *Frobenius or L^2-norm* $\|Q\|_2$ of Q is defined by

$$\|Q\|_2^2 = \mathrm{Tr}(Q^*Q).$$

Thinking of Q as an $n \times n$ complex matrix, the Frobenius norm is just the square root of the sum of the squares of the absolute values of the entries. There are other names for the Frobenius norm (e.g., Hilbert–Schmidt, Schur, and euclidean norm).

Exercise. Show that both the operator norm and the Frobenius norm are matrix norms; that is, they have the five properties listed above in the definition of matrix norm.

Lemma 2.

a) Let R_B denote the band-limiting operator defined in (4). Then the operator norm satisfies

$$\|R_B\| = 1.$$

b) If P_T denotes the time-limiting operator, then

$$\| P_T \| \le 1.$$

c) With R_B and P_T as defined above, we have

$$\| P_T R_B \| = \| R_B P_T \| \le 1.$$

Proof.

a) By the Plancherel theorem it is not hard to see that $\| R_B \| \le 1$, for if $f \in L^2(G)$, we have

$$\| R_B f \|_2^2$$

$$= \sum_{t \in G} |R_B f(t)|^2 = \sum_{t \in G} \frac{1}{|G|^2} \left| \sum_{\chi \in B} \hat{f}(\chi) \chi(t) \right|^2$$

$$= \frac{1}{|G|^2} \sum_{t \in G} \sum_{\chi \in B} \hat{f}(\chi) \chi(t) \sum_{\tau \in B} \overline{\hat{f}(\tau) \tau(t)} = \frac{1}{|G|} \sum_{\chi \in B} |\hat{f}(\chi)|^2 \le \| f \|_2^2.$$

We used the orthogonality relations for characters of G (the lemma after Theorem 2 in Chapter 10) to get the equality in the last line. Part a of the lemma follows from this and the fact that the last inequality can be equality when $\hat{f}$ is supported in B.

Parts b and c are proved in the next exercise. ∎

Exercise. Prove parts b and c of Lemma 2.

Hints.

b) Use the definition of P_T and that of the operator norm.

c) This holds for any pair of orthogonal projections. An *orthogonal projection* Q is a matrix such that $Q^* = Q$ and $Q^2 = Q$. It is easy to check that R_B and P_T are indeed orthogonal projections.

Lemma 3. Let $Q = R_W P_T$, the composition of band- and time-limiting operators. Then

$$\sqrt{\frac{1}{|G|}} \| Q \|_2 \le \| Q \| \le \| Q \|_2 = \sqrt{\frac{|W||T|}{|G|}}.$$

Proof.

- $\|Q\|_2 = \sqrt{\frac{|W||T|}{|G|}}$

To prove this equality, note that

$$(R_W P_T f)(x) = \frac{1}{|G|} \sum_{x \in W} \chi(y) \sum_{y \in T} f(y)\overline{\chi(x)}$$

$$= \frac{1}{|G|} \sum_{y \in T} f(y) \sum_{x \in W} \chi(xy^{-1}) = \sum_{y \in G} q_{T,W}(x, y) f(y).$$

Here we define the "kernel" $q_{T,W}(x, y)$ by

$$q_{T,W}(x, y) = \frac{1}{|G|} \delta_T(y) \sum_{x \in W} \chi(xy^{-1}), \quad \text{for } x, y \in G. \tag{5}$$

So $q_{T,W}(x, y)$ is the matrix of our band- and time-limiting operator Q corresponding to the basis of $L^2(G)$ consisting of delta functions δ_x, for $x \in G$. Note that $q_{T,W}(x, y)$ is the inverse DFT of δ_W at xy^{-1} times $\delta_T(y)$.

The square of the Frobenius norm of $Q = R_W P_T$ is

$$\|R_W P_T\|_2^2 = \sum_{\substack{x \in G \\ y \in G}} |q(x, y)|^2 = \sum_{\substack{x \in G \\ y \in G}} \frac{1}{|G|^2} \left| \delta_T(y) \sum_{x \in W} \chi(xy^{-1}) \right|^2$$

$$= \sum_{\substack{x \in G \\ y \in T}} \frac{1}{|G|} \sum_{x \in W} \chi(xy^{-1}) \frac{1}{|G|} \sum_{\rho \in W} \overline{\rho(xy^{-1})}.$$

Reverse the order of summation so that the one over $x \in G$ occurs first. This gives

$$\|R_W P_T\|_2^2 = \frac{1}{|G|^2} \sum_{y \in T} \sum_{\substack{\chi \in W \\ \rho \in W}} \rho(y)\overline{\chi(y)} \sum_{x \in G} \chi(x)\overline{\rho(x)}.$$

By the orthogonality relations, the inner sum over $x \in G$ is 0 unless $\chi = \rho$, in which case it is $|G|$. Therefore

$$\|R_W P_T\|_2^2 = \frac{1}{|G|} \sum_{y \in T} \sum_{\chi \in W} \chi(y)\overline{\chi(y)} = \frac{|T||W|}{|G|}.$$

This completes the proof of the equality.

- $\|Q\| \le \|Q\|_2$

Note that by an exercise above, we know that $\|Q\|^2$ is the maximum of λ, where λ runs through the eigenvalues of Q^*Q. It follows that if Ξ denotes the set of all eigenvalues of Q^*Q, then

$$\|Q\|^2 \leq \sum_{\lambda \in \Xi} \lambda = \mathrm{Tr}(Q^*Q) = \|Q\|_2^2.$$

- $\|Q\|_2 \leq \sqrt{|G|}\, \|Q\|$

  As before, if Ξ denotes the set of all eigenvalues of Q^*Q, we have

  $$\|Q\|_2^2 = \mathrm{Tr}(Q^*Q) = \sum_{\lambda \in \Xi} \lambda \leq |G| \max\{\lambda \mid \lambda \in \Xi\} = |G| \|Q\|^2.$$

  ∎

The lemma implies the following version of the uncertainty principle.

*Theorem 2. **Uncertainty – Version 2.*** With the definitions given by (3) and (4) of the time-limiting and band-limiting operators, we have

$$\|R_W P_T\|^2 \leq \frac{|W||T|}{|G|}.$$

Donoho and Stark [1989] note that the uncertainty principle has implications for recovery of missing frequencies in a signal. Let R_B be a band-pass operator. If s is the signal sent, then the received signal is $r = R_B s$. Let $W = B^c = \hat{G} - B$ be the frequencies that are not observed. Suppose that $s = P_T s$. Then

$$r = R_B P_T s = (I - R_W P_T)s. \tag{6}$$

We shall ignore noise in our discussion.

Theorem 5 of Donoho and Stark [1989] says that if $|T||W| < (1/2)|G|$, then the signal s can be reconstructed uniquely from r. Actually, we can improve on this somewhat.

Theorem 3. For any constant $c > 1$, if

$$|T||W| \leq \frac{1}{c}|G|,$$

then there is an inverse for the operator $(I - R_W P_T)$ and it is given by the geometric series

$$(I - R_W P_T)^{-1} = \sum_{k \geq 0} (R_W P_T)^k.$$

Proof. The matrix analogue of the geometric series converges if the operator norm satisfies $\|R_W P_T\| < 1$ (see Horn and Johnson [1991] or Lang [1983]). By Theorem 2 we know that

$$\|R_W P_T\| \le \sqrt{\frac{|T||W|}{|G|}} \le \sqrt{\frac{1}{c}} < 1.$$

∎

Moral. We can replace 2 in the hypothesis of Donoho and Stark's Theorem 5 with any constant $c > 1$. Donoho and Stark [1989, p. 915] find the hypothesis

$$|T||W| \le \frac{1}{2}|G|$$

disappointing. For even if $|W| = |G|/10$, we need $|T| < 5$. So $T = \text{support}(s)$ can have at most five elements. This means that the signal can be nonzero in at most five places. If we use the condition of Theorem 2 with $c > 1$ optimal, we find

$$|T| < \frac{10}{c}, \quad \text{for any } c > 1.$$

This means we can assume $c = 10/9$ and thus $|T| \le 9$. So the support of the signal can have at most nine elements – a slight improvement over five elements.

To learn more, we should do a few experiments. We can compute the eigenvalues of the matrices $q_{T,W}$ defined in (5).

Example 1. $G = \mathbb{Z}/15\mathbb{Z}$. Set $T = \{0, 3, 6, 9, 12(\text{mod } 15)\} \cong \mathbb{Z}/5\mathbb{Z}$, and set $W = (G/T)^{\widehat{}} \cong \{0, 5, 10(\text{mod } 15)\} \cong \mathbb{Z}/3\mathbb{Z}$. Note that the sets T and W do not satisfy the hypothesis of Theorem 3. Then compute the matrix entry $q_{T,W}(x, y)$ of the time- and band-limiting operator, for $x, y \in G$. We obtain from (5)

$$
\begin{aligned}
q_{T,W} &= \frac{1}{15}\delta_T(y) \sum_{W \in \{0,\, 5,\, 10(\text{mod } 15)\}} \exp\left\{\frac{2\pi i w(x - y)}{15}\right\} \\
&= \frac{1}{15}\delta_T(y)\left\{1 + \exp\left(\frac{2\pi i(x - y)}{3}\right) + \exp\left(\frac{4\pi i(x - y)}{3}\right)\right\} \\
&= \frac{1}{15}\delta_T(y)3\delta_T(x - y).
\end{aligned}
$$

Here we have used the orthogonality relations again. Therefore

$$q_{T,W}(x, y) = \frac{1}{5}\delta_T(y)\delta_T(x).$$

Now we can write the matrix $Q = q_{T,W}$ by rearrangement as

$$\begin{pmatrix} J & 0 \\ 0 & 0 \end{pmatrix},$$

where J is a 5×5 matrix, each entry being $1/5$. Thus the eigenvalues of q are one 1 and fourteen 0s. So the eigenvalues of $I - Q$ are one 0 and fourteen ones. Thus $I - Q$ is not invertible, but generalized inverses should work well enough on it.

Example 2. $G = \mathbb{Z}/15\mathbb{Z}$. This time we take $T = W = \{0, 5, 10(\mathrm{mod}\ 15)\}$. So this choice of T and W satisfies the hypothesis of Theorem 2. The matrix $I - Q$ is invertible; $Q = R_W P_T$. Let's compute the matrix $q_{T,W}$ using (5). We find that

$$\begin{aligned} q_{T,W}(s, t) &= \frac{1}{15}\delta_T(t) \sum_{w \in T} \exp\left\{\frac{2\pi i w(s-t)}{15}\right\} \\ &= \frac{1}{15}\delta_T(t)\left\{1 + \exp\left(\frac{2\pi i(s-t)}{3}\right) + \exp\left(\frac{4\pi i(s-t)}{3}\right)\right\} \\ &= \frac{1}{5}\delta_T(t)\delta_S(s-t), \end{aligned}$$

where $S = \{0, 3, 6, 9, 12(\mathrm{mod}\ 15)\}$. It follows that the matrix of Q has fifteen nonzero entries equal to $1/5$ (all in rows corresponding to 0, 5, 10). And we can rearrange the matrix of Q to look like

$$\begin{pmatrix} \frac{1}{5}I_3 & 0 \\ * & 0 \end{pmatrix}.$$

So the maximum eigenvalue of Q is $1/5$. It follows that the minimum eigenvalue of $I - Q$ is $4/5$ and thus $I - Q$ is invertible. From Lemma 3, we have, since $\|Q\|_2 = \sqrt{3/5}$,

$$\sqrt{\frac{3}{5}} = \|Q\|_2 \geq \|Q\| \geq \frac{1}{\sqrt{15}}\|Q\|_2 = \frac{1}{5}.$$

Exercise. Consider the possibility of reconstruction of a signal s from $(I - R_W P_T)s$ if $G = \mathbb{Z}/15\mathbb{Z}$, $T = \{-1, 0, 1(\mathrm{mod}\ 15)\}$, and $|W| \leq 4$.

According to Lemma 2, Smith [1990] notes that the question becomes:

$$\text{When is } \| P_T R_W \| = 1?$$

Otherwise $I - P_T R_W$ is invertible using the geometric series. We saw in Lemma 3 that

$$\frac{|T||W|}{|G|} < 1 \quad \text{implies that } \| P_T R_W \| < 1.$$

Theorem 4. (K. T. Smith [1990].)

a) $|T| + |W| > |G|$ implies $\| P_T R_W \| = 1$.
b) If G is $\mathbb{Z}/n\mathbb{Z}$ and $T = \{0, 1, 2, \ldots, k - 1 (\text{mod } n)\}$ then the converse holds.

Proof. See the Exercise below. ∎

Exercise. Prove Theorem 4. Then look at the three examples above from the point of view of this theorem.

Hints. a) Let $T = \{x_1, \ldots, x_k\}$. You must use the fact that there are constants c_j, $j = 1, \ldots, |T| = k$, not all 0, such that

$$\sum_{j=1}^{k} c_j \exp\left(\frac{-2\pi i x_j s}{n}\right) = 0, \quad \text{for all } s \in \hat{G} - W.$$

This implies that the function

$$f = \sum_{j=1}^{k} c_j \delta_{x_j}$$

has support in T and has Fourier transform with support in W.
b) Here you need to see that if

$$f = \sum_{j=0}^{k-1} c_j \delta_j,$$

its Fourier transform cannot vanish at $n - |W| \geq k$ points. Use the fact that a polynomial of degree k cannot vanish at more than k points.

We give one last version of uncertainty (following Donoho and Stark [1989], Smith [1990], and Wolf [1994]). First we need some definitions. Here ε and η are positive constants.

Definition. We say that $f : G \to \mathbb{C}$ is *ε-concentrated* on $T \subset G$ if

$$\| f - \delta_T f \|_2 \leq \varepsilon \| f \|_2.$$

We say that $F : \hat{G} \to \mathbb{C}$ is *η-concentrated* on $B \subset \hat{G}$ if

$$\| F - \delta_B F \|_2 \leq \eta \| F \|_2.$$

And we say that $f : G \to \mathbb{C}$ is *η-band-limited* to B if there is a function $f_B : G \to \mathbb{C}$ such that supp $(\hat{f}_B) \subset B$ and

$$\| f - f_B \|_2 \leq \eta \| f \|_2.$$

Theorem 5. Uncertainty – Version 3. Suppose that f is a nonzero function on G, which is ε-concentrated on $T \subset G$ as well as η-band-limited to $B \subset \hat{G}$. Then

$$\left(\frac{|T||B|}{|G|} \right)^{1/2} \geq \| P_T R_B \| \geq 1 - \varepsilon - \eta.$$

Proof. (Wolf [1994].) We have

$$\| f \|_2 - \| P_T R_B f \|_2 \leq \| f - P_T R_B f \|_2$$
$$\leq \| f - P_T f \|_2 + \| P_T f - P_T R_B f \|_2.$$

Note that $R_B f_B = f_B$, since f_B is supported in B. Then, since $\| P_T \| = 1$,

$$\| f \|_2 - \| P_T R_B f \|_2 \leq \varepsilon \| f \|_2 + \| P_T f - P_T f_B \| \leq \varepsilon \| f \|_2 + \eta \| f \|.$$

This implies the result of the theorem. ■

Part II
Finite Nonabelian Groups

The trouble is that groups behave in astonishingly subtle ways
that make them psychologically rather difficult to grasp. We
might say that they are adept at doing large numbers of
impossible things well before breakfast.

J. H. Conway [1980, p. 170]

In this part of the book we study Fourier analysis on finite nonabelian groups and
we should expect them to be able to be a bit more obnoxious before breakfast
than their nonabelian relatives. We will mostly deal with specific groups of 2×2
or 3×3 matrices over finite fields. Before meeting these groups for breakfast,
the reader will hopefully have tasted some undergraduate algebra books such
as Gallian [1990].

Brauer [1963] said: "Groups are the mathematical concept with which we
describe symmetry. One of the outstanding achievements of Greek mathematics
is the discovery of the 5 regular polyhedra, the Platonic solids. Each of them is
closely associated with a finite group." And we will find that these polyhedra
have much to do with some of the Cayley graphs that we associate to the affine
group of 2×2 invertible matrices over a finite field with lower row (0 1). Brauer
goes on to say: "We have to confess that it took mathematicians 2000 years to
achieve a mathematical formulation of the group concept. In its abstract form,
it was given by Cayley, after permutation groups had been used by Lagrange,
Cauchy, Abel, Galois, and others in their work on the solution of equations by
radicals."

The theory of group representations of finite nonabelian groups G was
created in 1897 by F. G. Frobenius [1849–1917]. He was inspired by a letter from R. Dedekind asking for a factorization of the group determinant
$\det(m(gh^{-1})_{g, h \in G})$ for a finite nonabelian group G. See the exercise at the end
of Chapter 15 for an example. Previously, group determinants of finite abelian

238   *Part II   Finite Nonabelian Groups*

groups had been used by number theorists such as Dedekind. The nonabelian group determinants were also destined to be precious to number theorists. They allow one to factor discriminants of normal extensions for example.

See Curtis [1979] and Lam [1998] for some of the history of representation theory. Besides Frobenius, others working on the theory of invariants were dealing in some way with the representations of the group $GL(n)$ of $n \times n$ matrices with nonzero determinant and complex entries (e.g., Cayley, Hermite, and Sylvester). In the beginning of the twentieth century, work of Burnside, Schur, Speiser, and others continued the theory. Emmy Noether [1983] viewed group representations as the study of modules over rings like $L^2(G)$ under convolution (also called the group ring).

There are applications to many things besides number theory, for example, signal processing, elementary particles, crystals, statistics, solution of equations, cryptography, codes, geometry, spectroscopy, radar, and communications networks. As Brauer said above, some of our favorite groups are the groups of symmetries of the five regular Platonic solids. See Weyl [1952]. Diaconis [1988] gives a large number of examples of the application of harmonic analysis on permutation groups to statistics. For example, he analyzes data from a survey which asked people to rank where they wanted to live: city, suburbs, or country [loc. cit., p. 143]. He finds, using Fourier analysis on S_3, the group of permutations of three objects, that "the best single predictor of f is what people rank last."

Galois saw that the Galois group of a polynomial equation $p(x) = 0$ tells much about the solutions (see Jacobson [1980], Lang [1983], van der Waerden [1991]). Assume the polynomial $p(x)$ has rational coefficients. Then the representations of its Galois group can be used to build the Artin L-function. The latter is a generalization of the Riemann zeta function that rules the arithmetic of the field extension of $\mathbb{Q}$ created by adjoining the roots of $p(x)$. See Heilbronn [1967] or Stark [1992] for some examples of the application of Fourier analysis on finite groups in algebraic number theory. Brauer himself was responsible for much of the work on Artin L-functions. For example, he used representation theory to show that the Artin L-function is meromorphic. It is still open whether it is in fact holomorphic for a nontrivial character of an irreducible representation (the Artin conjecture). The L-functions and zeta functions of graphs and graph coverings have an analogous but simpler theory. See Chapter 24.

The crystallographic groups tell us much about the structure of matter. See Kettle [1985], who manages to describe the group theory of interest to chemists using pictures and no matrices. He says in his introduction: "My experience is that for most students the physical picture becomes more and more hazy as the mathematics takes over. This is the reason for the structure of the present book.

The subject is presented pictorially – but accurately – so that the student is not referred to mathematical equations but rather to pictorial explanations."

Character tables by picture? It is an intriguing concept but here we will take a more analytic but still very concrete approach. We will usually take the point of view of analysis rather than algebra. Thus we discuss the Fourier transform on G rather than the Wedderburn structure theory of the group algebra. And we often view the Fourier inversion formula as the theorem of Peter and Weyl which was proved for continuous compact groups. This is an expression of a function on the finite group in a finite series of matrix entries of inequivalent irreducible unitary representations of G. And we will sometimes think of the Frobenius reciprocity law as a special case of that generalization of the Poisson summation formula, which is known as the Selberg trace formula (see Arthur [1989]). In short, we will always emphasize connections between the finite group theory and the continuous group theory.

Before beginning our discussion, we recall the warning of Dyson [1966], who said that Wigner, Weyl, and van der Waerden "developed the theory in a style of great generality and mathematical polish, which delayed for about 25 years its acceptance as a working tool by the majority of physicists, although the chemists during the same period found no great difficulty in adapting it to their needs."

We will attempt to avoid this pitfall, by concentrating on a few good examples – subgroups of $GL(2, \mathbb{F}_q)$ and $GL(3, \mathbb{F}_q)$. See also the book of Thomas and Wood [1980] for character tables for the groups of order less than 32 or Conway et al. [1985] for larger groups. Other references that the reader may wish to consult are: Curtis and Reiner [1966, 1981], Diaconis [1988], Dym and McKean [1972], Fulton and Harris [1991], Hamermesh [1962], Issacs [1976], James and Liebeck [1993], Lang [1984], Mackey [1976; 1978a,b], Serre [1977], Vilenkin [1968], and van der Waerden [1991]. We will mostly avoid the symmetric group S_n. There are many references which discuss the subject. The book of Grove [1997] makes use of the (free) computational group theory package GAP.

Chapter 15

Fourier Transform and Representations of Finite Groups

Rationalists, wearing square hats,
Think, in square rooms,
Looking at the floor,
Looking at the ceiling.
They confine themselves
To right-angled triangles.
     Wallace Stevens [1971] in
     "Six Significant Landscapes"

The finite groups we plan to analyze are all quite explicit subgroups of the *general linear group*

$$GL(n, \mathbb{F}_q) = \left\{ A \in \mathbb{F}_q^{n \times n} \mid \det(A) = |A| \neq 0 \right\} \tag{1}$$

for small values of n. Here $\mathbb{F}_q$ is the finite field with $q = p^r$ elements (for some prime p). We will study the *affine group*

$$\text{Aff}(q) = \left\{ \begin{pmatrix} y & x \\ 0 & 1 \end{pmatrix} \middle| x, y \in \mathbb{F}_q, y \neq 0 \right\} \tag{2}$$

as well as the *Heisenberg group*

$$\text{Heis}(q) = \left\{ \begin{pmatrix} 1 & x & z \\ 0 & 1 & y \\ 0 & 0 & 1 \end{pmatrix} \middle| x, y, z \in \mathbb{F}_q \right\}. \tag{3}$$

This latter group is a finite analogue of the real group which appears in quantum mechanics via the Heisenberg uncertainty principle (saying that one cannot measure position and momentum of a particle at the same time). There will

240

be applications of (2) to Ramanujan graphs, wavelets, and random number generators. The group (3) has similar applications as well as applications to radar imaging. See Chapters 17 and 18.

The special linear group $SL(n, \mathbb{F}_q)$ is the determinant one subgroup of $GL(n, \mathbb{F}_q)$. The projective special linear group $PLS(n, \mathbb{F}_q)$ is the quotient of $SL(n, \mathbb{F}_q)$ modulo its center. The *center* Z of a group G is the set of elements $z \in G$ such that $zx = xz$ for all $x \in G$.

A *simple group* is a nontrivial group whose only normal subgroups are G and $\{e\}$, where e is the identity element of G. Two of the five nonabelian simple groups of order less than 1,000 are $PSL(2, \mathbb{F}_7)$ of order 168 and $PSL(2, \mathbb{F}_{11})$ of order 660. It can be shown that $PSL(n, \mathbb{F}_q)$ is simple for $n \geq 2$, unless $n = 2$ and $q = 2$ or 3. See Scott [1987, p. 294]. Other simple groups come from analogues of the other simple real or complex Lie groups. But there are also so-called sporadic simple groups, the largest of which is called the monster.

The quest to find the list of all finite simple groups was one of the major quests of modern mathematics. Recently the classification of all the finite simple groups was obtained in 10,000 pages or so. Arbitrary finite groups are built up out of simple groups. So the simple groups may be viewed as analogues of the primes in number theory. See Conway [1980], Gallian [1990, pp. 365–9], and Gorenstein [1982, 1985] for some of the history. We will not have any more to say about this classification or the sporadic simple groups such as the monster.

Some of our favorite groups such as the affine and Heisenberg groups are not simple. They are solvable. And the Heisenberg group is nilpotent. See Chapters 17 and 18.

Even though the groups we are interested in are already matrix groups, if we insist on doing Fourier analysis on them, we must map them homomorphically into groups of matrices with complex entries. That is, we will be studying representations of finite groups G which are group homomorphisms into groups of complex $n \times n$ matrices. It is also possible to map G to other sorts of matrices (modular representations and l-adic representations come to mind) but we will never consider this possibility.

Exercise.

a) Show that Aff(3) is isomorphic to S_3, the symmetric group of all permutations of 3 objects.

b) Show that Heis(2) is isomorphic to D_4, the dihedral group of motions of a square. This group is generated by the rotation R and the flip F. Here $R^4 = I$, $F^2 = I$, $F^{-1}RF = R^{-1}$.

c) Show that $SL(2, \mathbb{F}_2) \cong S_3$.

Representations of G

Definition. A (finite-dimensional) *representation* of the finite group G is a group homomorphism $\pi : G \to GL(n, \mathbb{C})$. If $\pi(g)$ is a matrix with i, j entry $\pi_{i,j}(g)$, we call the functions $\pi_{i,j}$ the *matrix entries* of π.

The matrix entries are interesting special functions on G which are the analogues of the trigonometric functions on the circle or the exponential functions $\exp(2\pi i/n)$ on the discrete circle $\mathbb{Z}/n\mathbb{Z}$. That is, the matrix entries are replacements for the exponentials in the Fourier expansions of functions on G.

For real groups the matrix entries include many of the favorite special functions of mathematical physics (e.g., Legendre functions, Bessel functions, etc.) See Vilenkin [1968]. For finite groups one obtains special functions which are not so well known (e.g., Kloosterman sums). See Chapters 20 and 21 or Piatetski–Shapiro [1983] and Stanton [1981, 1984, 1990].

Many modern mathematicians identify $GL(n, \mathbb{C})$ and

$$GL(V) = \{T : V \to V \mid T \text{ is linear and invertible}\}, \qquad (4)$$

where V is an n-dimensional vector space over $\mathbb{C}$. This identification may throw out the baby special function with the bath water. Yes, the matrix entries do change with the basis of V. And that is rather murky bath water. But the baby special function can grow up to be a useful member of society hopefully. We will use the two definitions of representation interchangeably. It is often convenient to forget the matrix entries for a brief period of time. But we won't throw them out.

Exercise. Show that if you change the basis of the n-dimensional vector space V, you replace the matrix M of the linear transformation $T : V \to V$ with a matrix AMA^{-1}, where A either maps the new basis to the old one or vice versa.

Examples.

1) We always have the trivial representation which takes every element of G to the number 1. It may sound silly but we will need it, just as we need the constant function in the usual Fourier expansion of a periodic function.

2) Every undergraduate group theory course (e.g., see Gallian [1990]) talks about Cayley's theorem, which says that a finite group G is isomorphic to a subgroup of the symmetric group S_n, where $n = |G|$, the order of G. The symmetric group S_n consists of permutations (1-1, onto maps) of a set with n elements to itself. It is easily seen that this group is isomorphic to a group of matrices of 0s and 1s having at most one 1 in each row and column. This gives a representation of the group G called the regular representation, of

which we will have much to say. In some sense this is the mother of all representations.

Exercise. Prove Cayley's theorem and the fact that the symmetric group S_n is isomorphic to a group of 0, 1 matrices.

A chapter of a mathematics book often begins with a long parade of definitions. Sorry, but it appears that we are at the beginning of a chapter.

Definition. Two representations α and β of G into $GL(n, \mathbb{C})$ are said to be *equivalent* if you get one from the other by uniform change of basis, that is, there is a matrix $T \in GL(n, \mathbb{C})$ such that

$$T\alpha(g)T^{-1} = \beta(g), \quad \text{for all } g \in G.$$

Definition. A *unitary* representation π maps G into the unitary group $U(n)$, where

$$U(n) = \{A \in GL(n, \mathbb{C}) \mid {}^t\bar{A}A = I\}. \tag{5}$$

Equivalently, suppose that $\langle u, v \rangle = {}^t\bar{u}v$ is the standard Hermitian inner product on column vectors $u, v \in \mathbb{C}^n$. The unitary matrices are those that preserve this inner product, that is, $\langle Au, Av \rangle = \langle u, v \rangle$ for all $u, v \in \mathbb{C}^n$.

Exercise. Prove that the two definitions of unitary matrix given above are indeed equivalent.

Now we return to our parade of definitions.

Definition. A *subrepresentation* ρ of a representation $\pi : G \rightarrow GL(V)$ means that $\rho : G \rightarrow GL(W)$, where W is a subspace of V such that $\pi(g)W \subset W$ for all $g \in G$ and $\rho(g)$ is the restriction of $\pi(g)$ to W, that is,

$$\pi(g)|_W = \rho(g) \quad \text{for all } g \in G.$$

Equivalently, in matrix language, there is a basis of V such that the matrix of $\pi(g)$ has the block form:

$$\begin{pmatrix} \rho(g) & * \\ 0 & * \end{pmatrix}.$$

Definition. A representation π is *irreducible* if its only subrepresentations are π itself and 0.

The irreducible representations are the atoms from which all representations are built. You see this using Proposition 2 below.

Proposition 1. If G is a finite group, every representation of G is equivalent to a unitary representation.

Proof. First, recall that a matrix is unitary iff it leaves the standard inner product invariant. Next note that if a matrix M leaves invariant some other positive, definite, Hermitian inner product $c(u, v)$, for $u, v \in \mathbb{C}^n$, then M is conjugate to a unitary matrix. That is, there is a unitary matrix $U \in U(n)$ such that $M = A^{-1}UA$, for some $A \in GL(n, \mathbb{C})$. Let's prove this before finding $c(u, v)$.

Here a positive definite Hermitian inner product $c(u, v)$ means:

$c(u, v)$ is linear in u, holding v fixed;

$c(u, v) = c(v, u); \quad c(u, u) \geq 0; \quad c(u, u) > 0 \quad \text{if } u \neq 0.$

If $e_1, e_2, \ldots, e_n$ are the standard unit basis vectors of $\mathbb{C}^n$ and if $c(u, v)$ is a positive Hermitian inner product on $\mathbb{C}^n$, then $c(e_i, e_j) = c_{i,j}$ and the $n \times n$ matrix C with i, j entry $c_{i,j}$ is positive definite Hermitian. The spectral theorem (see Strang [1976]) says that $C = R^2$, where R is also positive Hermitian, for the spectral theorem will write $C = {}^t\bar{U}DU$, for some unitary $n \times n$ matrix U and some $n \times n$ diagonal matrix D with jth diagonal entry $d_j > 0, j = 1, \ldots, n$. Then we have $R = {}^t\bar{U}\sqrt{D}U$. Here $\sqrt{D}$ means the diagonal matrix with jth diagonal entry $\sqrt{d_j}$.

Since the matrix M leaves the inner product $c(u, v)$ invariant, it follows that

$${}^t\bar{M}CM = C,$$

which says

$${}^t\bar{M}\,{}^t\bar{R}RM = {}^t\bar{R}R.$$

Multiply the last equation on the right by R^{-1} and on the left by ${}^t\bar{R}^{-1} = R^{-1}$, which gives

$$R^{-1t}\bar{M}\,{}^t\bar{R}RMR^{-1} = R^{-1t}\bar{R}RR^{-1} = I.$$

This says that RMR^{-1} is unitary.

In order to find an inner product which is invariant under $\pi(g)$ for all $g \in G$, we use what is called "Weyl's unitary trick." This just means that we sum the standard inner product $\langle u, v \rangle$ over G; that is, define

$$c(u, v) = \sum_{g \in G} \langle \pi(g)u, \pi(g)v \rangle, \quad \text{for } u, v \in \mathbb{C}^n.$$

This completes the proof of the proposition. ∎

In the usual graduate abstract algebra course, much of which goes back to Emmy Noether, group representations are studied by applying Wedderburn theory to the group algebra. See Jacobson [1980], Lang [1984], or van der Waerden [1991], for example. Here we will instead use the Fourier analyst's approach to the subject. However, we still need to define the group algebra because it is just the algebra of all functions on G under convolution product.

Definition. The *group algebra* $\mathbb{C}[G]$ is a vector space over $\mathbb{C}$ of dimension $|G|$. Let $\{e_g \mid g \in G\}$ be a basis of the group algebra. Then we can write the elements of $\mathbb{C}[G]$ as

$$\sum_{g \in G} f(g)e_g, \quad \text{for } f(g) \in \mathbb{C}.$$

It is clear from this that we can identify elements of $\mathbb{C}[G]$ with functions $f : G \to \mathbb{C}$. That is, we can identify the group algebra with $L^2(G) = \{f : G \to \mathbb{C}\}$ as vector spaces over $\mathbb{C}$.

To make $\mathbb{C}[G]$ an algebra, we need a multiplication. This is defined by making use of multiplication in G and distributive laws. This corresponds to the convolution of functions in $L^2(G)$, for we have, using the formula $e_g e_h = e_{gh}$,

$$\sum_{g \in G} a(g)e_g \sum_{h \in G} b(h)e_h = \sum_{g \in G} \sum_{h \in G} a(g)b(h)e_{gh}$$

$$= \sum_{k \in G} \sum_{\substack{g, h \in G \\ gh = k}} a(g)b(h)e_k = \sum_{k \in G} \sum_{h \in G} a(kh^{-1})b(h)e_k$$

$$= \sum_{k \in G} (a * b)(k)e_k.$$

Here we define *convolution* $a * b$ by

$$(a * b)(x) = \sum_{t \in G} a(xt^{-1})b(t) = \sum_{y \in G} a(y)b(y^{-1}x). \tag{6}$$

Exercise. Show that the center of the algebra $L^2(G)$ under convolution consists of functions $f(g)$ such that $f(g) = f(xgx^{-1})$ for all $x, g \in G$. That is, prove that

$$f * h = h * f \quad \text{for all } h \in L^2(G)$$

iff f is constant on conjugacy classes

$$\{g\} = \{xgx^{-1} \mid x \in G\}. \tag{7}$$

Example. The Right and Left Regular Representations. As we noted earlier, the usual undergraduate algebra course includes Cayley's theorem, which says that a finite group G is isomorphic to a subgroup of the symmetric group S_n, for $n = |G|$. Here S_n is the group of permutations of the set $\{1, 2, 3, \ldots, n\}$. There is a unitary representation of S_n by the $n \times n$ permutation matrices of 0s and 1s. Composing the two representations gives the regular representation of G.

Equivalently, we define the *left regular representation L* of G by

$$L : G \to GL(L^2(G)),$$
$$g \mapsto L(g),$$

where

$$[L(g)a](x) = a(g^{-1}x),$$

for $a \in L^2(G), g, x \in G$.

Or define the *right regular representation R* of G by

$$[R(g)a](x) = a(xg)$$

for a function a in $L^2(G)$ and $x, g \in G$. As we said above L (or R) can be viewed as the mother of all representations of G. See Lemma 2 below.

Exercise.
a) Check that for the function L defined above, we do indeed have a group representation. That is, check $L(xy) = L(x)L(y)$ for all $x, y \in G$.
b) Write down the matrix entries for $L(g)$ using the basis of $L^2(G)$ consisting of delta functions $\delta_g(x) = 1$ if $x = g$ and 0 otherwise. Show that $L(g)\delta_h = \delta_{gh}$ and thus that the matrix of $L(g)$ has i, j entry $\delta_{g_i g_j^{-1}}(g)$, for $G = \{g_1, g_2, \ldots, g_n\}$. This means that the matrix of L is indeed a permutation matrix.

Exercise. Suppose that the finite group G acts on the finite set X; that is, given $g \in G$ and $x \in X$, we have $gx \in X$ such that if e is the identity of G, $ex = x$ for all $x \in X$, and for $g, h \in G$ and $x \in X$, we have $g(hx) = (gh)x$. Then we get a representation

$$\pi : G \to GL(V), \quad \text{where } V = \{f : X \to \mathbb{C}\}$$

defined by $\pi(g)f(x) = f(g^{-1}x)$ for all $g \in G$ and $x \in X$. Define the character of this representation to be $\chi_\pi(g) = \text{Tr}\,\pi(g)$. Show that

$$\chi_\pi(g) = \#\{x \in X \mid gx = x\}.$$

Fulton and Harris [1991] call this "the original fixed point formula."

Our goal is to decompose a representation π of G into indecomposable parts. Considering π to be a matrix, we want to replace π by an equivalent representation which has diagonal blocks that cannot be further decomposed. The following Proposition (due to Maschke in 1899) achieves this goal.

Proposition 2. **Complete Reducibility Theorem.**

1) Suppose that $\rho : G \to U(m)$ is a subrepresentation of the representation $\pi : G \to U(n)$. Then π is equivalent to the representation with block diagonal form:

$$(\rho \oplus \sigma)(g) = \begin{pmatrix} \rho(g) & 0 \\ 0 & \sigma(g) \end{pmatrix}.$$

2) By induction, $\pi(g)$ is equivalent to

$$(\pi_1 \oplus \cdots \oplus \pi_r)(g) = \begin{pmatrix} \pi_1(g) & & 0 \\ & \ddots & \\ 0 & & \pi_r(g) \end{pmatrix},$$

where each subrepresentation π_i is irreducible. We say, when this happens, that π is *completely reducible*.

Proof.

1) Since $\pi(g)$ is unitary, we know that $\pi(g)$ leaves invariant the standard inner product $\langle u, v \rangle = {}^t\bar{v}u$, for $u, v \in \mathbb{C}^n$. And ρ is a subrepresentation of π, meaning that there is a subspace W of $V = \mathbb{C}^n$ such that $\pi(g)W \subset W$ and the restriction $\pi(g)|_W = \rho(g)$ for all $g \in G$. Define the *orthogonal complement space* by

$$W^\perp = \{v \in V \mid \langle v, w \rangle = 0, \text{ for all } w \in W\}.$$

Then $\pi(g)(W^\perp) \subset W^\perp$ for all $g \in G$. Why?

This means that the restriction $\pi(g)|_{W^\perp} = \sigma(g)$ is a subrepresentation of π. So we have a basis of V coming from combining a basis of W with a basis of $W^\perp$. This basis yields the matrix of $\pi(g)$ given in the statement of the proposition.

2) This is a straightforward induction on n. ∎

Next we want to consider a famous lemma of Isaii Schur from 1905 which is basic to our subject. It implies the orthogonality relations for matrix entries in inequivalent irreducible representations, for example. And that gives the Fourier expansion of a function on G. So *Lemma 1 is the key to Fourier analysis on finite groups.*

Lemma 1. Schur's Lemma. Suppose that $\pi : G \to GL(V)$ and $\rho : G \to GL(W)$ are two representations of the finite group G. Define the space $I(\pi, \rho)$ of *intertwining operators* between π and ρ to be

$$I(\pi, \rho) = \{L : V \to W \mid L \text{ is linear and } L \circ \pi(g) = \rho(g) \circ L, \forall g \in G\}.$$

Suppose now that $L \in I(\pi, \rho)$. Then we have the following results:

1) Suppose the kernel of L is denoted ker $L = \{x \in V \mid Lx = 0\}$. Then the restriction of π to ker L is a subrepresentation of π, that is, $\pi(g)(\ker L) \subset \ker L$, for all $g \in G$. Denote the image of L by Im $L = LV = \{Lx \mid x \in V\}$. Then the restriction of ρ to Im L is a subrepresentation of ρ, that is, $\rho(g)(\operatorname{Im} L) \subset \operatorname{Im} L$, for all $g \in G$.
2) If π and ρ are both irreducible representations and L is not the zero map, then L is an isomorphism of V onto W. Thus π and ρ must be equivalent.
3) If π is irreducible and $L \in I(\pi, \pi)$, then there is a number $\lambda \in \mathbb{C}$ such that $Lv = \lambda v$, for all $v \in V$, that is, $L = \lambda I$, where I is the identity operator.
4) Suppose that both π and ρ are irreducible. Then

$$\dim_{\mathbb{C}} I(\pi, \rho) = \begin{cases} 1, & \text{if } \pi \text{ and } \rho \text{ are equivalent,} \\ 0, & \text{otherwise.} \end{cases}$$

Proof.

1) If $v \in \ker L$, then $\pi(g)v \in \ker L$, since

$$L(\pi(g)v) = \rho(g)(Lv) = \rho(g)(0) = 0, \quad \forall g \in G.$$

If $w \in \operatorname{Im} L$, then $w = Lv$ for some $v \in V$ and we have

$$\rho(g)w = \rho(g)(Lv) = L(\pi(g)v) \in \operatorname{Im} L.$$

2) Since $L \neq 0$, we know that $\ker L \neq V$. But using part 1 and the fact that π is irreducible, this means that $\ker L = \{0\}$. Again, since $L \neq 0$, we have $\operatorname{Im} L \neq \{0\}$. Again, by Part 1 and the fact that ρ is irreducible, this means that $\operatorname{Im} L = W$. Thus L is both 1-1 and onto.

3) Let λ be an eigenvalue of L, that is, a root of $\det(xI - L) = 0$, which must exist since all our vector spaces are over the complex numbers where any polynomial has a root. Next let V_λ denote the eigenspace corresponding to λ, that is,

$$V_\lambda = \{v \in V \mid Lv = \lambda v\} \neq \{0\}.$$

Then $\pi(g)(V_\lambda) \subset V_\lambda$, since $L(\pi(g)v) = \pi(g)(Lv) = \pi(g)(\lambda v) = \lambda \pi(g)v$, for all $v \in V_\lambda$. Since the restriction of $\pi(g)$ to V_λ gives a subrepresentation of $\pi(g)$ and π is irreducible, it follows that $V_\lambda = V$. Therefore $L = \lambda I$ on V as it is on V_λ.

4) This is the exercise below. ∎

Exercise. Prove the last part of Schur's lemma making use of the parts already proved. Given $K, L \neq 0$ in $I(\pi, \rho)$, look at the map $M = L^{-1} \circ K : V \to V$. Why does L^{-1} exist? Show that M is a nonzero element of $I(\pi, \pi)$. Then $M = \mu I$ for some $\mu \in \mathbb{C}$. This means $K = \mu L$ and that the dimension of the space $I(\pi, \rho)$ is at most 1.

Definition. Let $\pi : G \to GL(V)$ be a representation. We define the *degree* d_π *or dimension* of π to be the dimension of the vector space V.

Corollary. An irreducible representation $\pi : G \to GL(V)$ of an abelian group G must have degree one; that is, the representation space V of π is one dimensional.

Proof. Fix an element $g_0 \in G$ and consider the map

$$L_{g_0} : V \to V \quad \text{defined by } L_{g_0} v = \pi(g_0)v, \quad \text{for } v \in V.$$

Because the group G is abelian, the map L_{g_0} is in the space $I(\pi, \pi)$ of operators intertwining π with itself. Now Schur's lemma says that $L_{g_0} = \lambda_{g_0} I$, for some scalar $\lambda_{g_0} \in \mathbb{C}$. This means that the matrix of $\pi(g_0)$ is a scalar matrix, diagonal with all diagonal entries equal to λ_{g_0}. Therefore any subspace of V gives rise to a nontrivial subrepresentation of π. Thus V must be one dimensional in order for π to be irreducible. ∎

Now we can reconsider some of the work that we did in Chapter 2 where all our groups were abelian. First note that any two one-dimensional representations are inequivalent. We will need a complete list of inequivalent representations of a finite group G to do our Fourier analysis. Let's make another definition.

Definition. For a finite group G, let the *dual* $\hat{G}$ denote a complete set of inequivalent irreducible unitary representations of G.

Example. $G - \mathbb{Z}/n\mathbb{Z}$, under addition mod n.

This was one of our favorite examples in Part I. It is certainly abelian. Define $e_a(x) = \exp(2\pi i a x / n)$ for a, $x \in G$. Clearly e_a is a one-dimensional representation of G, also called a *character* of G. Then, as we shall see in an exercise below, $\hat{G} = \{e_a \mid a \in G\}$. That is, the e_a give all the irreducible unitary representations of $G = \mathbb{Z}/n\mathbb{Z}$, under addition. We can make $\hat{G}$ into a group and, in this case, it is isomorphic to the original group G. We say that finite abelian groups are "self-dual." In general, when G is nonabelian, this will not be the case. Note that the Fourier inversion formula in Theorem 1 of Chapter 1 involves a sum over G which is really $\hat{G}$ in general. That is, abelian groups are a little too special to get a complete idea of the nature of Fourier analysis for nonabelian groups.

Remarks on Constructing New Representations from Old Ones. Suppose that π and ρ are representations of G. Then one can form various other representations. For example, viewing representations as matrices, we have:

1) *the direct sum* of π and $\rho = \pi \oplus \rho =$ the matrix of part 1 of Proposition 2;
2) *the lift of a representation σ of G/N, where N is a normal subgroup of G* to a representation $\tilde{\sigma}$ of $G : \tilde{\sigma}(g) = \sigma(gN)$. Note that $N \subset \ker \tilde{\sigma}$.
3) *the tensor product* of π and ρ (which is an $n \times n$ matrix for $n = km$ if k is the dimension of π and m is the dimension of ρ), defined by forming $m \times m$ blocks such that the i, j block for $1 \leq i$, $j \leq k$ is $\pi_{i,j}(g)\rho(g)$;
4) *the dual representation* $\pi^d(g) = {}^t\pi(g)^{-1}$;
5) *the rth symmetric power* $S^r(\pi)$; see Fulton and Harris [1991, p. 472ff] or see Terras [1,Vol. I, p. 134] for an example;
6) *the rth exterior power* $\Lambda^r(\pi)$; see Fulton and Harris [loc. cit.];
7) our favorite construction for new representations from old, which starts with a representation ρ of a subgroup H of G and obtains a new representation $\pi = \mathrm{Ind}_H^G \rho$ of G. This is called the *representation induced by ρ on G*. We will consider this construction in formula (1) of Chapter 16.

Before we can do Fourier analysis on G, we need a few more tools. So we go back to the definition–theorem parade.

Definition. The *character* χ_π of a representation π of G is defined by $\chi_\pi(g) =$ Trace $\pi(g)$, for all $g \in G$.

Thus $\chi_\pi : G \to \mathbb{C}$. Of course if π is one dimensional, then the character of π is π.

Exercise. Show that for representations π and ρ of G we have the following formulas for characters:
1. $\chi_{\pi \oplus \rho} = \chi_\pi + \chi_\rho$;
2. $\chi_{\pi \otimes \rho} = \chi_\pi \cdot \chi_\rho$.

Theorem 1. **The Schur Orthogonality Relations.**

1) Suppose that $\pi : G \to U(n)$ and $\rho : G \to U(m)$ are two inequivalent irreducible unitary representations of the finite group G. Then, using the usual inner product on $L^2(G)$, we find that the matrix entries of π and ρ are pairwise orthogonal, that is,

$$\langle \pi_{ij}, \rho_{rs} \rangle = \sum_{g \in G} \pi_{ij}(g) \, \overline{\rho_{rs}(g)} = 0, \quad \text{for all } i, j, r, s.$$

2) If π is as above, then

$$\langle \pi_{ij}, \pi_{rs} \rangle = \frac{|G|}{d_\pi} \delta_{ir} \delta_{js},$$

where d_π is the dimension or degree of π and δ_{ij} is the Kronecker delta function. Thus the matrix entries of π are mutually orthogonal functions on G.
3) Let π and ρ be inequivalent, irreducible, unitary representations of G. Then the characters of π and ρ are also orthogonal. And we have

$$\langle \chi_\pi, \chi_\rho \rangle = \begin{cases} 0, & \text{if } \pi \text{ and } \rho \text{ are inequivalent,} \\ |G|, & \text{if } \pi \text{ and } \rho \text{ are equivalent.} \end{cases}$$

Proof.

1) We make use of an $n \times m$ complex matrix C that we pull out of a hat. Then define

$$P = \sum_{g \in G} \pi(g) C \rho(g^{-1}),$$

which is also an $n \times m$ matrix. The matrix P magically intertwines π and ρ, since

$$\pi(h)P = \sum_{g \in G} \pi(hg)C\rho(g^{-1}) = \sum_{u \in G} \pi(u)C\rho(u^{-1}h) = P\rho(h).$$

Here we made the substitution $u = hg$ and we used the fact that π and ρ are both group homomorphisms. Now we can use Schur's lemma to see that $P = 0$. Choose the matrix $C = E_{js}$, the matrix which is 1 in the js place and 0 everywhere else. This choice of C forces the ir entry of P to be

$$P_{ir} = \sum_{g \in G} \pi_{ij}(g) \overline{\rho_{rs}(g)} = 0.$$

2) Imitate the proof of 1 with a slight twist: Form the matrix

$$P = \sum_{g \in G} \pi(g)C\pi(g)^{-1}, \quad \text{for any } n \times n \text{ matrix } C.$$

By Schur's lemma, we know that $P = \lambda_C I$, for some scalar $\lambda_C \in \mathbb{C}$. Set $C = E_{ii}$ and take the trace to obtain

$$\sum_{g \in G} \text{Tr}(C) = \lambda_C \text{Tr}(I).$$

Therefore $\lambda_C = |G| \text{Tr}(C) d_\pi^{-1}$, if $d_\pi = $ degree of π. Now set $C = E_{js}$, as before, and obtain

$$P_{ir} = \sum_{g \in G} \pi_{ij}(g) \overline{\pi_{rs}(g)} = \frac{|G|}{d_\pi} \delta_{ir} \delta_{js}.$$

3) This is an exercise below. ∎

Our next problem is to see how to do a version of Theorem 1 when the representations π and ρ are not irreducible. We know that then they are direct sums of irreducible subrepresentations.

Proposition 3. Suppose that π and ρ are two representations of g and that we have the following direct sum decompositions:

$$\pi \cong n_1\pi_1 \oplus \cdots \oplus n_r\pi_r \quad \text{and} \quad \rho = m_1\pi_1 \oplus \cdots \oplus m_r\pi_r,$$

where the n_j and m_k are nonnegative integers and the π_j are irreducible representations of G. Then we have the following two formulas:

1) The inner product of the characters of π and ρ is

$$\langle \chi_\pi, \chi_\rho \rangle = |G| \sum_{j=1}^{r} n_j m_j,$$

which counts the number of irreducible components of π and ρ which agree.

2) Suppose that $I(\pi, \rho)$ is, as in Schur's lemma the space of intertwining operators. Then

$$\dim_{\mathbb{C}} I(\pi, \rho) = |G|^{-1} \langle \chi_\pi, \chi_\rho \rangle.$$

Proof. We leave this as an exercise.

Hints.
1) Use Part 3 of Theorem 1.
2) Show that both sides of the equality are additive in each variable; that is, show that if $\pi = \kappa \oplus \sigma$, the term in question is a sum of the term for κ plus that for σ. ∎

Corollary.
1) Two representations with the same character are equivalent and conversely two equivalent representations have the same character.
2) There are only a finite number of inequivalent irreducible unitary representations of the finite group G.
3) $1 \leq \langle \chi_\pi, \chi_\pi \rangle |G|^{-1} \in \mathbb{Z}$.
4) $\langle \chi_\pi, \chi_\pi \rangle = |G|$ if and only if π is irreducible.

Proof. Prove the corollary as an exercise.

Hints.
2) This follows from the orthogonality of the matrix entries since $L^2(G)$ is a finite-dimensional inner product space when G is finite.
3) See Part 1 of Proposition 3. ∎

Definition. If $\pi \in \hat{G}$ and ρ is any representation of G, define $m(\pi, \rho) =$ the *multiplicity of π in ρ* to be the number of times that π occurs in the direct sum decomposition of ρ given by the complete reducibility theorem.

Exercise. Suppose ρ is any representation of G and $\pi \in \hat{G}$. Show that

$$m(\pi, \rho)|G| = \langle \chi_\pi, \chi_\rho \rangle.$$

From the preceding corollary we see that the characters are extremely important. But they do not give us the complete story of Fourier analysis on a finite
group G. The fact that χ_π determines π up to equivalence does not mean that
there is an algorithm for writing down the matrix entries of π from χ_π.

The characters χ_π are so important that character tables of useful groups
are to be found in many sections of the usual science library (e.g., mathematics, physics, and chemistry). The chemists seem to know these tables by
heart for certain space groups, which are symmetry groups of certain chemical
substances. For example, Knox and Gold [1964, p. vii] say:

Group theory can no longer be regarded as an abstract mathematical technique of interest
only to the specialist. It provides a simple, concise, unifying approach to the large number
of problems in solid state physics that are best understood in terms of the translational
and rotational symmetry of crystal lattices. In addition, group theory provides a precise
mathematical language for the expression of the symmetries and the discussion of the
classification, degeneracy, and mixing of states of complicated systems. For the most
part, the only *exact* statements which can be made about a system are those which arise
as a direct consequence of its symmetry alone.

Exercise. Prove that the characters χ_π are constant on conjugacy classes $\{g\}$ in
G for all representations π of G.

What is a Character Table?

Definition. The *character table* of the finite group G is a matrix with columns
indexed by conjugacy classes $\{g\}$ of G and rows indexed by inequivalent, irreducible, unitary representations π in $\hat{G}$. The entry of the character table
corresponding to π and $\{g\}$ is $\chi_\pi(g)$; that is, the value of the character of π on
the conjugacy class. This is well defined by the exercise above.

Examples. $\mathbb{Z}/n\mathbb{Z}$ and S_3.

1) Suppose that G is the finite abelian group $\mathbb{Z}/n\mathbb{Z}$. Then all irreducible
   representations have degree 1 and there are n of them, parameterized by
   the elements $a \in \mathbb{Z}/n\mathbb{Z}$. And each conjugacy class has only one element
   $b \in \mathbb{Z}/n\mathbb{Z}$. Thus the character table is an $n \times n$ matrix. The representation $e_a(x) = \exp(2\pi i a x/n)$, for $a, x \in \mathbb{Z}/n\mathbb{Z}$, is its own character and the
   a, b entry of the character table is $\exp(2\pi i a b/n)$ for $a, b \in \mathbb{Z}/n\mathbb{Z}$. Thus the

character table in this case is actually the matrix of the Fourier transform on $G = \mathbb{Z}/n\mathbb{Z}$, using the natural basis of $L^2(G)$ consisting of delta functions corresponding to elements of $\mathbb{Z}/n\mathbb{Z}$. See Chapter 2.

2) Next let us consider the character table of the symmetric group S_3 of permutations of the set $\{1, 2, 3\}$. First note that the conjugacy classes of this group consist of three classes:

   a) class containing only the identity permutation;
   b) class of transpositions (12), (13), and (23);
   c) class of permutations (123) and (132).

Here we are using the standard notation for permutations. For example, (123) means the cyclic permutation that sends 1 to 2, 2 to 3, and 3 to 1. There are two obvious one-dimensional representations of S_3: the trivial representation, which equals its character $\chi_1(g) = 1$, for all $g \in S_3$, and the sign of the permutation representation, which equals its character

$$\chi_2(g) = \operatorname{sgn}(g)$$

$$= \begin{cases} +1, & g = \text{product of even number of transpositions,} \\ -1, & g = \text{product of odd number of transpositions.} \end{cases}$$

As we will soon learn, the character table is a square matrix. Thus there has to be one more irreducible representation with character χ_3. So far, then, we have the character table shown in Table II.1.

Theorem 3 below will tell us how to find x, y, z without figuring out what the third representation is. See the first exercise below. Later we will construct the third representation as an induced representation.

One beauty of character tables is that you can compute them without knowing too much about the group or the representations. In fact, group theorists knew the character table of the monster group (the largest of all the sporadic simple groups with order approximately 8×10^{53}) before they knew how to construct the group. See Conway [1980] for an example of how to construct a simple group when we only know the order is 60 plus the elementary properties of characters.

Table II.1. *First version of the character table for S_3*

$\hat{G}\backslash$conj.classes	$\{1\}$	$\{(12)\}$	$\{(123)\}$
χ_1	1	1	1
χ_2	1	-1	1
χ_3	x	y	z

Left Regular Representation

Next we prove a lemma which shows that the left regular representation contains every irreducible unitary representation within it.

*Lemma 2. **The Left Regular Representation is the Mother of All Representations.***

a) Let $L(g)$ denote the left regular representation of G. Then its character is

$$\chi_L(g) = \begin{cases} |G|, & \text{if } g = \text{the identity of } G, \\ 0, & \text{otherwise.} \end{cases}$$

b) Every irreducible representation $\pi \in \hat{G}$ is contained in L with multiplicity d_π, where d_π is the degree of π. That is, if $\hat{G} = \{\pi_1, \ldots, \pi_r\}$, L is similar to a direct sum of copies of all the π_j:

$$L \cong d_{\pi_1}\pi_1 \oplus \cdots \oplus d_{\pi_r}\pi_r.$$

c) If $d_\pi = $ degree of π,

$$\sum_{\pi \in \hat{G}} d_\pi^2 = |G|.$$

Proof.

a) The matrix of $L(g)$ is a permutation matrix corresponding to a permutation which fixes nothing and thus its trace is 0 unless g is the identity. Note that $L(g)\delta_h = \delta_{hg}$.

b) By Proposition 3, the multiplicity of π in L is found by computing the inner product of the characters of the two representations and dividing by $|G|$. Therefore we compute

$$\frac{1}{|G|}\langle \chi_L, \chi_\pi \rangle = \frac{1}{|G|}\sum_{y \in G} \chi_L(y)\overline{\chi_\pi(y)} = \frac{1}{|G|}|G|\chi_\pi(e) = d_\pi.$$

Here we have used part a to see that every term in the sum is zero except that coming from $y = $ the identity.

c) Compare the degree of L with the degree of the right-hand side of the formula for L given in b. ∎

The following exercise will be important in later sections (e.g., for Theorem 1 of Chapter 17).

Exercise. Suppose that S is a symmetric set of generators of the finite group G. That is, $s \in S$ implies $s^{-1} \in S$. Consider the Cayley graph $X(G, S)$ with vertices the elements of G and edges between $x \in G$ and xs for all $s \in S$. This is a connected undirected graph.

1) Show that the adjacency operator A on $X(G, S)$ has the form

$$Af = \sum_{s \in S} R(s), \quad \text{where } R \text{ is the right regular representation of } G.$$

2) Deduce from 1 and Lemma 2 above that the adjacency matrix A is similar to a block diagonal matrix

$$A \cong d_{\pi_1} M_{\pi_1} \oplus \cdots \oplus d_{\pi_r} M_{\pi_r}, \quad \text{where } M_\pi = \sum_{s \in S} \pi(s) \tag{8}$$

and

$$\hat{G} = \{\pi_1, \ldots, \pi_r\}.$$

Fourier Analysis on G

The next theorems are the big ones we've been looking for. They give us the main facts about the Fourier transform on a finite group as well as the basic information on character tables. It is a finite group analogue of the Peter–Weyl theorem [see Weyl, 1968] for compact groups as well as a nonabelian version of Theorem 1 in Chapter 2. That is, we find that the matrix entries $\pi_{ij}(x)$ of the $\pi \in \hat{G}$ are the analogues of the exponentials $\exp(2\pi i a x/n)$, $a, x \in \mathbb{Z}/n\mathbb{Z}$, when the cyclic group $\mathbb{Z}/n\mathbb{Z}$ is replaced by any finite group G. This is not hard to prove. You just need the preceding Theorem (Schur's orthogonality relations), Lemma 2, and some finite-dimensional linear algebra.

*Theorem 2. **Basic Facts about Fourier Analysis on a Finite Group.***

1) **Peter–Weyl Theorem.** As usual, let

$$\hat{G} = \{\text{inequivalent irreducible unitary representations } \pi \text{ of } G\}.$$

The matrix entries of the representations $\pi \in \hat{G}$ form a complete orthogonal set in $L^2(G)$.

2) **Plancherel Theorem.** For all $f, g \in L^2(G)$, the usual inner product is

$$\langle f, g \rangle = \sum_{x \in G} f(x)\overline{g(x)}.$$

Then defining $\|f\|_2^2 = \langle f, f \rangle$, we have

$$\|f\|_2^2 = |G|^{-1} \sum_{\substack{x \in \hat{G} \\ 1 \leq i, j \leq d_\pi}} d_\pi |\langle f, \pi_{i, j} \rangle|^2.$$

Here d_π is the degree of π.

3) **Fourier Inversion.** For any $f \in L^2(G)$, we have

$$f(x) = |G|^{-1} \sum_{\substack{\pi \in \hat{G} \\ 1 \leq i, j \leq n}} d_\pi \langle f, \pi_{i, j} \rangle \pi_{i, j}(x).$$

Proof of Theorem 2.

- *Part 1*

The Peter–Weyl theorem holds because the number of matrix entries for π running over all of $\hat{G}$ is equal to $|G|$ by part c of Lemma 2. Since the matrix entries are orthogonal (via Theorem 1), we then must have a complete orthogonal set in $L^2(G)$. Why? Well, the dimension of this inner product space is $|G|$ and we cannot have any more elements in a inner product space basis.

- *Parts 2 and 3*

Exercise. Prove parts 2 and 3 of Theorem 2.

Hints. Use the methods from Chapter 2. That is, use finite-dimensional inner product space theory and part 1. See your favorite linear algebra book (e.g., Lang, [1987] or Strang [1976]) for more information on finite-dimensional inner product or Hilbert spaces. You need to use Theorem 1 to see that the normalized matrix entries of the irreducibles $\pi \in \hat{G}$, namely $\pi_{ij}(d_\pi/|G|)^{1/2}$, form a complete orthonormal set. ∎

Next we want to show that $|\hat{G}|$ = the number of conjugacy classes in G. For this, it suffices to show that the characters χ_π, for $\pi \in \hat{G}$, form an orthogonal basis for the inner product space of class functions on G.

Definition. $f : G \to \mathbb{C}$ is a *class function* if

$$f(xgx^{-1}) = f(g) \quad \text{for all } x, g \in G.$$

To prove our claim, we prove a lemma.

Lemma 3. **Facts about Class Functions.**

a) The function f on G is a *class function* if and only if f is in the *center of the group algebra* $L^2(G)$ under convolution, that is, $f * h = h * f$, for all $h: G \to \mathbb{C}$.

b) If f is a class function, then for any representation $\pi \in \hat{G}$, the $d_\pi \times d_\pi$ matrix $C(\pi)$ of Fourier coefficients $\langle f, \pi_{ij} \rangle$ has the form $c_\pi I$, for some scalar $c_\pi \in \mathbb{C}$. Here I denotes the identity matrix.

c) If f is a class function, the Fourier inversion formula from Part 3 of Theorem 2 becomes

$$f(g) = \frac{1}{|G|} \sum_{\pi \in \hat{G}} \langle f, \chi_\pi \rangle \chi_\pi(g).$$

Proof of Lemma 3.

a) Let $\{\delta_a, a \in G\}$ be the basis of $L^2(G)$ of delta functions

$$\delta_a(x) = \begin{cases} 0, & x \neq a, \\ 1, & x = a, \end{cases}.$$

Then use the definition of convolution to see

$$(f * \delta_a)(x) = \sum_{y \in G} f(y)\delta_a(y^{-1}x) = f(xa^{-1})$$

and

$$(\delta_a * f)(x) = \sum_{y \in G} \delta_a(y) f(y^{-1}x) = f(a^{-1}x) = f(a^{-1}x a^{-1}a).$$

So f is a class function iff $f * \delta_a = \delta_a * f$, for all $a \in G$.

b) Using Part a, we see that if f is a class function then for any representation π of G, we have $f * \pi_{ij} = \pi_{ij} * f$. This means that if $C(\pi)$ is the matrix of Fourier coefficients $\langle f, \pi_{ij} \rangle$, the transpose of this matrix commutes with the matrix $\pi(x)$, for all $x \in G$. Why? Just write out the convolution equation:

$$({}^tC(\pi)\pi(x))_{i,j} = \sum_{y \in G} f(y) \sum_{k=1}^{d_\pi} \overline{\pi_{ki}(y)}\, \pi_{kj}(x)$$

$$= \sum_{y \in G} f(y)\pi_{ij}(y^{-1}x) = (f * \pi_{ij})(x) = (\pi_{ij} * f)(x)$$

$$= \sum_{y \in G} \pi_{ij}(xy^{-1})f(y) = (\pi(x)^t C(\pi))_{i,j}.$$

Therefore Part 3 of (Schur's) Lemma 1 proves Part b.

c) The Fourier inversion formula from part 3 of Theorem 2 says that any function on G, in particular our class function f, has an expansion:

$$f(g) = \frac{1}{|G|} \sum_{\pi \in \hat{G}} d_\pi \sum_{i,j=1}^{d_\pi} \langle f, \pi_{ij} \rangle \pi_{ij}(g)$$

$$= \frac{1}{|G|} \sum_{\pi \in \hat{G}} d_\pi c_\pi \sum_{i=1}^{d_\pi} \pi_{ii}(g) = \frac{1}{|G|} \sum_{\pi \in \hat{G}} d_\pi c_\pi \chi_\pi(g).$$

This says that f is a linear combination of characters. We know already that the characters are orthogonal as is the inner product $\langle \chi_\pi, \chi_\pi \rangle = |G|$. To prove the desired formula, just write

$$f(x) = \sum_{\pi \in \hat{G}} \lambda_\pi \chi_\pi(x), \quad \text{for some scalars } \lambda_\rho \in \mathbb{C}.$$

Then compute the inner product $\langle f, \chi_\rho \rangle = \lambda_\rho |G|$. ∎

Theorem 3. **Basic Facts about Characters.**

1) The character table of G is square; that is, $|\hat{G}|$ equals the number of conjugacy classes in G.
2) *Orthogonality Relations.* The matrix obtained from the character table by replacing each of the entries $\chi_\pi(g)$ by $\chi_\pi(g)\sqrt{\frac{|\{g\}|}{|G|}}$ is a unitary matrix.

- *Proof of Part 1 of Theorem 3.* This follows from Lemma 3.
- *Proof of Part 2 of Theorem 3.* It suffices to show that if we multiply the character table entry corresponding to $(\pi, \{g\})$ by $\sqrt{\frac{|\{g\}|}{|G|}}$, we make the character table (which we now know is square) into a matrix whose rows are orthonormal. This comes from the Schur orthogonality relations (part 3) of Theorem 1.
  Let $\pi, \rho \in \hat{G}$ and let Ξ denote the set of conjugacy classes of G. Then the orthonormality calculation is

$$\sum_{\{g\} \in \Xi} \frac{|\{g\}|}{|G|} \chi_\pi(g)\overline{\chi_\rho(g)} = \frac{1}{|G|} \sum_{x \in G} \chi_\pi(x)\overline{\chi_\rho(x)} = \begin{cases} 0, & \text{if } \pi \neq \rho, \\ 1, & \text{if } \pi = \rho. \end{cases}$$

∎

Exercise. Define the *centralizer* of g in G by $G_g = \{x \in G \mid xg = gx\}$. Prove that

$$|\{g\}| = \frac{|G|}{|G_g|}.$$

Exercise. Use the preceding results to determine the last row of the character table of S_3

Answer. $x = 2, y = 0, z = -1$.

Hints. First use part c of Lemma 2 to determine x. Then use part 2 of Theorem 3 to obtain two linear equations for the remaining unknowns y and z.

Exercise. Show that the orthogonality relations which give part 2 of Theorem 3 imply that we have dual orthogonality relations

$$\sum_{\pi \in \hat{G}} \chi_\pi(g)\chi_\pi(h) = \begin{cases} 0, & \text{if } g \text{ and } h \text{ are not conjugate,} \\ |G|/|\{g\}|, & \text{if } g \text{ and } h \text{ are conjugate.} \end{cases} \tag{9}$$

Remarks. There are many ways to prove Theorems 2 and 3. Look, for example, in Fulton and Harris [1991], James and Liebeck [1993], or van der Waerden [1991] for a more algebraic point of view.

In Theorem 2 we spoke of Fourier inversion without defining the Fourier transform. Here we remedy that problem.

Definition. Define the *Fourier transform* of $f : G \to \mathbb{C}$ by

$$\mathscr{F}f(\pi) = \hat{f}(\pi) = \sum_{g \in G} f(g)\pi(g). \tag{10}$$

Note that this Fourier transform $\hat{f}(\pi)$ is matrix valued; that is, it is a $d_\pi \times d_\pi$ matrix.

Remark. The Fourier transform defined above is not consistent with our earlier definition in formula (5) of Chapter 2. In order to make the two formulas agree, we should replace $\pi(y)$ by $\pi(y^{-1})$ or $\overline{\pi(y)}$. We won't worry about this inconsistency. It is more convenient for later computations to use this definition here.

*Theorem 4. **Basic Facts about the Fourier Transform on a Finite Group.***

1) *Convolution Property of Fourier Transform on G.*

Recall that we defined convolution of two functions f and g on G in the chapter on the group algebra as

$$(f * g)(x) = \sum_{y \in G} f(y)g(y^{-1}x).$$

Then

$$\mathscr{F}(f * g)(\pi) = \mathscr{F}f(\pi) \cdot \mathscr{F}g(\pi).$$

The multiplication on the right is matrix multiplication.

2) *Fourier Transform changes left regular representation to multiplication by* π. For $f \in L^2(G)$, define $[L(g)f](x) = f(g^{-1}x) = f^g(x)$, if $x, g \in G$. Then we have

$$\mathscr{F}[f^g](\pi) = \pi(g)\mathscr{F}f(\pi), \quad \text{for all } \pi \in \hat{G}.$$

3) *Explicit Similarity Transform from Part b of Lemma 2.* Using the appropriate basis, $\mathscr{F} \circ L(g) \circ \mathscr{F}^{-1}$ is a block diagonal matrix with diagonal blocks given by the $\pi(g)$ each listed d_π times for $\pi \in \hat{G}$.

- *Proof of Part 1 of Theorem 4.* The proof is the same as it was in Chapter 2 for the abelian case. We just need to apply the Fourier transform to the convolution and do the right change of variables:

$$\mathscr{F}(f * g)(\pi) = \sum_{x \in G} \pi(x) \sum_{u \in G} f(u)g(u^{-1}x)$$

$$= \sum_{y \in G} \pi(uy) \sum_{u \in G} f(u)g(y)$$

$$= \mathscr{F}f(\pi)\mathscr{F}g(\pi).$$

Here we made the substitution $y = u^{-1}x$ and we used the fact that π is a group homomorphism.

- *Proof of Part 2 of Theorem 4.* Use the definitions of the Fourier transform $\mathscr{F}$ and the left regular representation $L(g)$. For $f : G \to \mathbb{C}$ and $\pi \in \hat{G}$, we set $L(g)f = f^g$ and find that

$$\mathscr{F}[f^g](\pi) = \sum_{x \in G} [L(g)f](x)\pi(x) = \sum_{x \in G} f(g^{-1}x)\pi(x)$$

$$= \sum_{u \in G} f(u)\pi(gu) = \pi(g) \sum_{u \in G} f(u)\pi(u)$$

$$= \pi(g)\mathscr{F}(f)(\pi).$$

Here we set $u = g^{-1}x$ and used the fact that $f(u)$ is a scalar.

- *Proof of Part 3 of Theorem 4.* This follows from part 2. First define the vector space

$$L^2(\hat{G}) = \left\{ f : \hat{G} \to \overset{\text{disjoint}}{\underset{\pi \in \hat{G}}{\bigcup}} \mathbb{C}^{d_\pi \times d_\pi} \;\middle|\; f(\pi) \in \mathbb{C}^{d_\pi \times d_\pi}, \; \forall \pi \in \hat{G} \right\}. \quad (11)$$

Note that the Fourier transform is a 1-1 onto linear map taking $L^2(G)$ to $L^2(\hat{G})$.

Let k run through a basis of $L^2(G)$. In the formula of part 2, set $\mathscr{F}f = k$ and $k(\pi) = (k_1 \cdots k_{d_\pi})$, where the $k_j \in \mathbb{C}$ denote the columns of $k(\pi)$. Then we have, upon setting $M = \pi(g) \in \mathbb{C}^{d_\pi \times d_\pi}$,

$$\mathscr{F}L(g)\mathscr{F}^{-1}k(\pi) = \pi(g)k(\pi) = M(k_1 \cdots k_{d_\pi}) = (Mk_1 \cdots Mk_{d_\pi}).$$

If π is a one-dimensional representation, we are just multiplying the number $\pi(g)$ by $k(\pi)$. If π is d_π-dimensional, for k corresponding to that part of the basis of $L^2(G)$ coming from π, we are multiplying the matrix $k(\pi)$ by the matrix $M = \pi(g)$ on the left. The space of $d_\pi \times d_\pi$ matrices is a direct sum of d_π column spaces $\mathbb{C}^{d_\pi}$. On each of these spaces we are multiplying by M. So we can block diagonalize the linear map from the k_j to Mk_j by putting d_π copies of M down the diagonal, for each $\pi \in \hat{G}$. ∎

Exercise. Define an inner product on $L^2(\hat{G})$ by

$$\langle f, g \rangle = \frac{1}{|G|} \sum_{\pi \in \hat{G}} d_\pi \operatorname{Tr}(f(\pi)g(\pi)^*),$$

where, as usual, d_π denotes the degree of π. Show that the Plancherel theorem (part 2) of Theorem 2 says that the Fourier transform is an inner product space isometry from $L^2(G)$ to $L^2(\hat{G})$.

Exercise. Show that the Fourier inversion formula (part 3) of Theorem 2 is equivalent to

$$f(g) = \frac{1}{|G|} \sum_{\pi \in \hat{G}} d_\pi \operatorname{Tr}(\pi(g^{-1})\hat{f}(\pi)). \tag{12}$$

Last Remarks and Exercises

Example. In an exercise above, you used Theorem 1 to find that the character table of the symmetric group S_3 is that given in Table II.2.

At no time did you need to construct the representation corresponding to the third row of the table. However, we might as well do so, since it is easy and this group will come up in various guises throughout this book. At the moment, we do not have many ways to construct representations that are irreducible. The regular representation of S_3 has degree 6 and is certainly not irreducible. The

Table II.2. *The character table for S_3*

$\hat{G}$\conj.classes	{1}	{(12)}	{(123)}
χ_1	1	1	1
χ_2	1	−1	1
χ_3	2	0	−1

representation of S_3 by permutation matrices has degree 3:

$$1 \to I, \quad (12) \to \begin{pmatrix} 0 & 1 & 0 \\ 1 & 0 & 0 \\ 0 & 0 & 1 \end{pmatrix}, \quad (13) \to \begin{pmatrix} 0 & 0 & 1 \\ 0 & 1 & 0 \\ 1 & 0 & 0 \end{pmatrix},$$

$$(23) \to \begin{pmatrix} 1 & 0 & 0 \\ 0 & 0 & 1 \\ 0 & 1 & 0 \end{pmatrix}, \quad (123) \to \begin{pmatrix} 0 & 0 & 1 \\ 1 & 0 & 0 \\ 0 & 1 & 0 \end{pmatrix}, \quad (132) \to \begin{pmatrix} 0 & 1 & 0 \\ 0 & 0 & 1 \\ 1 & 0 & 0 \end{pmatrix}.$$

Of course, this representation can't be irreducible either. So how do we get the degree 2 representation π_3?

Note that we could describe the degree 3 representation above by writing

$$\pi(\sigma) \begin{pmatrix} v_1 \\ v_2 \\ v_3 \end{pmatrix} = \begin{pmatrix} v_{\sigma^{-1}(1)} \\ v_{\sigma^{-1}(2)} \\ v_{\sigma^{-1}(3)} \end{pmatrix},$$

for vectors

$$v = \begin{pmatrix} v_1 \\ v_2 \\ v_3 \end{pmatrix} \in \mathbb{C}^3.$$

Consider the subspace W of $\mathbb{C}^3$ defined by

$$W = \left\{ v = \begin{pmatrix} v_1 \\ v_2 \\ v_3 \end{pmatrix} \middle| v_1 + v_2 + v_3 = 0 \right\}.$$

Clearly if we define $\rho(g)$ to be the restriction of $\pi(g)$ to W, we get a subrepresentation ρ of π.

Exercise. Prove that the representation ρ of S_3 defined above is irreducible. Thus the character of ρ is that of the third row in the character table of S_3.

Hint. You can either give a direct proof or use character theory.

There are other ways of constructing the degree two irreducible representation of S_3 as we will see in the next chapter.

We can view S_3 as the group of motions of an equilateral triangle. And it can also be viewed as the affine group over the field with three elements as we saw in an earlier exercise.

For more character tables, see Fulton and Harris [1991], Jacobson [1980, 1985], James and Liebeck [1993], Thomas and Wood [1980], Serre [1973], or some of the other references mentioned earlier. We will find more character tables in the next chapters. There are a few more useful facts we should at least state. For proofs, see Jacobson [1980, 1985] or Serre [1973]. The proofs use some results from number theory – properties of rings of integers.

Definition. The *center* of G is the subgroup $Z = \{x \in G \mid xg = gx \text{ for all } g \in G\}$.

Theorem 5. (Schur). Suppose that π is an irreducible unitary representation of the finite group G and suppose that Z is the center of G. Then the degree of π divides

$$[G : Z] = |G|/|Z|.$$

Examples. More examples can be made by looking at direct products of groups. Suppose that H and G are groups. The *direct product* of G and H is the group

$$G \times H = \{(g, h) \mid g \in G, h \in H\}, \tag{13}$$

with coordinatewise multiplication defined by $(a, b)(g, h) = (ag, bh)$ for a, $g \in G$ and $b, h \in H$. It is an exercise to find that the inequivalent, irreducible, unitary representations of $G \times H$ are given by $\pi \otimes \rho$, for $\pi \in \hat{G}$ and $\rho \in \hat{H}$. See Serre [1973]. Thus the characters of direct products are just products of characters of the component groups. Later we will also look at semidirect products.

Exercise. Show that there are nonisomorphic groups with the same character table.
Hint. Consider the groups D_8 and Q_8, where the dihedral group D_8 is the group of symmetries of a square and the quaternion group Q_8 is the group of quarternions $\{\pm 1, \pm i, \pm j, \pm k\}$ with the usual multiplication $ij = k = -ji$ and so on. See James and Liebeck [1993, p. 283].

Exercise. Consider the alternating group A_5 of even permutations of five elements. Show that it can be viewed as the group of rotations of three-space that bring a regular icosahedron back to itself. Thus it is also called the icosahedral group. Find its character table.

Answer. See Chapter 21. Or see James and Liebeck [1993].

Exercise. Consider the *group determinant* $\det(g(y^{-1}x)_{x,y \in G})$ for some function $g : G \to \mathbb{C}$ in the special case that $G = S_3$. Note that the matrix $M = g(y^{-1}x)_{x,y \in G}$ is then the matrix of the operator $L_g f(x) = (f * g)(x)$ acting on $f \in L^2(G)$. List the elements of S_3 in the following order:

$$x_1 = (1), \quad x_2 = (12), \quad x_3 = (23), \quad x_4 = (13), \quad x_5 = (123), \quad x_6 = (132).$$

Write $g_i = g(x_i)$ and note that

$$M = \begin{pmatrix} g_1 & g_2 & g_3 & g_4 & g_6 & g_5 \\ g_2 & g_1 & g_6 & g_5 & g_3 & g_4 \\ g_3 & g_5 & g_1 & g_6 & g_4 & g_2 \\ g_4 & g_6 & g_5 & g_1 & g_2 & g_3 \\ g_5 & g_3 & g_4 & g_2 & g_1 & g_6 \\ g_6 & g_4 & g_2 & g_3 & g_5 & g_1 \end{pmatrix}.$$

Then use the decomposition of the left regular representation of S_3 as a direct sum of irreducibles to write

$$\det(M) = \prod_{\pi \in \hat{G}} \det(\hat{g}(\pi)).$$

The three factors on the right are

$$g_1 + g_2 + g_3 + g_4 + g_5 + g_6, \qquad g_1 - g_2 - g_3 - g_4 + g_5 + g_6,$$
$$g_1^2 + g_5^2 + g_6^2 - g_2^2 - g_3^2 - g_4^2 + g_2g_3 + g_2g_4 + g_3g_4 - g_1g_5 - g_1g_6 - g_5g_6.$$

Hint. See Curtis [1992]. You may want to wait until after we have considered the degree two representation of S_3 as an induced representation in the next chapter.

Chapter 16

Induced Representations

One of the highlights among the [Frobenius] papers published
after 1896 was a deep analysis of the relation between
characters of a group G and the characters of a subgroup H of
G .... As he stated in the introduction, an understanding of this
relationship is crucial for the practical computation of
representations and characters – a statement as true now as it
was then!

C. Curtis [1992, p. 51]

This chapter gives a way of constructing representations of finite groups from
those of subgroups called the method of induced representations. In 1898
Frobenius invented the method of induction by writing down the formula for
the character of the induced representation of a finite group. In 1927 Speiser
gave formulas for the matrix entries. Wigner obtained induced representations
of the (infinite) Lorentz group from the subgroup of translations in 1939. In
1940 Weil showed how to do induction for compact groups. In the 1950s
Mackey [1976], [1978a,b] developed the theory for noncompact, locally compact groups. Around the same time Selberg invented his trace formula and
applied it to the group $G = SL(2, \mathbb{R})$ and various discrete subgroups acting on
G/K, $K = SO(2)$. See Terras [1985, Vol. I]. We will see in Chapter 22 that
the finite analogue of Selberg's trace formula will allow us to prove some of the
results such as the Frobenius reciprocity law in an elegant way (see Chapter 22
and Arthur [1989]).

How do we define an induced representation?

Definition. Suppose that H is a subgroup of the finite group G and $\sigma : H \to$
$GL(W)$ is a representation of H. Then the *induced representation from H*
up to G denoted $\pi = \mathrm{Ind}_H^G \sigma$ is a group homomorphism $\pi : G \to GL(V)$,

where

$$V = \{f : G \to W \mid f(hg) = \sigma(h)f(g), \text{ for all } h \in H, g \in G\}. \qquad (1)$$

The representation $\pi(g)$ is then defined on $f \in V$ by

$$[\pi(g)f](x) = f(xg), \quad \text{for all } x, g \in G.$$

Thus we have defined the induced representation using the action of G by right translation of functions which are "σ-invariant."

Exercise. Check that π defined above is indeed a representation of G. If σ is a unitary representation, will π be unitary also?
Hints. To see if $\pi(g)$ is unitary, we need an inner product on V. Suppose $W = \mathbb{C}^r$, which we view as a space of complex column vectors. For $w \in W$ write $w^* = {}^t\bar{w}$. Then the inner product of the functions f and k in V is

$$\langle f, k \rangle = \sum_{x \in H \backslash G} k(x)^* f(x). \qquad (2)$$

You must decide if this inner product is well defined on cosets modulo H, and if it satisfies $\langle f, k \rangle = \langle \pi(g)f, \pi(g)k \rangle$, for all $g \in G$. For the first question, recall that for $h \in H$, $\sigma(h)$ is a unitary matrix, that is, $\sigma(h)^*\sigma(h) = I$, for all $h \in H$. Moreover, $f(hg) = \sigma(h)f(g)$ for $h \in H$ and $g \in G$.

Example 1. The *right regular representation* of G.
    The right regular representation is an induced representation with $H = \{e\}$, the subgroup with one element, the identity e of G, and σ being the trivial representation of this subgroup. Then the space $V = L^2(G)$. And $\pi(g)f(x) = f(xg)$ is just the right regular representation.

Example 2. The *two-dimensional representation* of the symmetric group S_3.
    Let H be the subgroup of S_3 generated by (123). Then H is a cyclic group of order three. Consider the representation σ of H defined by $\sigma(123)^x = \exp(2\pi i x/3)$. Let $\pi = \operatorname{Ind}_H^G \sigma$. We claim that the matrices of $\pi(g)$ have the following formulas if one chooses the correct basis:

$$\pi(12) = \begin{pmatrix} 0 & 1 \\ 1 & 0 \end{pmatrix} \quad \text{and} \quad \pi(123) = \begin{pmatrix} \exp(2\pi i/3) & 0 \\ 0 & \exp(4\pi i/3) \end{pmatrix}.$$

So let g_1 and g_2 be representatives in $G = S_3$ of the quotient $H \backslash G$, for $H = \langle(123)\rangle$, the cyclic subgroup generated by (123). Clearly we can take $g_1 = (1) =$

the identity and $g_2 = (12)$. Then we need a basis for the space

$$V = \{f : G \to \mathbb{C} \mid f(hg) = \sigma(h)f(g) \text{ for all } h \in H, g \in G\}.$$

The basis for V we choose is f_1, f_2 defined as follows:

$$f_i(x) = \left\{ \begin{array}{ll} \sigma(h), & \text{if } x = hg_i, \\ 0, & \text{otherwise.} \end{array} \right\} = \tilde{\sigma}\left(xg_i^{-1}\right),$$

where we define

$$\tilde{\sigma}(g) = \left\{ \begin{array}{ll} \sigma(g), & \text{if } g \in H, \\ 0, & \text{if } g \notin H. \end{array} \right.$$

So then we have for $g_1 = (1)$ and $g_2 = (12)$:

$$[\pi(12)f_i](x) = \tilde{\sigma}\left(x(12)g_i^{-1}\right) = \tilde{\sigma}\left(xg_j^{-1}\right) = f_j(x), \text{ if } j \neq i.$$

Then $\pi(12)$ does have the matrix stated above, since it interchanges f_1 and f_2.

Exercise.
a) Show that the formula above for $\pi(123)$ is correct.
b) Show that it follows from the formulas for $\pi(12)$ and $\pi(123)$ that the character χ_π does have the values claimed in the character table from Table II.2; that is, show that $\chi_\pi(12) = 0$ and $\chi_\pi(123) = -1$. For this last result, you need to recall that if $w = \exp(2\pi i/3)$, then $1 + w + w^2 = 0$. These are the orthogonality relations for $\mathbb{Z}/3\mathbb{Z}$.
c) Use the corollary to Proposition 3 of Chapter 15 to show that π is indeed an irreducible representation; that is, show that the inner product $\langle \chi_\pi, \chi_\pi \rangle = |G|$. This also follows from the fact that the character determines the representation up to equivalence and π does have the right character.

Hints.
a) First show that $[\pi(123)f_1](x) = \exp(2\pi i/3)f_1(x)$, by using the definitions of π and f_1 in a straightforward manner. Then note that if $x = hg_k$, for $h \in H$

$$[\pi(123)f_2](x) = \tilde{\sigma}(hg_k(123)(12)) = \sigma(132)f_2(hg_k).$$

With a little patience, we can use the method of the last example to determine the matrix entries of any induced representation. Suppose that H is a subgroup of the finite group G and that $\sigma : H \to GL(W)$ is a representation of H. We defined $\pi = \mathrm{Ind}_H^G \sigma$ to be a representation of G by linear transformations in V, where

$$V = \{f : G \to W \mid f(hg) = \sigma(h)f(g), \text{ for all } g \in G, h \in H\}.$$

Now set $\pi(g)f(x) = f(xg)$ for $g, x \in G$, and $f \in V$.
  As before set

$$\tilde{\sigma}(x) = \begin{cases} \sigma(x), & \text{if } x \in H, \\ 0, & \text{otherwise.} \end{cases}$$

Suppose that $g_i, i = 1, \ldots, m$ are a complete system of representatives for $H \backslash G$. And let $e_j, j = 1, \ldots, s$ form an orthonormal basis for $W \cong \mathbb{C}^s$. Then we claim that an orthonormal basis for V consists of functions:

$$f_{ij}(x) = \tilde{\sigma}\left(xg_i^{-1}\right)e_j, \quad \text{for } i = 1, \ldots, m \quad \text{and} \quad j = 1, \ldots, s. \tag{3}$$

Exercise. Prove this last statement. You need to check that using the inner product defined by (2) we have

$$\langle f_{ij}, f_{kr} \rangle = \delta_{ik}\delta_{jr}.$$

*Proposition 1. **Matrix Entries of an Induced Representation.*** Suppose that H is a subgroup of the finite group G. Assume that $\sigma : H \to GL(W)$ is a representation of H. Let $g_i, i = 1, \ldots, m$ denote a complete set of representatives for $H \backslash G$. And suppose that $e_j, j = 1, \ldots, s$ form an orthonormal basis for $W \cong \mathbb{C}^s$. Define

$$\tilde{\sigma}(x) = \begin{cases} \sigma(x), & \text{if } x \in H, \\ 0, & \text{otherwise.} \end{cases}$$

Then the matrix entries of $\pi = \mathrm{Ind}_H^G \sigma$ have the form

$$(\pi(g))_{kr, ij} = \tilde{\sigma}_{rj}\left(g_k g g_i^{-1}\right),$$

for $i, k \in \{1, \ldots, m\}$ and $r, j \in \{1 \ldots, s\}$. Here $m = |G/H|$ and $s = \dim_{\mathbb{C}} W$, the dimension of the vector space W. If σ is unitary, then so is the induced representation.

Proof. Using the definitions above, we find when $x = hg_k$, for $h \in H$,

$$[\pi(g)f_{ij}](x) = f_{ij}(xg) = \sigma(h)f_{ij}(g_k g) = \sigma(h)\tilde{\sigma}\left(g_k g g_i^{-1}\right)e_j.$$

Make the usual definition of the matrix entries of $\tilde{\sigma}$, that is,

$$\tilde{\sigma}(x)e_i = \sum_{j=1}^{s} \tilde{\sigma}(x)_{ji} e_j.$$

Then for $x = hg_k, h \in H$, we have

$$f_{ij}(xg) = \sigma(h) \sum_{r=1}^{s} \tilde{\sigma}_{rj}\left(g_k g g_i^{-1}\right)e_r = \sum_{r=1}^{s} \tilde{\sigma}_{rj}\left(g_k g g_i^{-1}\right)\sigma(h)e_r$$

$$= \sum_{r=1}^{s} \tilde{\sigma}_{rj}\left(g_k g g_i^{-1}\right)\sigma\left(xg_k^{-1}\right)e_r = \sum_{r=1}^{s} \tilde{\sigma}_{rj}\left(g_k g g_i^{-1}\right)f_{kr}(x).$$

In the last equality, we used formula (3) above.

This means that for general $x \in G$

$$f_{ij}(xg) = \sum_{k=1}^{m}\sum_{r=1}^{s} \tilde{\sigma}_{rj}\left(g_k g g_i^{-1}\right)f_{kr}(x),$$

which is just what we were trying to prove.

To see that the induced representation is unitary, refer to the first exercise in this chapter. ∎

Exercise. Give another proof of Proposition 1 as follows. Let M be an $m \times m$ matrix and $e_1, \dots, e_m$ an orthonormal set in $\mathbb{C}^s$ with respect to the usual inner product $\langle v, w \rangle = w^* v$. As usual, we view the elements of $\mathbb{C}^s$ as column vectors. Then the matrix entries of M are inner products

$$m_{ij} = \langle Me_j, e_i \rangle = {}^t\overline{e_i} M e_j.$$

Thus you can compute the matrix entries of the induced representation $\pi = \operatorname{Ind}_H^G \sigma$ as

$$\langle \pi(g)f_{ij}, f_{kr} \rangle = \sum_{x \in H\backslash G} \overline{{}^t f_{kr}(x)}\, f_{ij}(xg), \quad \text{with } f_{ij} \text{ defined by (3)}.$$

Corollary. **Frobenius Formula for Characters of Induced Representations.**
Suppose that H is a subgroup of the finite group G and $\sigma : H \to GL(W)$ is a representation of H. Let $\pi = \operatorname{Ind}_H^G \sigma$. Then the following formula holds relating the characters of the two representations:

$$\chi_\pi(g) = \frac{1}{|H|}\sum_{x \in G} \tilde{\chi}_\sigma\left(xgx^{-1}\right) = \sum_{a \in H\backslash G} \tilde{\chi}_\sigma\left(aga^{-1}\right).$$

Here we define

$$\tilde{\chi}_\sigma(x) = \begin{cases} \chi_\sigma(x), & \text{if } x \in H, \\ 0, & \text{otherwise.} \end{cases}$$

The sum over $a \in H \backslash G$ means a sum over a complete set of representatives in G for the quotient $H \backslash G$.

Exercise. Prove the corollary above.

The Affine Group

An important example for the rest of this book is the following.

Example. The Affine Group over a Finite Field $\mathbb{F}_q$. Here the affine group is defined as in Chapter 15 by

$$\text{Aff}(q) = \left\{ \begin{pmatrix} y & x \\ 0 & 1 \end{pmatrix} \middle| x, y \in \mathbb{F}_q, y \neq 0 \right\}. \tag{4}$$

Our goal is to find the representations of this group.

To find the representations of $\text{Aff}(q)$ which aren't of degree one, we will need a nontrivial character ψ of the additive group of the finite field $\mathbb{F}_q$. For example, set

$$\psi(x) = \exp(2\pi i \text{Tr}(x)/p), \quad \text{if } q = p^r, \text{ where } p = \text{prime},$$

and the trace of $x = \text{Tr}(x) = x + x^p + x^{p^2} + \cdots + x^{p^{r-1}}$.

Definition. Let N be the subgroup of $\text{Aff}(q)$ consisting of matrices of the form

$$\begin{pmatrix} 1 & x \\ 0 & 1 \end{pmatrix}.$$

Then N is isomorphic to the additive group of the field $\mathbb{F}_q$. Thus ψ gives a representation of N via

$$\psi \begin{pmatrix} 1 & x \\ 0 & 1 \end{pmatrix} = \psi(x).$$

Proposition 2. **Irreducible Representations of the Affine Group over $\mathbb{F}_q$.** A complete list of the representations in $\hat{G}$, for $G = \text{Aff}(q)$, is given by the two types of representations below.

I. The *one-dimensional representations* have the form

$$\chi \begin{pmatrix} y & x \\ 0 & 1 \end{pmatrix} = \chi(y),$$

where χ is a character of the multiplicative group of $\mathbb{F}_q$.

II. The $(q-1)$-*dimensional irreducible unitary representation* π is defined by

$$\pi = \text{Ind}_N^G \psi, \quad \text{with } N \text{ and } \psi \text{ as above.}$$

Proof. As usual, there is not any trouble seeing that the one-dimensional representations are irreducible and inequivalent. To see that the $(q-1)$-dimensional representation is irreducible, we can compute the inner product:

$$\langle \chi_\pi, \chi_\pi \rangle = \sum_{x \in G} \chi_\pi(x) \overline{\chi_\pi(x)}.$$

If the inner product is $|G| = q(q-1)$, then π is irreducible.

In order to do our computation, it will help to know the *conjugacy classes* of $G = \text{Aff}(q)$. We claim there are three sorts of conjugacy classes.

Conjugacy Classes of The Finite Affine Group

Type 1. $\{I\} = $ the class of the identity;
Type 2.

$$\left\{ \begin{pmatrix} 1 & x \\ 0 & 1 \end{pmatrix} \, \middle| \, x \neq 0, x \in \mathbb{F}_q \right\},$$

one class, with $q-1$ elements;
Type 3.

$$\left\{ \begin{pmatrix} y & x \\ 0 & 1 \end{pmatrix} \, \middle| \, x \in \mathbb{F}_q \right\},$$

$y \neq 1, q-1$ classes, each with q elements.

See the exercise below for hints on the proof.

We must next use the Frobenius character formula to compute the value of χ_π on each conjugacy class. We get:

Type 1. degree $\pi = q - 1$;
Type 2.

$$\chi_\pi \begin{pmatrix} 1 & x \\ 0 & 1 \end{pmatrix} = \sum_{b \neq 0} \tilde{\psi}\left(\begin{pmatrix} b & 0 \\ 0 & 1 \end{pmatrix}\begin{pmatrix} 1 & x \\ 0 & 1 \end{pmatrix}\begin{pmatrix} b^{-1} & 0 \\ 0 & 1 \end{pmatrix}\right)$$

$$= \sum_{b \neq 0} \psi(bx) = -1,$$

since the last sum runs over all nontrivial characters of the additive group and $x \neq 0$;
Type 3.

$$\chi_\pi \begin{pmatrix} y & x \\ 0 & 1 \end{pmatrix} = \sum_{b \neq 0} \tilde{\psi}\left(\begin{pmatrix} b & 0 \\ 0 & 1 \end{pmatrix}\begin{pmatrix} y & x \\ 0 & 1 \end{pmatrix}\begin{pmatrix} b^{-1} & 0 \\ 0 & 1 \end{pmatrix}\right) = 0,$$

since the argument of $\tilde{\psi}$ is never in N.

Now we can compute the desired inner product to show the irreducibility of π. We obtain

$$\frac{1}{|G|} \sum_{x \in G} \chi_\pi(x)\overline{\chi_\pi(x)} = \frac{1}{q(q-1)}\{1 \cdot (q-1)^2 + (q-1)(-1)^2 + 0\} = 1.$$

This completes the proof of Proposition 2. ∎

We have now found the character table for $G = \text{Aff}(q)$, since we know the conjugacy classes and the irreducible representations and the values of the irreducible characters on the conjugacy classes. See Table II.3.

Table II.3. *Character table for the affine group* $\text{Aff}(q)$:
$\chi_j(y) = \exp[(2\pi i u j)/(q-1)]$, *if* $y = g^u$, *where* $\mathbb{F}_q^*$ *is generated by* g

	{I}	$\left\{\begin{pmatrix} 1 & x \\ 0 & 1 \end{pmatrix}, x \in \mathbb{F}_q^*\right\}$	$\left\{\begin{pmatrix} y & x \\ 0 & 1 \end{pmatrix}, x \in \mathbb{F}_q\right\}, y \neq 1$
# classes	1	1	$q - 2$
# elements of class	1	$q - 1$	q
$\chi_j, 1 \leq j \leq q - 1$	1	1	$\chi_j(y)$
χ_π	$q - 1$	-1	0

Exercise.

1) Check that the conjugacy classes of Aff(q) are as stated in the proof of the previous proposition.

2) Check that all nontrivial additive characters of $\mathbb{F}_q$ have the form $\psi_b(x) = \psi(bx)$, as b runs over the nonzero elements of $\mathbb{F}_q$.

3) Check that Table II.3 really is the character table of Aff(q).

Hints.

1) First note that

$$\begin{pmatrix} b & a \\ 0 & 1 \end{pmatrix}^{-1} \begin{pmatrix} y & x \\ 0 & 1 \end{pmatrix} \begin{pmatrix} b & a \\ 0 & 1 \end{pmatrix} = \begin{pmatrix} y & b^{-1}[a(y-1)+x] \\ 0 & 1 \end{pmatrix}.$$

3) Check that we found *all* the representations of Aff(q) and that they are indeed inequivalent and unitary.

Next we want to write down the matrix of the induced representation $\pi = \operatorname{Ind}_N^G \psi$ from the character table shown in Table II.3.

*Proposition 3. **Matrix of the ($q-1$)-Dimensional Representation of the Finite Affine Group.*** For $x, y \in \mathbb{F}_q$, with $y \neq 0$, $y = g^t$, and for g a generator of the multiplicative group $\mathbb{F}_q^*$, we have

$$\pi \begin{pmatrix} y & x \\ 0 & 1 \end{pmatrix} = D(x)W^t,$$

where $D(x)$ is the $(q-1) \times (q-1)$ diagonal matrix

$$D(x) = \begin{pmatrix} \psi(x) & 0 & 0 & \cdot & \cdot & 0 \\ 0 & \psi(gx) & \cdot & \cdot & \cdot & 0 \\ \cdot & \cdot & \cdot & \cdot & \cdot & \cdot \\ \cdot & \cdot & \cdot & \cdot & \cdot & \cdot \\ \cdot & \cdot & \cdot & \cdot & \cdot & \cdot \\ 0 & 0 & & & & \psi(g^{q-2}x) \end{pmatrix},$$

$$\psi(x) = \exp\left(\frac{2\pi i \operatorname{Tr}(x)}{p}\right), \quad q = p^r, \quad \operatorname{Tr} = \text{trace down to } \mathbb{F}_p,$$

and W is the $(q-1) \times (q-1)$ shift matrix

$$W = \begin{pmatrix} 0 & 1 & 0 & \cdot & \cdot & 0 \\ 0 & 0 & 1 & \cdot & \cdot & 0 \\ \cdot & & & \cdot & & \cdot \\ \cdot & & & & \cdot & \cdot \\ \cdot & & & & & \cdot \\ 1 & 0 & 0 & \cdot & \cdot & 0 \end{pmatrix}.$$

Proof. We use Proposition 1 above. Note that

$$\begin{pmatrix} y & x \\ 0 & 1 \end{pmatrix} = \begin{pmatrix} 1 & x \\ 0 & 1 \end{pmatrix}\begin{pmatrix} y & 0 \\ 0 & 1 \end{pmatrix}.$$

Thus, since π is a representation, we need only compute

$$\pi\begin{pmatrix} g & 0 \\ 0 & 1 \end{pmatrix} \quad \text{and} \quad \pi\begin{pmatrix} 1 & x \\ 0 & 1 \end{pmatrix}.$$

Since ψ is one dimensional, we only need two indices for our matrix entries of π, rather than four. Proposition 1 gives

$$\pi(h)_{k,j} = \tilde\psi\left(g_k h g_j^{-1}\right),$$

where the coset representatives for $N\backslash G$ can be taken to be

$$g_{k+1} = \begin{pmatrix} g^k & 0 \\ 0 & 1 \end{pmatrix}, \quad k = 0, \ldots, q-2.$$

Here g is always a generator of the multiplicative group of $\mathbb{F}_q$.
We first consider the $(k+1, j+1)$ matrix entry of

$$\pi\begin{pmatrix} 1 & x \\ 0 & 1 \end{pmatrix},$$

which is found as follows:

$$\tilde\psi\left(\begin{pmatrix} g^k & 0 \\ 0 & 1 \end{pmatrix}\begin{pmatrix} 1 & x \\ 0 & 1 \end{pmatrix}\begin{pmatrix} g^{-j} & 0 \\ 0 & 1 \end{pmatrix}\right) = \tilde\psi\left(\begin{pmatrix} g^k & g^k x \\ 0 & 1 \end{pmatrix}\begin{pmatrix} g^{-j} & 0 \\ 0 & 1 \end{pmatrix}\right)$$

$$= \tilde\psi\left(\begin{pmatrix} g^{k-j} & g^k x \\ 0 & 1 \end{pmatrix}\right) = \begin{cases} 0, & \text{for } k \neq j, \\ \psi(g^k x), & \text{for } k = j. \end{cases}$$

This says that the off-diagonal entries of

$$\pi \begin{pmatrix} 1 & x \\ 0 & 1 \end{pmatrix}$$

vanish and the $(k+1)$st diagonal entry is $\psi(g^k x)$, $k = 0, 1, \ldots, q-2$. That is exactly the formula for $D(x)$.

To find the $(k+1, j+1)$ matrix entry of

$$\pi \begin{pmatrix} g & 0 \\ 0 & 1 \end{pmatrix},$$

we again use Proposition 1 and obtain

$$\tilde{\psi}\left(\begin{pmatrix} g^k & 0 \\ 0 & 1 \end{pmatrix} \begin{pmatrix} g & 0 \\ 0 & 1 \end{pmatrix} \begin{pmatrix} g^{-j} & 0 \\ 0 & 1 \end{pmatrix} \right) = \tilde{\psi}\left(\begin{pmatrix} g^{k+1-j} & 0 \\ 0 & 1 \end{pmatrix} \right)$$

$$= \begin{cases} 1, & \text{if } k+1 \equiv j \pmod{(q-1)}, \\ 0, & \text{otherwise.} \end{cases}$$

This is exactly the formula for the $(k+1, j+1)$ matrix entry of W. ∎

Exercise. Let $q = p^r$ and define $\mathrm{Aff}(\mathbb{Z}/q\mathbb{Z})$ to be the matrices

$$\begin{pmatrix} y & x \\ 0 & 1 \end{pmatrix},$$

such that $x, y \in \mathbb{Z}/q\mathbb{Z}$, and p does not divide y. Find the representations of $\mathrm{Aff}(\mathbb{Z}/q\mathbb{Z})$. See Angel et al. [1995].

In the next few chapters we will find that the affine group has applications to expander graphs, random number generators, and wavelets.

Properties of Induced Representations

Our next goal is to prove some of the properties of induced representations such as the Frobenius reciprocity law and investigate connections with the Selberg trace formula.

*Theorem 1. **The Frobenius Reciprocity Law.*** Suppose that H is a subgroup of the finite group G and that we have two representations: $\sigma : H \to GL(W)$ and $\pi : G \to GL(V)$. Define $\mathrm{Res}^G_H \pi$ to be the representation of H obtained by restricting π to H. Then when $I(\pi, \rho)$ denotes the space of intertwining operators, as before, we have

$$\dim_{\mathbb{C}} I\left(\pi, \operatorname{Ind}_H^G \sigma\right) = \dim_{\mathbb{C}} I\left(\sigma, \operatorname{Res}_H^G \pi\right),$$

which means that the multiplicity of π in $\rho = \operatorname{Ind}_H^G \sigma$ is equal to the multiplicity of σ in $\operatorname{Res}_H^G \pi = \pi|_H$. Then, as in Proposition 3 of Chapter 15, this means that the number of irreducible components of π and ρ which agree is the same as the number of irreducible components of σ and $\operatorname{Res}_H^G \pi$ which agree.

Proof. According to Proposition 3 of Chapter 15, we must do a computation involving characters. We will have to use the Frobenius formula for the character of the induced representation of course. The calculation is as follows:

$$
\begin{aligned}
\frac{1}{|G|} \langle \chi_\pi, \chi_\rho \rangle_G &= \frac{1}{|G|} \sum_{g \in G} \chi_\pi(g) \frac{1}{|H|} \sum_{s \in G} \overline{\tilde{\chi}_\sigma(sgs^{-1})} \\
&= \frac{1}{|G|} \frac{1}{|H|} \sum_{g \in G} \sum_{s \in G} \chi_\pi(sgs^{-1}) \overline{\tilde{\chi}_\sigma(sgs^{-1})} \\
&= \frac{1}{|G|} \frac{1}{|H|} \sum_{z \in G} \sum_{s \in G} \chi_\pi(z) \overline{\tilde{\chi}_\sigma(z)} \\
&= \frac{1}{|H|} \sum_{s \in H} \chi_\pi(s) \overline{\chi_\sigma(s)} = \frac{1}{|H|} \langle \chi_\pi|_H, \sigma \rangle_H.
\end{aligned}
$$

This is exactly the formula we needed to prove. ∎

Example. Let H be any subgroup of G. Suppose that σ is the trivial representation of H, that is, $\sigma(h) = 1$ for all $h \in H$. Then the Frobenius reciprocity formula says that if π is an irreducible representation of G, the multiplicity of π in $\operatorname{Ind}_H^G 1$ is the same as the multiplicity of 1 in $\operatorname{Res}_H^G \pi$. That is, the multiplicity of $\pi : G \to GL(V)$ in $\operatorname{Ind}_H^G 1$ is $\dim_{\mathbb{C}} V^H$, where V^H is the space of H-fixed vectors in V; that is,

$$V^H = \{v \in V \mid \pi(h)v = v, \text{ for all } h \in H\}.$$

So π occurs in $\operatorname{Ind}_H^G 1$ if and only if the trivial representation occurs in $\operatorname{Res}_H^G \pi$.

Thus when $H = \{e\}$, the multiplicity of π in the right regular representation is the degree of π since in this case $V^H = V$. This gives another proof of Part b of Lemma 2 in Chapter 15 – a result which was important in our proof of Theorem 2 of that same chapter – the basic result on Fourier analysis on G. It showed that the matrix entries of the $\pi \in \hat{G}$ span $L^2(G)$. Of course our proof of Part b of Lemma 2 was the same as our proof of the Frobenius reciprocity law.

Proposition 4. **Induction in Stages.** Suppose we have a chain of subgroups: $K \subset H \subset G$. And suppose that we have a representation $\sigma : K \to GL(U)$. Let $\rho = \operatorname{Ind}_K^H \sigma$ and let $\pi = \operatorname{Ind}_H^G \rho$. Then $\pi \cong \pi' = \operatorname{Ind}_K^G \sigma$. That is, the two representations π and $\pi' = \operatorname{Ind}_K^G \sigma$ are equivalent, which means that

$$\operatorname{Ind}_K^G \sigma \cong \operatorname{Ind}_H^G \left(\operatorname{Ind}_K^H s \right).$$

Proof. We can once more use the Frobenius character formula and the fact that characters determine representations up to equivalence. So we calculate first the character of the one-step induction π':

$$\chi_{\pi'}(g) = \frac{1}{|K|} \sum_{x \in G} \tilde{\chi}_\sigma (xgx^{-1}).$$

Next we compute the character of the two-step induction π:

$$\chi_\pi (g) = \frac{1}{|H|} \sum_{u \in G} \tilde{\chi}_\rho (ugu^{-1}) = \frac{1}{|H|} \sum_{u \in G} \frac{1}{|K|} \sum_{y \in H} \tilde{\chi}_\sigma (yugu^{-1}y^{-1})$$

$$= \frac{1}{|H|} \sum_{x \in G} \frac{1}{|K|} \sum_{y \in H} \tilde{\chi}_\rho (xgx^{-1}) = \chi_{\pi'}(g),$$

which proves the proposition. ∎

Notes. One can ask now whether all irreducible representations $\pi \in \hat{G}$ can be realized as induced representations. This does not seem to be true even for such a basic group as $GL(2, \mathbb{F}_q)$. However, Serre: [1977, p. 78, Theorem 20] proves a *theorem of Brauer*, which implies that each irreducible character of G is a linear combination with integer coefficients of monomial characters, where a *monomial character* is induced from a degree 1 character of some subgroup. However, the integer coefficients can be negative. Serre writes: [1977, p. 79] "Theorem 20 plays an essential role in many applications of representation theory: to a large extent, it gives a reduction of questions pertaining to an arbitrary character χ to the case where χ has degree 1 (hence comes from a character of a cyclic group). It is by this method, for example, that Brauer proved the Artin L-functions are *meromorphic* in the entire complex plane." See H. Heilbronn's article on Artin L-functions in Cassels and Fröhlich [1967, Chapter 8].

Although we will not be able to get all representations of the groups we are interested in by induction, there are generalizations of induction which can be used to obtain all the irreducible unitary representations of Lie groups like $SL(2, \mathbb{R})$. See Gurarie [1992, p. 308], Lang [1985, p. 183], and Mackey [1978b, p. 252]. Of course, here we will be interested in $SL(2, \mathbb{F}_q)$. The holomorphic induced representations for such groups are sketched in Gelfand, Graev, and Piatetski–Shapiro [1990, p. 158]. See also Piatetski–Shapiro [1983] and Naimark and Stern [1982].

Exercise. Show that

$$\operatorname{Ind}_H^G(\sigma \oplus \tau) = \operatorname{Ind}_H^G \sigma \oplus \operatorname{Ind}_H^G \tau.$$

Exercise.

a) Let

$$S = \bigcup_{i=1}^{s} C_i,$$

where C_i is a conjugacy class in the finite group G. Suppose S is symmetric; that is, $x \in S$ implies $x^{-1} \in S$. Consider the Cayley graph $X(G, S)$. Show that the eigenvalues of the adjacency matrix of this graph have the form

$$\lambda_\pi = \frac{1}{d_\pi} \sum_{s \in S} \chi_\pi(s), \quad \text{where } \pi \in \hat{G}, \text{ and } d_\pi = \text{degree of } \pi.$$

b) Work out the eigenvalues of $S(G, S)$ for a specific example such as $G = \operatorname{Aff}(q)$.

Chapter 17
The Finite $ax + b$ Group

The wavelet transform achieves the Holy Grail of complexity
theory (or simplicity theory). The transform is an $O(n)$
computation. But does it separate the true signal from noise?
G. Strang and T. Nguyen [1996, p. xiv]

Here we aim to consider the applications of the affine group Aff(q) of 2×2
matrices of the form

$$\begin{pmatrix} a & b \\ 0 & 1 \end{pmatrix}, \quad \text{for } a, b \in \mathbb{F}_q, \quad \text{with } a \neq 0.$$

This group acts on vectors $^t(x\ 1)$ via

$$\begin{pmatrix} a & b \\ 0 & 1 \end{pmatrix}\begin{pmatrix} x \\ 1 \end{pmatrix} = \begin{pmatrix} ax + b \\ 1 \end{pmatrix}$$

and thus it is sometimes called the finite $ax + b$ group.

$G = \text{Aff}(q)$ is an example of a *solvable* group. This means there is a sequence

$$\{I\} = G_0 \subset G_1 \subset \cdots \subset G_n = G$$

of subgroups of G with G_{i-1} normal in G_i and G_i/G_{i-1} abelian for $1 \leq i \leq n$.

Exercise. Show that Aff(q) is a solvable group.

The word "solvable" comes from the fact that a polynomial equation $f(x) = 0$
is solvable in radicals iff the Galois group is solvable.

We say that G is a *nilpotent* group if the G_i are as above except that G_i/G_{i-1} must also be in the center of G/G_{i-1}. In the next chapter we will consider an example of a nilpotent group.

Exercise. Define the *commutator (or derived) subgroup G'* of the finite group G to be the subgroup of G generated by all commutators $x^{-1}y^{-1}xy$, for $x, y \in G$. Show that G is solvable iff the sequence of subgroups

$$G \supset G' \supset G'' \supset \cdots \supset G^{(i)} \supset \cdots$$

terminates in the identity in a finite number of steps.
Hint. See Hall [1959, p. 139].

It is also useful to note that Aff(q) is the semidirect product of two of its subgroups. See Hall [1959, pp. 88–90] for more information.

Definition. We say that the group G is the *semidirect product $G = A \propto H$*, if A and H are two subgroups of G such that A is normal, $G = A \cdot H$, and $A \cap H = \{e\}$, where e is the identity of G.

If we assume, in addition that A is abelian, we can construct the irreducible representations of G by the *little group method of Mackey and Wigner* [see Serre, 1973, p. 62]. First note that since A is abelian its irreducible representations χ have degree 1. Moreover, $h \in H$ acts on $\chi \in \hat{A}$ via

$$(h\chi)(a) = \chi(h^{-1}ah), \quad \text{for } a \in A,$$

since A is normal. Let $H_\chi = \{h \in H \mid h\chi = \chi\}$. Set $G_\chi = A \cdot H_\chi$. Extend $\chi \in \hat{A}$ to a representation of G by $\chi(ah) = \chi(a)$, for $a \in A$, $h \in H_\chi$. Then let $\rho \in \hat{H}_\chi$. Obtain a representation of G_χ by $\rho(ah) = \rho(h)$, $a \in A$, $h \in H_\chi$. Then the irreducible representations of G are $\theta_{\chi,\rho} = \text{Ind}_{G_\chi}^{G}(\chi \otimes \rho)$, as χ runs over representatives of $\hat{A}$ modulo the equivalence by the action of H.

Exercise.
a) Prove the preceding statement.
b) Apply the little group method to obtain all the irreducible representations of Aff(q) with

$$A = \left\{ \begin{pmatrix} 1 & x \\ 0 & 1 \end{pmatrix} \right\} \quad \text{and} \quad H = \left\{ \begin{pmatrix} y & 0 \\ 0 & 1 \end{pmatrix} \right\}.$$

Compare with Proposition 2 of Chapter 16.

Hint.

a) See Serre [1977, p. 62].
b) Write

$$\begin{pmatrix} y & x \\ 0 & 1 \end{pmatrix} = (y, x).$$

We know that $\hat{A}$ is given by exponentials

$$\chi_a(1, x) = \exp\left(\frac{2\pi i \, \mathrm{Tr}(ax)}{p}\right).$$

Then $(y, 0) \in H$ acts on χ_a via.

$$(y, 0)\chi_a(1, x) = \chi_{y^{-1}a}(1, x).$$

Thus, there are only two representatives of $\hat{A}$ mod H, namely χ_0 and χ_1. We find that

$$G_0 = G \quad \text{and} \quad G_1 = A.$$

From χ_0, we get the 1-dimensional representations of the affine group. From $\rho \in \hat{A} = \hat{G}_1$, we get the $(q - 1)$-dimensional representation of the affine group as $\mathrm{Ind}_A^G \rho$.

We will consider three topics: Cayley graphs, random walks, and wavelets.

Application #1. Cayley Graphs

Recall the definition of Cayley graph in Chapter 3. First let's consider a small symmetric generating set for $G = \mathrm{Aff}(q)$, namely

$$S = \left\{ \begin{pmatrix} g & 0 \\ 0 & 1 \end{pmatrix}, \begin{pmatrix} g^{-1} & 0 \\ 0 & 1 \end{pmatrix}, \begin{pmatrix} 1 & 1 \\ 0 & 1 \end{pmatrix}, \begin{pmatrix} 1 & -1 \\ 0 & 1 \end{pmatrix} \right\},$$

where g generates $\mathbb{F}_q^*$. (1)

In order to compute the eigenvalues of the adjacency matrix for the Cayley graph $X(\mathrm{Aff}(\mathbb{F}_q), S)$, we need a theorem.

Just as in Theorem 2 of Chapter 3, we can describe the spectrum of the adjacency operator of the Cayley graph $X(G, S)$ using Fourier analysis on the group $G = \mathrm{Aff}(q)$. We simply need to recall the exercise after Lemma 2 of

Chapter 15, which writes the adjacency operator A of $X(G, S)$ as a sum of right regular representations R of G:

$$Af = \sum_{s \in S} Rf(s).$$

Then Lemma 2 of Chapter 15 block diagonalizes the right regular representation. Or, equivalently, we can use part 3 of Theorem 4 in Chapter 15 to see that the Fourier transform gives the explicit similarity transform needed to block diagonalize A. The discussion is completed by recalling the list of inequivalent irreducible representations of Aff(q). See Propositions 2 and 3 of Chapter 16. We have thus proved the following theorem. There are other ways to prove Theorem 1. See, for example, Chung [1989], Diaconis [1988, p. 49], Li [1992], and Lovász [1975].

*Theorem 1. **Partial Diagonalization of the Adjacency Operators of Cayley Graphs of Aff(q).*** The adjacency operator A of the Cayley graph $X(G, S)$ for $G = $ Aff(q) with symmetric generating set S is similar to a block diagonal matrix

$$A \cong \begin{pmatrix} R & 0 \\ 0 & \widetilde{M} \end{pmatrix}.$$

Here R is a $q \times q$ diagonal matrix with diagonal entries indexed by the 1-dimensional representations of Aff(q) coming from characters χ of the multiplicative group $\mathbb{F}_q^*$. The corresponding diagonal entry of R is

$$R_\chi = \sum_{\left(\begin{smallmatrix} y & x \\ 0 & 1 \end{smallmatrix}\right) \in S} \chi(y).$$

The matrix $\widetilde{M}$ is a block diagonal matrix with $q - 1$ identical blocks down the diagonal, each block given by the matrix

$$M_\pi = \sum_{s \in S} \pi(s).$$

Here π is the $(q - 1)$-dimensional representation of Aff(q) defined by

$$\pi \begin{pmatrix} y & x \\ 0 & 1 \end{pmatrix} = D(x)W^t, \quad \text{where } y = g^t, \quad \text{with } g \text{ a generator of } \mathbb{F}_q^*,$$

and

$$
D(x) = \begin{pmatrix}
\psi(x) & 0 & 0 & \cdot & \cdot & 0 \\
0 & \psi(gx) & \cdot & \cdot & \cdot & 0 \\
\cdot & \cdot & \cdot & & & \cdot \\
\cdot & \cdot & & \cdot & & \cdot \\
\cdot & \cdot & & & \cdot & \cdot \\
0 & 0 & \cdot & \cdot & \cdot & \psi(g^{q-2}x)
\end{pmatrix},
$$

$$
W = \begin{pmatrix}
0 & 1 & 0 & 0 & 0 & 0 \\
0 & 0 & 1 & 0 & 0 & 0 \\
\cdot & \cdot & \cdot & & & \cdot \\
\cdot & \cdot & & \cdot & & \cdot \\
\cdot & \cdot & & & \cdot & \cdot \\
1 & 0 & 0 & \cdot & \cdot & 0
\end{pmatrix},
$$

for $\psi(x) = \exp(2\pi i\, \mathrm{Tr}(x)/p)$ when $q = p^r$ and $\mathrm{Tr}(x) = x + x^p + \cdots + x^{p^{r-1}}$.

Corollary. Assume that p is a prime with $p \geq 5$. If S is given by formula (1), the 1-dimensional eigenvalues of the Cayley graph $X(\mathrm{Aff}(p), S)$ have the form

$$
2\left\{ 1 + \cos\frac{2\pi a}{p-1} \right\} = 4\,\cos^2\frac{\pi a}{p-1} \geq 0, \quad \text{for } a = 0, 1, \ldots, p-2.
$$

The eigenvalue 0 occurs at $a = (p-1)/2$.

Using Matlab to generate the $(p-1)$-dimensional eigenvalues of $X\,(\mathrm{Aff}(p), S)$ as well as the histograms of the complete set of eigenvalues, we obtain Figure II.1. This should be compared with the figures in Lafferty and Rockmore [1992] who consider spectra of adjacency operators of Cayley graphs associated to $G = SL(2, \mathbb{F}_q)$ and generating sets such as

$$
S = \left\{ \begin{pmatrix} 1 & \pm 1 \\ 0 & 1 \end{pmatrix}, \pm \begin{pmatrix} 0 & 1 \\ -1 & 0 \end{pmatrix} \right\}.
$$

Exercise. Do the 1-dimensional eigenvalues of $X(\mathrm{Aff}(p), S)$ with S as in formula (1) satisfy the Ramanujan bound $|\lambda| \leq 2\sqrt{q}$?

Exercise. Draw a figure analogous to Figure II.1 for the next 25 primes.

In a later chapter, we will consider other Cayley graphs for the affine group with larger sets of generators. However, we invite the reader to consider inventing her or his own generating sets.

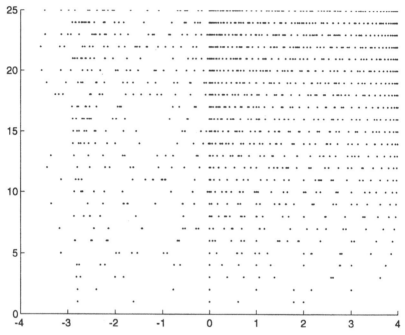

Figure II.1. Spectra of $X(\text{Aff}(p), S)$ with S as in formula (1) for 25 consecutive primes starting at $p = 5$.

Application #2. Random Number Generators

We discussed random number generators in Chapter 6. In particular, we noted that Chung, Diaconis, and Graham [1987] considered the generator

$$X_{j+1} \equiv a_j X_j + b_j \pmod{n}, \tag{2}$$

where n is an odd prime, $a_j = 1$, and b_j takes values ± 1 with equal probability. They showed that in this case convergence to uniform takes something like cn^2 steps. They also show that "the situation is changed drastically" if we replace the condition $a_j = 1$ with $a_j = 2$. They take the values of b_j in $\{0, 1, -1\}$ (each with equal probability). Then the upper bound on the number of steps to uniform is $c \log n \log \log n$ steps. At the end of Chapter 6 we gave a short summary of these results.

Here we follow some work of Diaconis [1988, p. 35] and Diaconis and Shahshahani [1986], who consider random number generators of the form

$$X_{k+1} \equiv a_k X_k + b_k \pmod{n}.$$

We take $(a_k b_k)$ to be the top row of an element of the affine group acting on vectors

$$Y_k = \begin{pmatrix} X_k \\ 1 \end{pmatrix}.$$

Clearly one can study this random number generator using similar methods to those of Chapter 6, except that now we need to think about the affine group.

One needs a generalization of Lemma 2 of Chapter 6 – the upper bound lemma.

*Lemma 1. **The Upper Bound Lemma.*** Suppose that p is a probability function on a finite group G and that u is the uniform probability on G. Then

$$\|p - u\|^2 \leq \sum_{1 \neq \pi \in \hat{G}} d_\pi \operatorname{Tr}(\hat{p}(\pi)\hat{p}(\pi)^*).$$

Proof. The proof is left as an exercise.
Hint. Imitate the proof of Lemma 2 in Chapter 6. ∎

It also makes sense to replace b_k in (2) with $b_k X_{k-1}$. Then the vector Y_k is replaced by

$$\begin{pmatrix} X_k \\ X_{k-1} \end{pmatrix}.$$

Diaconis and Shahshahani [1986, p. 191] use this result but not on the affine group, instead on $GL(2, \mathbb{F}_p)$, to show that the joint distribution of X_k, X_{k-1} converges to uniform in $cp^2 \log p$ steps. They expect that the truth is $c(\log p)^3$ or less steps.

Exercise. Consider the random walk on $\operatorname{Aff}(p)$, p prime, with probability function defined by

$$p \begin{pmatrix} a & b \\ 0 & 1 \end{pmatrix} = \frac{1}{9} \delta_{S_1}(a) \, \delta_{S_2}(b),$$

where $S_1 = \{1, g, g^{-1}\}$, g is a primitive root (mod p), and $S_2 = \{0, 1, -1\}$. Show that

$$\|p^{(k)} - u\|_1^2 \leq p^2 \exp\left(\frac{-\beta k}{p^2}\right), \quad \text{for } k > cp^2 \log p. \tag{3}$$

Conclude that $cp^2 \log p$ steps suffice to make $p^{(k)}$ close to uniform on $\operatorname{Aff}(p)$.

Hint. See Diaconis [1988, p. 35] for more information. Another reference is Maria Zack's thesis [1989].

Questions about the Last Exercise. The last result would seem to say that the idea of using the affine group to improve the speed of the random number generator is a bust, since this estimate is even worse than the one at the end of Chapter 6. (But perhaps a saving grace is that we can actually generate pairs of random numbers (a, b) this way with $a \not\equiv 0 \pmod{p}$?) However, Hildebrand [1993] has shown that, under restrictive hypotheses on the prime p, $c(\log p)^2$ steps suffice to reach uniform for the random process $X_{n+1} \equiv a_n X_n + b_n \pmod{p}$, assuming that one does not take $a_n = 1$ always or $b_n = 0$ always or that both a_n and b_n take only one value. Thus an improvement is actually attained.

Application #3. Wavelets

Wavelets are a new way to speed up signal processing. They are closely related to filter banks. Most of the work concerns real continuous wavelets. Some references are Ingrid Daubechies [1992], M. Holschneider [1995a,b], G. Strang and T. Nguyen [1996], and Strichartz [1994]. Here we will only consider a finite analogue of the real wavelet. Our main reference is Flornes et al. [1994].

Let $G = \mathrm{Aff}(p)$, for prime p.

Notation. We will write

$$\begin{pmatrix} a & b \\ 0 & 1 \end{pmatrix} = (a, b), \quad \text{for } a, b \in \mathbb{F}_p, a \neq 0. \tag{4}$$

We have a representation $\pi : \mathrm{Aff}(p) \to GL(V)$, $V = L^2(\mathbb{F}_p)$ defined as follows:

$$\pi(a, b)f(x) = f(a^{-1}(x - b)) = f_{a,b}(x), \quad \text{for } x \in \mathbb{F}_p. \tag{5}$$

We can also rewrite this in terms of the translation and dilation operators defined at the end of Chapter 2. Recall that these were given by

$$T_b f(x) = f(x + b) \quad \text{and} \quad D_a f(x) = f(ax), \quad \text{for } a, b \in \mathbb{F}_p, a \neq 0. \tag{6}$$

Then

$$\pi(a, b) = T_b D_{a^{-1}}. \tag{7}$$

Exercise.

1) Prove that $\pi(a, b) = T_b D_{a^{-1}}$ does give a representation of the affine group.
2) Show that this representation is unitary.
3) Show that if we restrict this representation π to the subspace

$$E = \{f \in L^2(\mathbb{F}_p) \mid \langle f, 1 \rangle = 0\}, \tag{8}$$

we get an irreducible representation. Here the inner product is $\langle f, 1 \rangle = \sum_{x \in \mathbb{F}_p} f(x)$.

4) Finally show that this representation is the same as the $(p-1)$-dimensional induced representation that we called π in Proposition 2 of Chapter 16.

Definition. Take $\psi \in V = L^2(\mathbb{F}_p)$ such that $\langle \psi, 1 \rangle = 0$. Call ψ a "wavelet." Then the *wavelet transform* of $f \in V$ is defined for a, b in $\mathbb{F}_p, a \neq 0$, by

$$\mathscr{W}f(a, b) = \sum_{x \in \mathbb{F}_p} f(x) \overline{\psi(a^{-1}(x - b))} = \langle f, \psi_{a,b} \rangle, \tag{9}$$

where $\psi_{a,b}$ is defined by (5).

*Theorem 2. (**Inversion #1**).* Assume our wavelet ψ is in E defined by (8) above; that is, $\langle \psi, 1 \rangle = 0$. Given $f \in E$, we have the following inversion formula for the wavelet transform defined by (9):

$$f(x) = \frac{1}{c_\psi} \sum_{(a, b) \in G} \mathscr{W}f(a, b) \psi_{a,b}(x),$$

where $c_\psi = p \langle \psi, \psi \rangle$.

Proof. It suffices to prove the result for a basis of E as defined in (8). So take $e_d(x) = \exp(2\pi i dx/p)$, for $d \in \mathbb{F}_p^*$. Then the sum on the right-hand side of the formula in our theorem is

$$\Sigma = \sum_{(a, b) \in G} \sum_{y \in \mathbb{F}_p} e_d(y) \overline{\psi_{a,b}(y)} \psi_{a,b}(x)$$

$$= \sum_{(a, b) \in G} \langle e_d, \psi_{a,b} \rangle \psi_{a,b}(x) = \frac{1}{p} \sum_{(a, b) \in G} \langle \hat{e}_d, \hat{\psi}_{a,b} \rangle \psi_{a,b}(x).$$

Here we have used the Plancherel theorem in Chapter 2.

Next note that by the exercise on the translation and dilation operators at the end of Chapter 2, we have, upon writing $\mathscr{F}$ rather than ^ for the DFT on the additive group of $\mathbb{F}_p$:

$$\langle \hat{e}_d, \hat{\psi}_{a,b} \rangle = \langle \mathscr{F}e_d, \mathscr{F}T_{b^{-1}}D_{a^{-1}}\psi \rangle = \langle \mathscr{F}e_d, \overline{e_b}D_a\mathscr{F}\psi \rangle.$$

Now, using Table I.1 of Chapter 2, we have $\mathscr{F}e_d = p\delta_d$. Thus Σ becomes

$$\Sigma = \sum_{(a,b)\in G} \langle \delta_d, \overline{e_b}D_a\mathscr{F}\psi \rangle \psi_{a,b}(x) = \sum_{(a,b)\in G} e_b(d)\overline{D_a\mathscr{F}\psi(d)}\psi_{a,b}(x)$$

$$= \sum_{(a,b)\in G} e_b(d)\overline{\hat{\psi}(ad)}\psi(a^{-1}(x-b)).$$

Now change variables via $c = a^{-1}(x-b)$. This gives

$$\Sigma = \sum_{(a,c)\in G} \exp\left(\frac{2\pi i d(x-ac)}{p}\right)\overline{\hat{\psi}(ad)}\,\psi(c)$$

$$= \exp\left(\frac{2\pi i dx}{p}\right)\sum_{a\in\mathbb{F}_p^*}\overline{\hat{\psi}(ad)}\sum_{c\in\mathbb{F}_p}\exp\left(\frac{-2\pi i dac}{p}\right)\psi(c)$$

$$= \exp\left(\frac{2\pi i dx}{p}\right)\sum_{a\in\mathbb{F}_p^*}\overline{\hat{\psi}(ad)}\,\hat{\psi}(ad).$$

Since, by hypothesis on the wavelet $\hat{\psi}(0) = 0$, we have

$$\Sigma = e_d(x)\langle \hat{\psi}, \hat{\psi} \rangle = e_d(x)p\langle \psi, \psi \rangle.$$

This completes the proof. ∎

Next we change our assumption that $\hat{\psi}(0) = 0$, and prove another synthesis theorem.

*Theorem 3. (**Inversion #2**).* Let ψ be a wavelet such that

$$(p-1)|\hat{\psi}(0)|^2 = \sum_{k=1}^{p-1}|\hat{\psi}(k)|^2.$$

Then the map $f \mapsto \mathscr{W}f$ is an isometry on $f \in L^2(\mathbb{F}_p)$ up to the constant c_ψ and the inversion formula for the wavelet transform $\mathscr{W}f$ defined by (9) is

$$f(x) = \frac{1}{c_\psi}\sum_{(a,b)\in G}\mathscr{W}f(a,b)\psi_{a,b}(x),$$

where $c_\psi = (p-1)|\hat{\psi}(0)|^2.$

Proof. Again it suffices to check the formula for $e_d(x)$, for all $d \in \mathbb{F}_p$, but now the third to last line of our preceding proof will not work. However, everything up to that point is valid. So we have

$$\Sigma = \sum_{(a,b) \in G} \mathscr{W}e_d(a,b)\,\psi_{a,b}(x) = e_d(x) \sum_{a \in \mathbb{F}_p^*} \overline{\hat{\psi}(ad)}\,\hat{\psi}(ad).$$

There are two cases.

Case 1. $d = 0$. Then $\Sigma = e_d(x)(p-1)|\hat{\psi}(0)|^2$ and we are done.

Case 2. $d \neq 0$. Here we use the condition on the wavelet to see that the theorem holds. ∎

See Flornes et al. [1994] for illustrations of the uses of the finite wavelet transform.

Some Information about the Wavelet Transform on the Real Line

Taking some mother wavelet ψ such that $\int_{\mathbb{R}} \psi(t)dt = 0$, the continuous wavelet transform is

$$\mathscr{W}f(a,b) = |a|^{-1/2} \int_{\mathbb{R}} f(t)\psi\left(\frac{t-b}{a}\right)dt.$$

Examples of useful wavelets are:

Mexican hat: $\psi(t) = (1-t^2)e^{-t^2/2}$;

$$\text{Haar: } \psi(t) = \begin{cases} 1, & 0 \leq t < 1/2; \\ -1, & 1/2 \leq t < 1; \\ 0, & \text{otherwise.} \end{cases}$$

Then one forms a family of functions $\psi_{j,k}(x) = 2^{j/2}\psi(2^j x - k)$. One wants to expand a function f on $\mathbb{R}$ in a Haar series

$$f = \sum_{j \in \mathbb{Z}} \sum_{k \in \mathbb{Z}} \langle f, \psi_{j,k} \rangle \psi_{j,k}(x).$$

There are problems as usual. Of course f must satisfy some hypothesis such as square integrability. And the convergence of the Haar series may not be pointwise. Nor does it necessarily allow term-by-term integration. See Strichartz [1994] for more information.

The Haar wavelet has been around for a long time but has not been used much because it is discontinuous. Thus the approximations it gives are bad for smooth functions.

The Daubechies wavelets are much better for this sort of thing. They are compactly supported and differentiable a finite number of times. Although they are not so easily defined, their graphs are becoming almost as familiar as those of sines and cosines – at least to readers of wavelet books. They involve an associated scaling function ϕ. One needs the equations

$$\psi(x) = \phi(2x) - \phi(2x - 1),$$
$$\phi(x) = \phi(2x) + \phi(2x - 1).$$

These are high- and low-pass filters. They lead to a convergent product for the Fourier transform of the scaling function. See Strichartz [1994, p. 147].

Chapter 18
The Heisenberg Group

Nowadays the textbooks speak without exception of
Heisenberg's matrices, Heisenberg's commutation law, and
Dirac's field quantization.
   In fact, Heisenberg knew at that time very little of matrices
and had to study them.

M. Born [1978, p. 219]

The history of physics in the first half of the twentieth century
is a history of argument on the heroic scale, of scientific
disputes carried on year after year in letters, conferences,
late-night encounters in which champions never abandoned a
position until pure intellectual defeat was at last undeniable.
But for some reason men willing to argue forever whether it
was nature or only our knowledge which was uncertain in
Heisenberg's uncertainty principle, soon dropped the more
easily answered question of what Heisenberg did during the
war, and why he did it.

T. Powers [1993, p. 470]

The finite *Heisenberg group* is

$$\mathrm{Heis}(q) = \left\{ \begin{pmatrix} 1 & x & z \\ 0 & 1 & y \\ 0 & 0 & 1 \end{pmatrix} \middle| x, y, z \in \mathbb{F}_q \right\}. \tag{1}$$

It is a nilpotent group, which is defined near the beginning of Chapter 17.
   The real Heisenberg group is extremely important in many areas of applied
mathematics – from Heisenberg's uncertainty principle to radar cross-ambiguity
functions. Some call Heis $(\mathbb{R})$ the Heisenberg–Weyl group. Some references are:

Blahut et al. [1991], Grünbaum et al. [1992], Perelomov [1986], and Schempp [1986a,b]. The Heisenberg group over a finite field $\mathbb{F}_q$ or a finite ring $\mathbb{Z}/q\mathbb{Z}$ has just begun to see some applications. We will discuss these at the end of this chapter. See also Schempp [1986b] and the article of L. Auslander, I. Gertner, and R. Tolimieri in Grünbaum et al. [1992].

Exercise. Prove that a group is nilpotent if and only if its lower central series terminates. The lower central series is found as follows: Let $[H, K]$ be the subgroup generated by the commutators $[h, k] = h^{-1}k^{-1}hk$, for $h \in H, k \in K$. Set $G_1 = G$ and $G_i = [G, G_{i-1}]$.
Hint. See Hall [1959, Ch. 10].

   Our first goal is to find the irreducible unitary representations of Heis(q). We could attempt to use the little group method of the preceding chapter. But instead we will just write them down, following Schempp [1986b].

Exercise. Use the little group method to find the irreducible unitary representations of the group

$$E = \left\{ \begin{pmatrix} y & x & u \\ 0 & 1 & 0 \\ 0 & 0 & 1 \end{pmatrix} \middle| \; x, y, u \in \mathbb{F}_q, y \neq 0 \right\}.$$

This group is of interest thanks to Nancy Allen's thesis [1996], which creates some Ramanujan graphs similar to those of the next section.

Conjugacy Classes of the Heisenberg Group

Note that

$$\begin{pmatrix} 1 & x & z \\ 0 & 1 & y \\ 0 & 0 & 1 \end{pmatrix} \begin{pmatrix} 1 & a & c \\ 0 & 1 & b \\ 0 & 0 & 1 \end{pmatrix} \begin{pmatrix} 1 & x & z \\ 0 & 1 & y \\ 0 & 0 & 1 \end{pmatrix}^{-1}$$

$$= \begin{pmatrix} 1 & x & z \\ 0 & 1 & y \\ 0 & 0 & 1 \end{pmatrix} \begin{pmatrix} 1 & a & c \\ 0 & 1 & b \\ 0 & 0 & 1 \end{pmatrix} \begin{pmatrix} 1 & -x & xy-z \\ 0 & 1 & -y \\ 0 & 0 & 1 \end{pmatrix}$$

$$= \begin{pmatrix} 1 & a & c+(bx-ay) \\ 0 & 1 & b \\ 0 & 0 & 1 \end{pmatrix}.$$

It follows that we have two types of conjugacy classes in Heis(q):

I.

$$\left\{ \begin{pmatrix} 1 & 0 & c \\ 0 & 1 & 0 \\ 0 & 0 & 1 \end{pmatrix} \right\},$$

q classes each having one element, and

II.

$$\left\{ \begin{pmatrix} 1 & a & c \\ 0 & 1 & b \\ 0 & 0 & 1 \end{pmatrix} \middle| c \in \mathbb{F}_p \right\}, \quad (a, b) \neq (0, 0),$$

$q^2 - 1$ classes, each having q elements.

Now we seek the representations of Heis(q). There are

Two Types of Representations of Heis(q)

I. q^2 one-dimensional representations defined, for $a, b \in \mathbb{F}_q$, by

$$\mu_{a,b} \begin{pmatrix} 1 & x & z \\ 0 & 1 & y \\ 0 & 0 & 1 \end{pmatrix} = \psi(ax + by), \quad \text{where } \psi(x) = \exp\left(\frac{2\pi i \operatorname{Tr}(x)}{p} \right). \quad (2)$$

II. $q - 1$ representations which are q dimensional defined for nonzero $s \in \mathbb{F}_q$ by setting

$$A = \left\{ \begin{pmatrix} 1 & 0 & z \\ 0 & 1 & y \\ 0 & 0 & 1 \end{pmatrix} \middle| y, z \in \mathbb{F}_q \right\}$$

and

$$\pi_s = \operatorname{Ind}_A^G \psi_s, \quad \text{where } \psi_s \begin{pmatrix} 1 & 0 & z \\ 0 & 1 & y \\ 0 & 0 & 1 \end{pmatrix} = \psi(sz) \quad (3)$$

$$= \exp\left(\frac{2\pi i \operatorname{Tr}(sz)}{p} \right).$$

Table II.4. *Character table for the Heisenberg group*
$Heis(q)$, $q = p^r$, $p = \text{prime}$

	$\left\{\begin{pmatrix} 1 & 0 & c \\ 0 & 1 & 0 \\ 0 & 0 & 1 \end{pmatrix}\right\}$	$\left\{\begin{pmatrix} 1 & a & * \\ 0 & 1 & b \\ 0 & 0 & 1 \end{pmatrix}\right\}$
# Classes	q	$q^2 - 1$
# Elements in class	1	q
$\mu_{f,g}$, $f, g \in \mathbb{F}_q$	1	$\psi(fa + gb)$
π_s, $s \in \mathbb{F}_q$, $s \neq 0$	$q\psi(sc)$	0

Note that A is a normal abelian subgroup of G and G/A is abelian.
This gives us the character table shown in Table II.4.

Exercise.
a) Check that the representations of $Heis(q)$ defined above are irreducible.
b) Check that the representations of $Heis(q)$ defined above are pairwise inequivalent.
c) Check that we found all the irreducible unitary representations of $Heis(q)$.

Exercise. Show that the matrix entries for the induced representation defined in (3) are given as follows. First note

$$\begin{pmatrix} 1 & x & z \\ 0 & 1 & y \\ 0 & 0 & 1 \end{pmatrix} = \begin{pmatrix} 1 & 0 & z \\ 0 & 1 & 0 \\ 0 & 0 & 1 \end{pmatrix} \begin{pmatrix} 1 & 0 & 0 \\ 0 & 1 & y \\ 0 & 0 & 1 \end{pmatrix} \begin{pmatrix} 1 & x & 0 \\ 0 & 1 & 0 \\ 0 & 0 & 1 \end{pmatrix}.$$

Then show

i)

$$\pi_s \begin{pmatrix} 1 & 0 & z \\ 0 & 1 & 0 \\ 0 & 0 & 1 \end{pmatrix} = \psi(sz)\, I, \quad \text{where } I \text{ is the } q \times q \text{ identity matrix.}$$

ii)

$$\pi_s \begin{pmatrix} 1 & 0 & 0 \\ 0 & 1 & y \\ 0 & 0 & 1 \end{pmatrix} = D(sy), \quad \text{where } D(sy) \text{ is the diagonal } q \times q \text{ matrix:}$$

$$D(sy) = \begin{pmatrix} \psi(0) & 0 & 0 & \cdots & 0 \\ 0 & \psi(syb_1) & 0 & \cdots & 0 \\ \vdots & \vdots & \vdots & \cdots & \vdots \\ 0 & 0 & 0 & \cdots & 0 \\ 1 & 0 & 0 & \cdots & \psi(syb_{q-1}) \end{pmatrix},$$

where $\mathbb{F}_q = \{0, b_1, \ldots, b_{q-1}\}$.

iii)

$$\pi_s \begin{pmatrix} 1 & x & 0 \\ 0 & 1 & 0 \\ 0 & 0 & 1 \end{pmatrix} = W(x),$$

where $W(x)$ is a $q \times q$ permutation matrix. Here $W(x)_{a,b} = \delta_b(a + x)$.

iv) For example, if $q = p = $ prime show that $W(x) = W^x$, where

$$W = \begin{pmatrix} 0 & 1 & 0 & \cdots & 0 \\ 0 & 0 & 1 & \cdots & 0 \\ \vdots & \vdots & \vdots & \cdots & \vdots \\ 0 & 0 & 0 & \cdots & 1 \\ 1 & 0 & 0 & \cdots & 0 \end{pmatrix}.$$

v) Consider the example $\mathbb{F}_q = \mathbb{F}_4 = \mathbb{F}_2(\alpha) = \{0, 1, \alpha, \alpha + 1\}$, where $\alpha^2 + \alpha + 1 = 0$. Show that in this case representing $x \in \mathbb{F}_4$ as $x = x_0 + x_1\alpha$, where $x_i \in \{0, 1\}$, we have

$$\pi_s \begin{pmatrix} 1 & x & 0 \\ 0 & 1 & 0 \\ 0 & 0 & 1 \end{pmatrix} = W_1^{x_0} W_\alpha^{x_1}, \quad \text{where } W_1 = \begin{pmatrix} 0 & 1 & 0 & 0 \\ 1 & 0 & 0 & 0 \\ 0 & 0 & 0 & 1 \\ 0 & 0 & 1 & 0 \end{pmatrix} \quad \text{and}$$

$$W_\alpha = \begin{pmatrix} 0 & 0 & 1 & 0 \\ 0 & 0 & 0 & 1 \\ 1 & 0 & 0 & 0 \\ 0 & 1 & 0 & 0 \end{pmatrix}.$$

Exercise. Find the analogue of Table II.4 if we replace the finite field $\mathbb{F}_q$ with the finite ring $\mathbb{Z}/q\mathbb{Z}$.

Next we want to consider a few applications of the Heisenberg group. We will only provide sketches of the results.

Application #1. Random Walks on the Heisenberg Group

Maria Zack [1989] envisions applications of these results to random number generators. The idea is similar to that of the previous chapter. Let p be an odd prime and let (a_i, b_i, c_i) be independent, identically distributed, random variables on $\mathbb{F}_p^3$. Consider the simple random number generator

$$X'_n = X'_{n-1} + b_n \pmod p$$

and then the interlaced random number generator

$$X_n = X_{n-1} + a_n X'_{n-1} + c_n \pmod p.$$

One can analyze the properties of this random number generator via random walks on the Heisenberg group since

$$\begin{pmatrix} X_n \\ X'_n \\ 1 \end{pmatrix} = \begin{pmatrix} 1 & a_n & c_n \\ 0 & 1 & b_n \\ 0 & 0 & 1 \end{pmatrix} \begin{pmatrix} X_{n-1} \\ X'_{n-1} \\ 1 \end{pmatrix}.$$

Zack finds that if the probabilities are all supported on $\{0, 1, -1\}$, then the behavior is very similar to that of the corresponding random walk on the affine group.

Diaconis and Saloff-Coste [1994, 1995] consider generalizations of this random walk. But in the special case of interest here, they look at the Cayley graph $X(G, S)$, where $G = \text{Heis}(p)$ and S is a symmetric generating set of G such as

$$S = \{I, X^{\pm 1}, Y^{\pm 1}\}, \quad \text{where } X = \begin{pmatrix} 1 & 1 & 0 \\ 0 & 1 & 0 \\ 0 & 0 & 1 \end{pmatrix}, \quad Y = \begin{pmatrix} 1 & 0 & 0 \\ 0 & 1 & 1 \\ 0 & 0 & 1 \end{pmatrix}. \quad (4)$$

Note that here we are allowing the graph to have loops. In fact Diaconis and Saloff-Coste allow $\mathbb{F}_p$ to be replaced by $\mathbb{Z}/n\mathbb{Z}$.

Let β be the second largest eigenvalue of $M = |S|^{-1}A$ in absolute value, where A is the adjacency matrix of the Cayley graph $X(G, S)$. And let γ be the diameter of the Cayley graph. For the generating set above $\gamma \approx p$. Then for each integer $N \geq 2$ with $|S| \leq N$, there are positive constants a, A, depending on N, such that

$$a\gamma^{-2} \leq -\log \beta \leq A\gamma^{-2}.$$

This allows one to bound the time for the random walk associated to the graph to reach uniform in a similar way to that which we used in Chapter 6. The result is that the time to uniform is $\approx p^2$ for S as above.

Exercise. Study the spectrum of the adjacency matrix of $X(\text{Heis}(p), S)$ with S as in (4) above for some small primes p. Produce a figure similar to Figure II.1 of Chapter 17.

Application #2. Signal Processing and Radar

Schempp [1986a,b] considers applications of the finite and real Heisenberg groups to radar systems. (Radar stands for radio detection and ranging). See also the paper of W. Miller in Blahut et al. [1991, pp. 66–168]. A radar transmitter sends out pulses of electromagnetic energy of wavelength a few centimeters. These pulses have large amplitude and short duration. An object (such as an airplane or storm) in the path of the beam scatters the radiation and some will return to the antenna. The modern system uses the same antenna for transmission as well as reception.

Consider the signal pulse to be $f(t)e^{2\pi i \omega t}$ at time t. Then $\|f\|_2^2$ is the total input signal energy. Assuming that the received signal is reflected from a stationary object, the echo signal is delayed in time. To find the distance to the object, one needs to know the time x at which the echo arrives at the receiver. The distance is $cx/2$ if c is the velocity of the pulse. If the object is moving, the carrier frequency ω of the echo signal also differs from that of the transmitted pulse thanks to the Doppler effect. If the Doppler frequency shift can be measured, one can compute the component of the object's velocity in the direction of the antenna.

The mixed *radar cross-ambiguity function* is

$$H(f, g)(x, y) = \int_{\mathbb{R}} f(t + x/2)\overline{g(t - x/2)}\, e^{2\pi i y t}\, dt.$$

This function is the basis for the search for the optimal waveform to use in order to distinguish two echo signals by means of their arrival times x and the Doppler shifts y of their carrier frequencies from a common reference value.

$H(f, g)(x, y)$ is a matrix entry for the *Schrödinger representation* of the Heisenberg group. For $s \in \mathbb{R}$, this representation $\pi_s(x, y, z)$ acts on a function f in the Hilbert space $L^2(\mathbb{R})$ by

$$[\pi_s(x, y, z)f](t) = \exp(2\pi i s(z + t y))f(t + x).$$

Exercise. Show that if π_s is the induced representation of the finite Heisenberg group defined by (3) above, then it acts on functions on $\mathbb{F}_q$ in the same way as the Schrödinger representation above.

The radar uncertainty principle says that there is an ambiguity in determining the distance and relative velocity of the object. This is related to the uncertainty principle in quantum mechanics, which leads Schempp to make use of the Heisenberg group. Using the Hermite functions, which give a Hilbert basis of $L^2(\mathbb{R})$, he computes the radar auto-ambiguity function $H(f, f; x, y)$ in terms of Laguerre–Weber functions $L_m(x)$.

Auslander, Gertner, and Tolimieri, writing in Grünbaum et al. [1992, pp. 21–35], consider finite analogues of this theory for the Heisenberg group over $\mathbb{Z}/n\mathbb{Z}$, where n need not be prime. They consider finite Heisenberg group wavelets coming from a mother wavelet ψ in $L^2(\mathbb{Z}/n\mathbb{Z})$ via

$$\psi_{ab}(t) = \psi(t + a) \exp\left(\frac{2\pi i b t}{n}\right).$$

The *finite cross-ambiguity function* of $f, g \in L^2(\mathbb{Z}/n\mathbb{Z})$ is

$$H(f, g)(x, y) = \sum_{t \in \mathbb{Z}/n\mathbb{Z}} f(t) g_{xy}(t) = \langle f, g_{xy} \rangle, \quad \text{with } g_{xy} \text{ as above.} \qquad (5)$$

Exercise. Prove the following properties of the finite cross-ambiguity function $H(f, g)$ defined by Equation (5):

1)

$$\|H(f, g)\|_2^2 = n \|f\|_2^2 \|g\|_2^2.$$

2)

$$H(f, g)(x, y) = \frac{1}{n} \exp\left(\frac{-2\pi i x y}{m}\right) H(\hat{f}, \hat{g})(-y, x),$$

where $\hat{f}$ denotes the Fourier transform of f.

3) The finite cross-ambiguity function is a matrix entry of a representation of the finite Heisenberg group Heis $(\mathbb{Z}/n\mathbb{Z})$.

Application #3. Ramanujan Graphs

Perla Myers [1995 and preprint], Archie Medrano [1997], and Michelle DeDeo [1998] have examined various Cayley graphs for the Heisenberg group. Some of these graphs are Ramanujan. For example, consider the following exercise.

Exercise.

a) Consider the Cayley graph $X(\text{Heis}(p), S)$ with

$$S = \{(x, y, 0) \mid x, y \in \mathbb{F}_p, x \cdot y = 0, x \text{ and } y \text{ not both } 0\},$$

where

$$(x, y, z) = \begin{pmatrix} 1 & x & z \\ 0 & 1 & y \\ 0 & 0 & 1 \end{pmatrix}.$$

Is this a connected graph? Is it a Ramanujan graph?

b) Compare it with the graph $X(\mathbb{F}_p^3, S')$, where

$$S' = \{(x, y, 0) \mid x, y \in \mathbb{F}_p, x \cdot y = 0, x \text{ and } y \text{ not both } 0\}.$$

Chapter 19

Finite Symmetric Spaces – Finite Upper Half Plane H_q

> Analysis on Symmetric spaces, or more generally
> homogeneous spaces $[G/K]$ of semisimple Lie groups, is a
> subject that has undergone a vigorous development in recent
> years, and has become a central part of contemporary
> mathematics. This is only to be expected, since homogeneous
> spaces and group representations arise naturally in diverse
> contexts ranging from Number Theory and Geometry to
> Particle Physics and Polymer Chemistry.
> R. Gangolli and V.S. Varadarajan [1988, Preface]

In this chapter we want to consider finite analogues of the symmetric spaces G/K mentioned above. These are related to the concept of Gelfand pair (G, K). See also Gelfand [1988], Helgason [1968, 1984], Selberg [1989], Terras [1985 and 1988], and Vilenkin [1968]. In particular, we will study a finite analogue of the Poincaré upper half plane $H \cong G/K$, with $G = SL(2, \mathbb{R})$ and $K = SO(2)$. We will replace $\mathbb{R}$ with the finite field $\mathbb{F}_q$. Such quotients have been considered by Angel et al. [1992], Celniker et al. [1993], Soto-Andrade [1987], and Terras [1991, 1996]. Other quotients G/B, for B the Borel subgroup of upper triangular matrices, have been considered by Krieg [1990] and Stanton [1990], for example.

One goal is to obtain a Fourier transform on $K \backslash G / K$ called the spherical transform (see Corollary 1 of Theorem 2 in Chapter 20 and formula (15) of Chapter 23). The spherical transform has many of the properties of the Fourier transform on abelian groups since it is complex valued rather than matrix valued. The spherical transform of a K-bi-invariant function f on G is the inner product of f with a spherical function. Thus we will need to study these spherical functions in Chapter 20. The Selberg trace formula for K-bi-invariant functions on G will also be considered in Chapters 22 and 23.

Harmonic analysis on finite symmetric spaces has not had many treatments so far, especially in comparison with the continuous symmetric spaces such as

$\mathbb{R}^n$, the sphere, or the Poincaré upper half plane. This makes the theory more difficult simply because the spherical functions are not always so familiar. Every undergraduate Fourier analysis course covers the spherical functions for the sphere $SO(3)/SO(2)$ – the Legendre polynomials. But the spherical functions for $GL(2, \mathbb{F}_q)/K$, where K, defined in (3) below, is an analogue of the group of rotations, are not well known. Another difficulty is that many different languages have been used to discuss some of these things, for example, the languages of association schemes as in Bannai and Ito [1984], Hecke algebras as in Krieg [1990], and the language of permutation groups as in Neumann [1977].

Some additional references for this chapter are: Angel [1993, 1996], Angel et al. [1992, 1994, 1995], Brouwer et al. [1989], Nancy Celniker [1994], Celniker et al. [1993], Evans [1994, 1995], Harish-Chandra [1984], Li [1996a,b], Lubotzky [1994], Piatetski-Shapiro [1983], Poulos [1991], Soto-Andrade [1987], Valette [1997], and Elinor Velasquez [1991].

The first question one might ask is: What makes a quotient G/K a symmetric space? In Terras [1985 and 1988], we had several approaches. For example, in Terras [1988, Vol. II, p. 260] we sought a Riemannian manifold with a geodesic-reversing isometry at each point. See Helgason [1968] for further discussion of this approach. One could also model Selberg's *weakly symmetric space* (see Selberg [1989] or Terras [1988, Vol. II, pp. 18 and 28]). Vilenkin [1968] would say that we are seeking a *massive* subgroup K of G, that is, such that $\pi \in \hat{G}$ occurs in the induced representation $\mathrm{Ind}_K^G 1$ with multiplicity at most one. The most important property of our symmetric space will be that (G, K) form a *Gelfand pair*, that is, the convolution algebra $L^2(K \backslash G/K)$ is commutative.

Before saying anything else about definitions, let's give an example.

Example. A Finite Analogue of Poincaré's Upper Half Plane. References for this example are: Angel [1996], Angel et al. [1992, 1994, 1995], Celniker et al. [1993], Evans [1994, 1995], Katz [1993, 1995], Kuang [1995 and 1997], Li [1995, 1996a,b, in press], Soto-Andrade [1987], Terras [1991, 1996, in press], and Valette [1997] as well as the theses of Angel [1993], Celniker [1994], Poulos [1991], and Velasquez [1991].

Before defining the finite upper half plane, let's quickly review the basics of the infinite one. For more information, see Terras [1985, Ch. 3].

Real Poincaré Upper Half Plane

The *Poincaré upper half plane* is

$$H = \{z = x + iy \in \mathbb{C} \mid y > 0\},$$

with the noneuclidean arc length element $ds^2 = y^{-2}(dx^2 + dy^2)$ and corresponding Laplacian

$$\Delta u = y^2 \left(\frac{\partial u^2}{\partial x^2} + \frac{\partial u^2}{\partial y^2} \right).$$

Both ds and Δ are invariant under the action of g in $SL(2, \mathbb{R})$ acting on $z \in H$ by fractional linear transformation; that is,

$$\text{if } g = \begin{pmatrix} a & b \\ c & d \end{pmatrix}, \quad \text{then } gz = \frac{az + b}{cz + d}.$$

The geodesics or curves minimizing arc length in H are circles and lines orthogonal to the real axis. They give rise to a noneuclidean geometry. There is a geodesic-reversing isometry at each point of H. The geodesic-reversing isometry at i is $z \mapsto -1/z$. The subgroup K of $G = SL(2, \mathbb{R})$ fixing the point i is the group $K = SO(2, \mathbb{R})$ of 2×2 rotation matrices

$$\begin{pmatrix} \cos t & \sin t \\ -\sin t & \cos t \end{pmatrix}, \quad t \in \mathbb{R}.$$

Thus we can identify H with G/K. Thanks to the existence of the geodesic-reversing isometries, H is a symmetric space. Note that K is the subgroup of G fixed by the involution which sends x to ${}^t x^{-1}$.

The eigenfunctions of Δ which are invariant under various subgroups of $SL(2, \mathbb{R})$ have been intensively studied. The simplest such eigenfunction is the power function $p_s(z) = y^s$. Eigenfunctions of Δ which are invariant under K are the spherical functions and they are obtained from the classical Legendre functions. Those invariant under the subgroup N of matrices of the form

$$\begin{pmatrix} 1 & x \\ 0 & 1 \end{pmatrix}$$

are Bessel functions. Those invariant under $SL(2, \mathbb{Z})$ (and of at most polynomial growth) are Maass wave forms. These are much more mysterious except for the Eisenstein series.

In Terras [1985, Ch. 3] we discuss Fourier analysis on H as well as the solution of the heat equation, the central limit theorem. And we consider the theory of L-functions attached to Maass wave forms f which are eigenfunctions of Hecke operators

$$T_n f(z) = n^{-1/2} \sum_{\substack{ad=n, d>0 \\ b \bmod d}} f\left(\frac{az + b}{d} \right).$$

That is, we assume $T_n f(z) = u_n f$, for some $u_n \in \mathbb{C}$, and consider L-functions $L_f(s)$ which are infinite products:

$$\prod_{p \text{ prime}} (1 - u_p p^{-s} + p^{-2s})^{-1}.$$

Finite Upper Half Plane

The finite upper half plane is modeled on the real one above as follows. Suppose that $\mathbb{F}_q$ is a finite field of odd characteristic p, $q = p^r$. Let $\delta \in \mathbb{F}_q$ be a nonsquare.

Definition. The *finite upper half plane* is

$$H_q = \{z = x + y\sqrt{\delta} \mid x, y \in \mathbb{F}_q, y \neq 0\}. \tag{1}$$

You can view H_q as a subset of the quadratic extension $\mathbb{F}_q(\sqrt{\delta})$ of $\mathbb{F}_q$. Here $\sqrt{\delta}$ is our substitute for $i = \sqrt{-1}$ in $\mathbb{C}$.

Actually we should call H_q a union of an upper and a lower half plane (or a double cover of an upper half plane). This won't matter much. Since there is no notion of positivity in $\mathbb{F}_q$, it is just as natural to draw H_3, for example, as the points

$$-1 \overset{\bullet}{+} 2i \quad \overset{\bullet}{2i} \quad 1 \overset{\bullet}{+} 2i$$
$$-1 \overset{\bullet}{+} i \quad \overset{\bullet}{i} \quad 1 \overset{\bullet}{+} i$$

as it is to view H_3 as the points

$$-1 \overset{\bullet}{+} i \quad \overset{\bullet}{i} \quad 1 \overset{\bullet}{+} i$$
$$-1 \overset{\bullet}{-} i \quad \overset{\bullet}{-i} \quad 1 \overset{\bullet}{-} i$$

where $i = \sqrt{2} = \sqrt{-1}$, in the field with nine elements.

We will use the standard notation from complex analysis. For example, if $z = x + y\sqrt{\delta} \in H_q$ or more generally if $z \in \mathbb{F}_q(\sqrt{\delta})$, we will write $x = \text{Re } z = real part$ of z and $y = \text{Im } z = imaginary part$ of z. The *conjugate* of z is $\bar{z} = x - y\sqrt{\delta}$, which can also be viewed as the *Frobenius automorphism* z^q. The map z to z^q is the generator of the Galois group of $\mathbb{F}_q(\sqrt{\delta})$ over $\mathbb{F}_q$. The *norm* of z is $Nz = z\bar{z}$. The *trace* of z is $z + \bar{z}$.

The *general linear group* $GL(2, \mathbb{F}_q)$ consists of matrices

$$g = \begin{pmatrix} a & b \\ c & d \end{pmatrix},$$

with $a, b, c, d \in \mathbb{F}_q$ and $\det g = ad - bc \neq 0$. The matrix g acts on z in H_q

by *fractional linear transformation:*

$$gz = \frac{az+b}{cz+d} \in H_q. \tag{2}$$

Exercise. Show that if

$$g = \begin{pmatrix} a & b \\ c & d \end{pmatrix}$$

and $z \in H_q$, then

$$\mathrm{Im}\,\frac{az+b}{cz+d} = \frac{\mathrm{Im}\,z\,\det g}{N(cz+d)} \quad \text{and}$$

$$\mathrm{Re}\,\frac{az+b}{cz+d} = \frac{acNz + bd + \mathrm{Re}\,z(ad+bc)}{N(cz+d)}.$$

Conclude that gz is in H_q.

Exercise.

1) Show that the center Z of $GL(2, \mathbb{F}_q)$ consists of all matrices of the form aI, $a \in \mathbb{F}_q$, where I is the identity matrix. Such matrices fix all points $z \in H_q$. Thus we could restrict consideration to the *projective general linear group* $PGL(2, \mathbb{F}_q) \cong GL(2, \mathbb{F}_q)/Z$.

2) Show that the action (by fractional linear transformation) of the following subgroup A of $GL(2, \mathbb{F}_q)$ is transitive on H_q:

$$A = \mathrm{Aff}(q) = \left\{ \begin{pmatrix} a & b \\ 0 & 1 \end{pmatrix} \middle| a, b \in \mathbb{F}_q, a \neq 0 \right\}, \quad \text{the affine group.}$$

That is, show that for a fixed point, for example, $\sqrt{\delta}$, which we think of as the origin, for any $z \in H_q$, there is a $g \in A$ so that $z = g\sqrt{\delta}$. Is the same true for the special linear group

$$G = SL(2, \mathbb{F}_q) = \{ g \in GL(2, \mathbb{F}_q) \mid \det g = 1 \}?$$

It follows from the preceding exercise that we can identify H_q with the quotient G/K, where $G = GL(2, \mathbb{F}_q)$ and K is the subgroup of G which fixes $\sqrt{\delta}$. That is,

$$K = \left\{ \begin{pmatrix} a & b\delta \\ b & a \end{pmatrix} \middle| a, b \in \mathbb{F}_q, a^2 - \delta b^2 \neq 0 \right\}, \tag{3}$$

which is isomorphic to the multiplicative group $\mathbb{F}_q(\sqrt{\delta})^*$. The isomorphism is given by

$$\begin{pmatrix} a & b\delta \\ b & a \end{pmatrix} \mapsto a + b\sqrt{\delta}.$$

Exercise. Check the last statements. Note that they imply that since $|H_q| = q(q-1)$ and $|K| = q^2 - 1$, we have

$$|GL(2, \mathbb{F}_q)| = q(q-1)(q^2 - 1).$$

You can also prove this last formula by counting the number of ways to obtain two linearly independent column vectors from $\mathbb{F}_q^2$.

What is a Finite Symmetric Space?

Now that we have an example, let us go back to the question: What is a finite symmetric space? The main property of interest is that of the following definition.

Definition. Suppose that K is a subgroup of the finite group G. Then (G, K) is a *Gelfand pair* iff $L^2(K \backslash G / K)$ is a commutative algebra under convolution. We will then call the quotient space G/K a *symmetric space*.

Note. We will view $L^2(K \backslash G / K)$ as the space of K-bi-invariant functions $f : G \to \mathbb{C}$; that is, $f(kgk') = f(g)$ for all $g \in G$ and $k, k' \in K$.

Next we list a few criteria for Gelfand pairs.

Definition. We say that G and its subgroup K satisfy *Gelfand's criterion* iff there is a group isomorphism $\tau : G \to G$ such that $s^{-1} \in K\tau(s)K$ for all $s \in G$.

The following is a finite analogue of the criterion given by Selberg in his 1956 paper. See Selberg [1989, p. 427].

Definition. We say that $X = G/K$ satisfies *Selberg's criterion* if there is a 1-1 map $\mu : X \to X$ such that $\mu(eK) = eK$, for $e =$ the identity of G, and for every $x, y \in X$ there is an $m \in G$ so that $mx = \mu y$ and $my = \mu x$.

Another definition of symmetric space can be found in Loos [1969, Vol. I, p. 63].

See Krieg [1990, p. 29] for an example of a Gelfand pair which does not satisfy Gelfand's criterion. In the example, $G = GL(2, \mathbb{F}_q)$ and $K = \text{Aff}(q)$. The connection with the book of Krieg on Hecke algebras is that the algebra

$L^2(K\backslash G/K)$ of K-bi-invariant functions on G under convolution is a Hecke algebra. Hecke introduced the algebra of Hecke operators acting on spaces of modular forms around 1937. We discussed these algebras for groups like $GL(n, \mathbb{Z})$ in Terras [1988, Vol. II] using methods of Tamagawa and Shimura and others such as Maass. Soon people realized that one could deal with Hecke algebras in other situations such as that of finite groups. Some of this work is discussed in Krieg [1990]. There are both group theoretical and combinatorical efforts under way. We will not say too much about the generalities here but stick to concrete examples.

Theorem 1. Both the Gelfand and Selberg criteria above imply that (G, K) is a Gelfand pair, that is, that $L^2(K\backslash G/K)$ is a commutative algebra under convolution.

Proof. First we assume Gelfand's criterion. If f is in $L^2(K\backslash G/K)$, which means that f is a K-bi-invariant function on G, we set $\check{f}(x) = f(x^{-1})$, for $x \in G$. The Gelfand criterion says that $\check{f}(x) = f(\tau x) = f^\tau(x)$.

First recall formula (6) of Chapter 15 defining convolution. Note that $f, g \in L^2(K\backslash G/K)$ implies that $f * g \in L^2(K\backslash G/K)$. Now since τ is a group isomorphism, $(f * g)^\tau = f^\tau * g^\tau$. See the exercise below. It follows that $(f * g)^\vee = \check{f} * \check{g}$.

Therefore

$$(\check{f} * \check{g})(t) = (f * g)^\vee(t) = \sum_{s \in G} f(t^{-1}s^{-1})g(s)$$

$$= \sum_{b \in G} \check{f}(b)\, \check{g}(tb^{-1}) = (\check{g} * \check{f})(t).$$

Thus $\check{f} * \check{g} = \check{g} * \check{f}$. Since $\check{f} = f$, we are done. Why?

Now is the time to show that the Selberg criterion implies the same sort of setup as the Gelfand criterion. To do this, we make use of *point-pair invariants* on $X = G/K$, that is, functions

$$k : X \times X \to \mathbb{C} \quad \text{such that } k(gx, gy) = k(x, y), \quad \text{for all}$$
$$x, y \in X \text{ and } g \in G. \tag{4}$$

Given a function $f \in L^2(K\backslash G/K)$, we define a point-pair invariant $K_f(a, b) = f(b^{-1}a)$, for all $a, b \in G$. See the exercise below.

Suppose that $G/K = X$ satisfies Selberg's criterion. Then we have a 1-1 map $\mu : G/K \to G/K$ which will lift to $\mu : G \to G$ via $\mu(x) = \mu(xK)$, for

$x \in G$. And there is some $m \in G$ with

$$K_f(\mu x, \mu y) = K_f(my, mx) = K_f(y, x) = K_{\vee f}(x, y),$$

for all $x, y \in G$. (5)

Here we need the fact that K_f is a point-pair invariant, as well as the definition of K_f.

It follows from (5) that

$$f((\mu y)^{-1} \mu x) = f(x^{-1} y). \qquad (6)$$

Now if e is the identity of G, we know that $\mu(eK) = eK = K$. Thus if we set $y = K = eK$ in (6), we obtain

$$f(\mu x) = f(x^{-1}). \qquad (7)$$

Then (6) and (7) imply that

$$f((\mu y)^{-1} \mu x) = f(x^{-1} y) = f(\mu(y^{-1} x)). \qquad (8)$$

Thus μ acts like the isomorphism τ of Gelfand's criterion on K-bi-invariant functions. It is as good as a group homomorphism as far as functions on $K \backslash G / K$ are concerned. And it acts like $s \to s^{-1}$ on such functions. Thus the proof that Gelfand's criterion implies that (G, K) is a Gelfand pair completes the proof that Selberg's criterion also implies the same conclusion. ∎

Exercise. Check the following statements.

1) If $\tau : G \to G$ is a group isomorphism, then

$$(f * g)^\tau = f^\tau * g^\tau, \quad \text{where } f^\tau(x) = f(\tau x), \quad \text{for all } x \in G.$$

2) If $f \in L^2(K \backslash G / K)$, then

$$K_f(a, b) = f(b^{-1} a) = K_f(ga, gb), \quad \text{for all } a, b, g \in G.$$

Exercise. Suppose that H is a normal subgroup of G. When is G/H a symmetric space?

Example. Finite Upper Half Planes Are Symmetric Spaces. To show that H_q satisfies Selberg's criterion with $\mu(gK) = g^{-1}K$, suppose that $z, w \in H_q$ and $z = g\sqrt{\delta}$, $w = h\sqrt{\delta}$, setting $\mu z = z^* = g^{-1}\sqrt{\delta}$ and $w^* = h^{-1}\sqrt{\delta}$. We need to find m in $GL(2, \mathbb{F}_q)$ so that $mz = w^*$ and $mw = z^*$. First, note that we can move the points z, w to $\sqrt{\delta}$, $p = g^{-1}h\sqrt{\delta}$:

$$g^{-1}z = \sqrt{\delta} \quad \text{and} \quad g^{-1}w = p.$$

Write $t = g^{-1}h$. Now we want $n \in G$ so that

$$n\sqrt{\delta} = p^* = t^{-1}\sqrt{\delta} \quad \text{and} \quad np = (\sqrt{\delta})^* = \sqrt{\delta}.$$

Clearly $n = t^{-1}$ does the job.

Sometimes the easiest criterion to use is that found in the following corollary.

Corollary 1. Suppose that the K double cosets of G are stable under inversion, that is, $(KsK)^{-1} = KsK$, for all $s \in G$. Then $L^2(K\backslash G/K)$ is a commutative algebra under convolution.

Proof. This is a special case of Gelfand's criterion with $\tau =$ the identity. ∎

Exercise. Show that the finite upper half plane satisfies the criterion of Corollary 1.

What is so good about having $L^2(K\backslash G/K)$ be a commutative algebra? It means that we have the following corollary.

Corollary 2. Suppose that G/K satisfies the criterion of the preceding corollary. Then there is a common basis of $L^2(K\backslash G/K)$ consisting of orthogonal eigenfunctions of the operators of convolution by δ_c, where C runs through the double cosets $K\backslash G/K$.

Proof. This is elementary linear algebra, once you know that the operators are Hermitian or self-adjoint with respect to the inner product on $L^2(K\backslash G/K)$. See the exercise below. ∎

It is the eigenfunctions of the preceding corollary that are the spherical functions for the finite symmetric space G/K. We will have much more to say about them in Chapter 20.

Exercise. Complete the proof of Corollary 2 above by showing that the convolution operators

$$f \mapsto \delta_c * f, \quad \text{for } f, g \in L^2(K\backslash G/K) \quad \text{and} \quad C \in K\backslash G/K$$

satisfy $\langle \delta_c * f, g \rangle = \langle f, \delta_c * g \rangle$. Here we are thinking that $\langle , \rangle$ is the inner product from $L^2(G)$. You need to use the fact that $C = C^{-1}$.

Before moving on to the consideration of spherical functions on finite symmetric spaces G/K, let us try to explain our fascination with them. We first met one in Chapter 10 – the Krawtchouk polynomial. In that case G was the semidirect product of F_2^n and S_n.

Of course, the first spherical functions were Laplace spherical harmonics (see Terras [1985, Vol. I, Ch. 2]) – eigenfunctions of the Laplacian on the sphere. Fourier expansions in terms of Laplace spherical harmonics are necessary for the study of phenomena with spherical symmetry such as the magnetic field of the sun or the hydrogen atom.

As in Helgason [1968, 1984], Terras [1988, Vol. II, pp. 65–70], and Wawrzyńczyk [1984], there are many ways to characterize spherical functions on G/K. The main definition requires them to be eigenfunctions of the Laplacian and all the G-invariant differential operators on G/K. Thus we need to ask: What is the analogue of the Laplacian for a finite symmetric space? We saw in Chapters 3 and 7 that the adjacency operators for graphs are natural analogues of the Laplacian. Now one can also view these adjacency operators as convolution with δ_c, for $C = KsK$, for some $s \in G$, or as mean-value operators (as in Theorem 4 of Chapter 10). What does this mean for our favorite example, the finite upper half plane?

Finite Noneuclidean Geometry

To obtain a finite analogue of noneuclidean geometry on the finite upper half plane H_q, we need a notion of distance. This will not be a metric but instead a distance with values in $\mathbb{F}_q$. Here we imitate Stark [1987] who defined a p-adic analogue of the Poincaré distance, using a p-adic point-pair invariant.

Definition. The "*distance*" between two points $z, w \in H_q$ is given by the point-pair invariant as in (4) defined by

$$d(z, w) = \frac{N(z - w)}{\operatorname{Im} z \operatorname{Im} w} = \frac{(x - u)^2 - \delta(y - v)^2}{yv}, \qquad (9)$$

for $z = x + y\sqrt{\delta}$ and $w = u + v\sqrt{\delta}$, with $x, y, u, v \in \mathbb{F}_q$, $yv \neq 0$.

Note that $d(z, w) \in \mathbb{F}_q$. So we are not talking about a metric here. There is no possibility of a triangle inequality. However, it is certainly analogous to the Poincaré arc length $ds^2 = y^{-2}(dx^2 + dy^2)$ if we think of w in (9) as $z + \Delta z$ so that $\operatorname{Im} w \cong \operatorname{Im} z$. And one should take q such that $\delta = -1$ is a nonsquare in $\mathbb{F}_q$ to be close to the real case.

This is the distance we used in Angel et al. [1992, 1994], Celniker et al. [1993], and Terras [1991, 1996]. Soto-Andrade [1987] uses a slightly different point-pair invariant:

$$\Delta(z, w) = \frac{N(z - w)}{N(z - \bar{w})}, \quad \text{for } z, w \in H_q.$$

Exercise.
1) Show that with the definition above, $d(z, w)$ is indeed a point-pair invariant, that is, $d(gz, gw) = d(z, w)$, for all $g \in G = GL(2, \mathbb{F}_q)$ and all $z, w \in H_q$.
2) Do the same for $\Delta(z, w)$, defined above. Give a formula relating $\Delta(z, \sqrt{\delta})$ and $d(z, \sqrt{\delta})$.

Next we want to describe how to associate $q - 2$ graphs to H_q.

Definition. Define a *graph* $X_q(\delta, a)$, for $a \in H_q$, as follows. Let the vertices of the graph be the points of H_q. Draw an edge between two vertices z, w in H_q iff $d(z, w) = a$.

Example. $q = 3$. *The Octahedron.* We can take $\delta = -1$ for $q = 3$, since $-1 \equiv 2 \pmod 3$ is not a square in $\mathbb{F}_3$. We will write $i = \sqrt{-1}$ in $\mathbb{F}_9$. To understand the graph $X_3(-1, 1)$, we must find the points $z = x + iy$, with $x, y \in \mathbb{F}_3, y \neq 0$, satisfying

$$d(z, i) = \frac{N(z - i)}{y} = 1.$$

This is equivalent to solving $x^2 + (y - 1)^2 = y$. Solutions are the points $\pm 1 \pm i$. These are the points adjacent to i. To find the points adjacent to

$$z = \begin{pmatrix} y & x \\ 0 & 1 \end{pmatrix} i,$$

just apply the matrix

$$\begin{pmatrix} y & x \\ 0 & 1 \end{pmatrix}$$

to the points $\pm 1 \pm i$. The graph that you get is the octahedron. See Figure II.2.

Exercise. 1) Check that we drew the graph $X_3(-1, 1)$ correctly.
Hint. The points adjacent to

$$1 + i = \begin{pmatrix} 1 & 1 \\ 0 & 1 \end{pmatrix} i$$

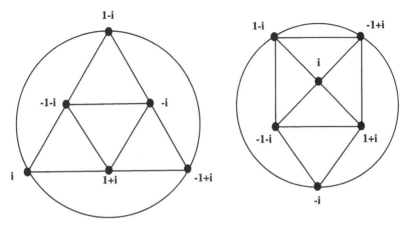

Figure II.2. H_3. The octahedron (two views).

are

$$\begin{pmatrix} 1 & 1 \\ 0 & 1 \end{pmatrix} (\pm 1 \pm i).$$

These are the points $-1 \pm i$ and $\pm i$, just as drawn.

2) Find the adjacency operator of the graph and its eigenvalues. You can use Matlab or Mathematica to do this easily.

Answer. The eigenvalues λ of the adjacency operator are $4, -2, -2, 0, 0, 0$. Note that the second largest satisfies $|\lambda| \leq 2\sqrt{3} \cong 3.46$. So the graph $X_3(-1, 1)$ is Ramanujan, as defined in Chapter 3.

The octahedron graph can be associated to the fundamental domain of a discrete group $\Gamma(3)$ acting on the Poincaré upper half plane, with $\Gamma(n)$ defined as the congruence group of level n:

$$\Gamma(n) = \{\gamma \in PSL(2, \mathbb{Z}) \mid \gamma \equiv I \pmod{n}\},$$

where the projective special linear group is $PSL(2, \mathbb{Z}) = SL(2, \mathbb{Z})/\pm I$. The congruence is induced from that defined for 2×2 integral matrices. It means that the off-diagonal entries are congruent to 0 mod n while the diagonal entries are congruent to 1 mod n.

Now $\Gamma(3)$ is a subgroup of index 12 in the modular group $\Gamma(1) = PSL(2, \mathbb{Z})$. A fundamental domain $D_3 \cong \Gamma(3)\backslash H$ can be taken as a union of twelve copies of the standard fundamental domain $D_1 \cong \Gamma(1)\backslash H$. See Figure II.3. Terras

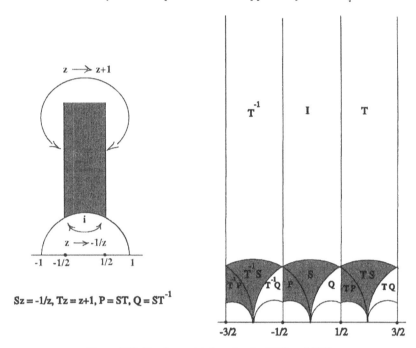

Figure II.3. Fundamental domains for $\Gamma(1)$ and $\Gamma(3)$.

[1985, Vol. I, Ch. 3] gives more details on the fundamental domains. Many examples of fundamental domains can be found in Klein and Fricke [1966, pp. 353 ff].

It is easily checked that you get the octahedron graph from D_3 by letting the vertices correspond to a union of two adjacent copies of D_1 (one shaded and one unshaded). Thus we obtain six vertices. Then connect vertices if the corresponding domains have a boundary edge in common. Here you must remember that some of the edges of D_3 are identified by generators of $\Gamma(3)$. For example, the vertical edges are identified by $z \mapsto z + 3$. The sides at the bottom of D_3 are identified as in Figure II.3. When we glomp together the shaded and unshaded copies of D_1, we are actually getting the fundamental domain for

$$\Gamma_v(3) = \left\{ \gamma \in \Gamma(1) \,\middle|\, \gamma \cong \pm I, \text{ or } \pm \begin{pmatrix} 0 & 1 \\ -1 & 0 \end{pmatrix} \pmod{3} \right\}.$$

This group has index 6 in $\Gamma(1) = SL(2, \mathbb{Z})$. See Terras [loc. cit.] or Rankin [1977] for more details.

Exercise. Check out the identification described above between graphs associated to $D_3 \cong \Gamma(3)\backslash H$ (where H is the Poincaré upper half plane) and the octahedron.

Note. The fact that graphs obtained this way coincide for small primes p does not mean this will happen for large primes. See Chapter 5 and Medrano et al. [1996] for euclidean analogues.

There are many connections between the theory of graphs and Riemann surfaces. See Besseon [1988], Brooks [1986, 1991, 1997], Buser [1988 and 1992], and Sunada [1988].

Example. $q = 5$. *Graphs on the Dodecahedron.* For $q = 5$, there are three connected nonisomorphic graphs $X_5(\delta, a)$ obtained by taking $\delta = 2$ and $a = 1, -1, 2$. See Figure II.4. At first (see Terras [1991]), we considered only the graph $X_5(2, 1)$. However, Poulos [1991] discovered that there are other interesting graphs. Stark was the one who observed that the graphs all sit on the dodecahedron in a natural way. The points are labeled in the following order:

$$\sqrt{2}, 1 + \sqrt{2}, 2 + \sqrt{2}, 3 + \sqrt{2}, 4 + \sqrt{2}, 2\sqrt{2}, 1 + 2\sqrt{2}, 2 + 2\sqrt{2},$$
$$3 + 2\sqrt{2}, 4 + 2\sqrt{2}, 3\sqrt{2}, 1 + 3\sqrt{2}, 2 + 3\sqrt{2}, 3 + 3\sqrt{2},$$
$$4 + 3\sqrt{2}, 4\sqrt{2}, 1 + 4\sqrt{2}, 2 + 4\sqrt{2}, 3 + 4\sqrt{2}, 4 + 4\sqrt{2}.$$

In Figure II.4, we draw all three graphs $X_5(2, a), a = 1, -1, 2$, on the dodecahedron. We represent the finite upper half plane by a dodecahedron. The edges of the graph $X_5(2, a)$ that are incident on a fixed vertex are shown as solid lines. The remaining edges can be obtained by applying the symmetries of the affine group.

You can reformulate these concepts in the language of permutation groups. See Neumann [1977].

Exercise.

1) Let $X = G/K$. Show that there is a one-to-one correspondence between the orbits* of G in $X \times X$ and the orbits of K in X. Here $g \in G$ acts on $(x, y) \in X \times X$ via $g(x, y) = (gx, gy)$. The correspondence comes about as follows. Let Δ denote a G-orbit in $X \times X$. Define for $a \in X$,

$$\Gamma = \Gamma(a) = \{b \in X \mid (a, b) \in \Delta\}.$$

---

* If a group G acts on a set X, an *orbit* of $p \in X$ is the set of points $\{gp \mid g \in G\}$.

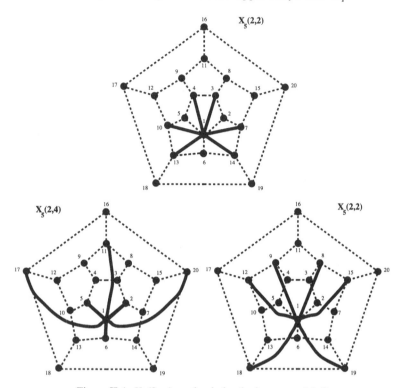

Figure II.4. $X_5(2, a)$ on the dodecahedron, $a = \pm 1, 2$.

Then Γ is a K-orbit in X. Note that this gives a 1-1 correspondence between K-orbits in X and our point-pair invariants on X.

2) Then define an adjacency matrix B_Δ by

$$(B_\Delta)_{a,b} = \begin{cases} 1, & \text{if } (a, b) \in \Delta, \\ 0, & \text{otherwise.} \end{cases}$$

Show that the B_Δ forms a basis of the space of all operators on $L^2(G/K)$ commuting with the action of G. This is called the *centralizer ring* V. Show that the graph with B_Δ as its adjacency operator is our finite upper half plane graph.

Hint. See Neumann [1977].

Most of the following theorem is due to Poulos [1991] for the case $q = p$, an odd prime. For the general case, see Celniker [1994] and Angel [1993, 1996]. In fact, Angel also works out the case of characteristic 2. Many of these results are also discussed in Angel et al. [1992, 1994], Celniker et al. [1993], and Terras [1991, 1996].

Theorem 2. **Properties of Finite Upper Half Plane Graphs.** (From Poulos [1991].) Assume that $q = p^r$, where p is an odd prime. Suppose that δ is a nonsquare in $\mathbb{F}_q$. Let $a \in \mathbb{F}_q$.

1) The graph $X_q(\delta, a)$ is a $(q + 1)$-regular graph provided that $a \neq 0$ or 4δ.
2) The graphs $X_q(\delta, a)$ and $X_q(\delta c^2, ac^2)$ are isomorphic.
3) Let τ be an element of the Galois group $\mathrm{Gal}(\mathbb{F}_q/\mathbb{F}_p)$. Then the graphs $X_q(\delta, a)$ and $X_q(\tau(\delta), \tau(a))$ are isomorphic.
4) The graph $X_q(\delta, a)$ is connected, provided that $a \neq 0, 4\delta$. In fact the graph $X_q(\delta, a)$ is a Cayley graph for the affine group

$$\mathrm{Aff}(q) = \left\{ \begin{pmatrix} y & x \\ 0 & 1 \end{pmatrix} \,\middle|\, x, y \in \mathbb{F}_q, y \neq 0 \right\},$$

using the generators

$$S_q(\delta, a) = \left\{ \begin{pmatrix} y & x \\ 0 & 1 \end{pmatrix} \,\middle|\, x, y \in \mathbb{F}_q, y \neq 0, x^2 = ay + \delta(y - 1)^2 \right\}.$$

5) The K-double cosets for $G = GL(2, \mathbb{F}_q)$,

$$K = \left\{ \begin{pmatrix} a & b\delta \\ b & a \end{pmatrix} \,\middle|\, a, b \in \mathbb{F}_q, a^2 - \delta b^2 \neq 0 \right\},$$

are represented by the sets $S_q(\delta, a)$, for $a \in \mathbb{F}_q$.

Proof. 1) It suffices, since $GL(2, \mathbb{F}_q)$ acts transitively on H_q, to show that $q + 1$ points are adjacent to the origin $\sqrt{\delta}$. These are the points $z = x + y\sqrt{\delta}$ such that $N(z - \sqrt{\delta}) = ay$. Equivalently z satisfies

$$x^2 = ay + \delta(y - 1)^2. \tag{10}$$

Another version of this equation can be obtained using the fact that $\bar{z} = z^q$. Thus we find that (10) is the equation

$$(z - \sqrt{\delta})^{q+1} = a(2\sqrt{\delta})^{-1}(z - z^q).$$

This is a polynomial of degree $q + 1$ in z over $\mathbb{F}_q(\sqrt{\delta})$. Such a polynomial has at most $q + 1$ solutions in $\mathbb{F}_q(\sqrt{\delta})$. Of course, we seek solutions in H_q, that is, such that the imaginary part does not vanish, which is a further restriction. If

$y = 0$, then δ would have to be a square, contradicting the choice of δ. So, if we have solutions, they are automatically in H_q.

To get more information on the solutions of (10), recall that the multiplicative group $\mathbb{F}_q(\sqrt{\delta})^*$ is cyclic with generator γ. And our equation (10) is seen by the exercise below to be equivalent to the equation

$$N\{z + c\} = r, \tag{11}$$

where $c = (a/2\delta - 1)\sqrt{\delta}$ and $r = a(1 - a/4\delta)$. Therefore we are solving the equation $w^{q+1} = r$, for $w = z + c$.

Exercise. Show that formula (10) is equivalent to formula (11).

As long as $r \neq 0$, which is the case when $a \neq 0$ or 4δ, Equation (11) is solvable because the norm is a group homomorphism mapping $\mathbb{F}_q(\sqrt{\delta})^*$ onto $\mathbb{F}_q^*$.

To see that the norm is onto, note that the kernel of the norm has $q + 1$ elements, for the kernel consists of elements of the form γ^e, such that

$$(q + 1)e \equiv 0 \,(\mathrm{mod}\; q^2 - 1).$$

Thus $q - 1$ must divide e. And $e = (q - 1)j, j = 0, 1, 2, \ldots, q$. Therefore the image of the norm map has $(q^2 - 1)/(q - 1) = q + 1$ elements, which is exactly as many elements as $\mathbb{F}_q^*$. This discussion implies that the equation $w^{q+1} = r$ has $q + 1$ solutions w given any nonzero $r \in \mathbb{F}_q$.

2) This result was found by Stark. The mapping defined by $T_c(x + y\sqrt{\delta}) = cx + y\sqrt{\delta}$ is an isomorphism when viewed as a mapping of the affine group $\mathrm{Aff}(q)$. Thus once we have proved part 3, we will have a graph isomorphism by T_c from $X_q(\delta, a)$ to $X_q(c^2\delta, c^2a)$, since g and h in $\mathrm{Aff}(q)$ correspond to an edge in the graph $X_q(\delta, a)$ iff $d(g\sqrt{\delta}, h\sqrt{\delta}) = a$, which is equivalent to saying $h^{-1}g \in S_q(\delta, a)$. Also $T_c(g\sqrt{\delta})$ and $T_c(h\sqrt{\delta})$ are connected in the graph $X_q(c^2\delta, c^2a)$ iff

$$T_c(h\sqrt{\delta})^{-1}\, T_c(g\sqrt{\delta}) = T_c(h^{-1}g\sqrt{\delta}) \in S_q(c^2\delta, c^2a).$$

Clearly if we set

$$z = h^{-1}g\sqrt{\delta} = x + y\sqrt{\delta},$$

then

$$T_c(h^{-1}g\sqrt{\delta}) = cx + y\sqrt{\delta}.$$

Furthermore

$$z \in S_q(\delta, a) \quad \text{iff } x^2 = ay + \delta(y-1)^2.$$

If we multiply this last equation by c^2, we obtain the equation for $T_c(h^{-1}g\sqrt{\delta}) = cx + y\sqrt{\delta}$, showing that it lies in $S_q(c^2\delta, c^2a)$.

3) Do the proof as an exercise.

4) First note that if $g, h \in \mathrm{Aff}(q)$, the points $g\sqrt{\delta}$ and $h\sqrt{\delta}$ are adjacent in the graph $X_q(\delta, a)$ iff $h^{-1}g \in S_q(\delta, a)$. Thus $g = hs$ for some $s \in S_q(\delta, a)$. It follows that our graph is a Cayley graph, once we have proved that the set $S_q(\delta, a)$ does generate $\mathrm{Aff}(q)$ and is stable under inversion; that is, $s \in S_q(\delta, a)$ implies $s^{-1} \in S_q(\delta, a)$.

To see that $S_q(\delta, a)$ is stable under inversion, note that if $s \in S_q(\delta, a)$, we have

$$a = d(s\sqrt{\delta}, \sqrt{\delta}) = d(\sqrt{\delta}, s^{-1}\sqrt{\delta}) = d(s^{-1}\sqrt{\delta}, \sqrt{\delta}),$$

which implies that $s^{-1} \in S_q(\delta, a)$.

It is harder to see that the graph is connected, that is, that the set $S_q(\delta, a)$ generates $\mathrm{Aff}(q)$. This was proved by Poulos [1991] in the case $q = p$ and modified by Angel [1993] for general q. Let $S = S_q(\delta, a)$ and let G be the subgroup of $\mathrm{Aff}(q)$ generated by S. Since S has $q + 1$ elements, it follows that it must contain an element of the form

$$\begin{pmatrix} y & x \\ 0 & 1 \end{pmatrix}, \quad \text{where both } x \text{ and } y \text{ are nonzero in } \mathbb{F}_q.$$

Since S is stable under inversion,

$$\begin{pmatrix} y^{-1} & -y^{-1}x \\ 0 & 1 \end{pmatrix} \in S.$$

Therefore we have

$$\begin{pmatrix} y & x \\ 0 & 1 \end{pmatrix}\begin{pmatrix} y^{-1} & y^{-1}x \\ 0 & 1 \end{pmatrix} = \begin{pmatrix} 1 & 2x \\ 0 & 1 \end{pmatrix} \in G.$$

It follows that

$$\begin{pmatrix} y & x \\ 0 & 1 \end{pmatrix}^{-1}\begin{pmatrix} 1 & v \\ 0 & 1 \end{pmatrix}\begin{pmatrix} y & x \\ 0 & 1 \end{pmatrix} = \begin{pmatrix} 1 & y^{-1}v \\ 0 & 1 \end{pmatrix} \in G, \quad \text{for } v = 2x.$$

Next suppose that g generates the multiplicative group $\mathbb{F}_q^*$. Let a be the greatest common divisor of all the exponents $e \geq 1$ such that

$$\begin{pmatrix} g^e & x \\ 0 & 1 \end{pmatrix} \in S, \quad \text{for some } x.$$

We know that an $e \geq 1$ exists because $|S| = q + 1$.

Suppose that $a > 1$. We will deduce a contradiction. For each exponent e, we have at most two values of x such that

$$\begin{pmatrix} g^e & x \\ 0 & 1 \end{pmatrix} \in S.$$

So the number of matrices

$$\begin{pmatrix} g^e & x \\ 0 & 1 \end{pmatrix} \in S$$

is less than or equal to $2(q - 1)/a \leq q - 1$. This contradicts the fact that $|S| = q + 1$. It follows then that $a = 1$; that is, the g.c.d of all the exponents e must be 1. This means that

$$\begin{pmatrix} g & x \\ 0 & 1 \end{pmatrix} \in S,$$

for some x, and thus for every $y \in \mathbb{F}_q^*$ the matrix

$$\begin{pmatrix} y & x \\ 0 & 1 \end{pmatrix} \in G,$$

for some x.

Finally, we must show $G = \text{Aff}(q)$. To do this, we note that, from our earlier computations, we have elements

$$s = \begin{pmatrix} 1 & v \\ 0 & 1 \end{pmatrix} \in G,$$

for some $v \neq 0$. Then for all $y \in \mathbb{F}_q^*$, we have an $x \in \mathbb{F}_q$, so that

$$r = \begin{pmatrix} y & x \\ 0 & 1 \end{pmatrix} \in G.$$

This implies

$$r^{-1}sr = \begin{pmatrix} 1 & y^{-1}x \\ 0 & 1 \end{pmatrix} \in G.$$

Thus for every $b \in \mathbb{F}_q$, the matrix

$$\begin{pmatrix} 1 & b \\ 0 & 1 \end{pmatrix}$$

lies in G, since we can always solve the equation $b = y^{-1}x$, for y, when we are given nonzero x and b in $\mathbb{F}_q$. So for every $y \in \mathbb{F}_q^*$, the matrix

$$\begin{pmatrix} y & 0 \\ 0 & 1 \end{pmatrix} = \begin{pmatrix} 1 & -x \\ 0 & 1 \end{pmatrix} \begin{pmatrix} y & x \\ 0 & 1 \end{pmatrix} \quad \text{is in } G.$$

But then G must contain every matrix

$$\begin{pmatrix} a & b \\ 0 & 1 \end{pmatrix} = \begin{pmatrix} 1 & b \\ 0 & 1 \end{pmatrix} \begin{pmatrix} a & 0 \\ 0 & 1 \end{pmatrix}, \quad \text{for all } a, b \in \mathbb{F}_q \text{ with } a \neq 0.$$

This completes our proof. See Angel [1996, pp. 76–8] for a generalization.
5) (From Poulos [1991].) First recall that K is the subgroup of $GL(2, \mathbb{F}_q)$ in Equation (3), which fixes the point $\sqrt{\delta}$. See an earlier exercise. Then note that if $k \in K$ and $kz = z$, for some $z \in H_q$, either $k = aI$ for some $a \in \mathbb{F}_q^*$ or $z = \pm\sqrt{\delta}$. See the exercise below. It is also easy to see that

$$S_q(\delta, 0) = \{\sqrt{\delta}\} \quad \text{and} \quad S_q(\delta, 4\delta) = \{-\sqrt{\delta}\}.$$

When $a \neq 0, 4\delta$, we want to show that for any $z_a \in S_q(\delta, a)$, we have

$$S_q(\delta, a) = K z_a.$$

First note that certainly

$$K z_a \subset S_q(\delta, a),$$

since any $k \in K$ fixes $\sqrt{\delta}$ and $d(kz_a, \sqrt{\delta}) = d(z_a, k^{-1}\sqrt{\delta}) = d(z_a, \sqrt{\delta}) = a$.
Next note that, if $Z =$ the center of $GL(2, \mathbb{F}_q)$, that is,

$$Z = \{aI \mid a \in \mathbb{F}_q\},$$

we have a mapping

$$K/Z \twoheadrightarrow K z_a$$

given by mapping the coset kZ to the point kz_a. This mapping is certainly well-defined and onto. Thus we have

$$|K z_a| = |K|/|Z| = \frac{q^2 - 1}{q - 1} = q + 1.$$

Since $|S_q(\delta, a)| = q + 1$, it follows that $K z_a = S_q(\delta, a)$. It follows that the $S_q(\delta, a)$, $a \in \mathbb{F}_q$, are all the K-double cosets of $GL(2, \mathbb{F}_q)$. ■

Notes.
1) Once we have shown that the eigenvalue $q + 1$ of the adjacency matrix of our graph $X(q, \delta)$ has multiplicity one we will have another proof that the graph is connected.
2) Since $S - S_q(\delta, a), a \in \mathbb{F}_q$, represent the K-double cosets of G and $S = S^{-1}$, we know that (G, K) is a Gelfand pair by Gelfand's criterion with $\tau =$ identity.
3) The exercise before Theorem 2 gives another proof of part 5 of the theorem.

Exercise. Prove that if K is as defined in (3), then if $kz = z$ for $z \in H_q$, either $z = \pm\sqrt{\delta}$ or $k = aI$, for $a \in \mathbb{F}_q$ and I the identity matrix.

Exercise. Use some of the reasoning in the proof of part 1 of Theorem 2 to prove Hilbert's Theorem 90, which says that w is an element of $\mathbb{F}_q(\sqrt{\delta})^*$ such that $Nw = 1$ iff $w = a\bar{a}^{-1} = q^{1-q}$, for some $a \in \mathbb{F}_q(\sqrt{\delta})^*$.
Hint. See Piatetski-Shapiro [1983, p. 33].

As in Chapters 4 and 5, there are many combinatorial questions that one can ask about the finite upper half plane graphs. For example, one can ask: What are their girths? For the answer (3 or 4), see Celniker [1994] and Poulos [1991]. One can also ask about the diameter (≤ 4) and the chromatic number. See the same references plus Evans [1994, 1995].

We will consider next the questions about graphs which were posed in Chapters 4 and 5. In particular we wonder:

Are these Graphs $X_q(\delta, a)$ Ramanujan?

Recall that we defined Ramanujan graph in Chapter 3. It is a connected $(q+1)$-regular graph such that if A denotes the adjacency operator, we have the inequality

$$|\lambda| \leq 2\sqrt{q},$$

for all eigenvalues λ of A such that $|\lambda| \neq q + 1$.

As we saw in Chapter 1, this question arises in connection with the problem of creating good expander graphs. The constructions of Ramanujan graphs given by Lubotzky et al. [1988] required Deligne's proof of the Weil conjectures and his proof of the Ramanujan bound for Fourier coefficients of modular forms. However, Davidoff and Sarnak (preprint) have an elementary proof of a weaker inequality. See also Lubotzky [1994] and Sarnak [1990].

We will see that in order to show that our graphs $X_q(\delta, a)$ are Ramanujan, one needs to estimate some exponential sums, some of which are the sort that can be handled using elementary methods as in Schmidt [1976]. Those corresponding to discrete series or cuspidal representations of $GL(2, \mathbb{F}_q)$ are a little more intransigent. Soto-Andrade [1987] gave a useful formula for the eigenvalues that allowed Katz [1993] to estimate them – proving that the finite upper half plane graphs are Ramanujan. Later Winnie Li [1996b] found a proof that does not involve étale cohomology (but instead uses the theory of L-functions for the ideles). Li (in press) obtains the finite upper half plane graphs as quotients of Morgenstern's function field analogues of the Lubotzky, Phillips, and Sarnak.

Before the work of Katz, Nancy Celniker [1994] and others investigated the question on various computers. Celniker [1994] showed, in particular, that all the graphs $X_q(\delta, a)$ for $q < 100$ or so are indeed Ramanujan.

There is much more to be said about the eigenvalues than just that they satisfy the Ramanujan bound. One can ask, for example: When does the eigenvalue 0 occur? What are the multiplicities of the eigenvalues? What algebraic number field contains the eigenvalues? See Poulos [1991]. In Table II.5 we list the eigenvalues of the various graphs for $q = 5, 7$, and 9. We will have more to say about the eigenvalues and eigenfunctions in the remainder of this chapter.

Exercise. Check Table II.5 using Matlab or Mathematica or your favorite linear algebra software.
Hint. Theorem 1 in Chapter 17 will shrink the size of the matrices whose spectra are required.

Table II.5. *Eigenvalues for H_3, H_5, H_7*

$X_3(2, 1)$	4, 0	$(-2, 0)$ repeated twice
$X_5(2, 1)$	6, $-2, 2, 2$	$(2, 1, -2, -3)$ repeated 4 times
$X_5(2, 2)$	6, $-2, -2, -2$	$(3, 1, -2, -2)$ repeated 4 times
$X_5(2, 4)$	6, 2, 0, 0	$(2, 0, 0, -4)$ repeated 4 times
$X_7(3, 1)$	8, $-4, 2, 2, 2, 2$	$(-4, -3.4142, 2, 2, 2, -.5858)$ 6 times
$X_7(3, 2)$	8, 3, 3, $-1, -1, 0$	$(-4, 3, \pm2.8284, -1, 0)$ 6 times
$X_7(3, 3)$	8, $-3, -3, -1, -1, 0$	$(4,-3, \pm2.8284, -1, 0)$ 6 times
$X_7(3, 4)$	8, 4, $\pm2, \pm2$	$(4, -3.4142, \pm2, -2, -.5858)$ 6 times
$X_7(3, 6)$	8, $-4, -4, 0, 0, 0$	$(4.8284, -4, -.8284, 0, 0, 0)$ 6 times

Exercise. Compute the eigenvalues for $p = 11$. You will find that there are two graphs that have the same eigenvalues. Are these graphs isomorphic? *Hint.* Take $\delta = 2$, $a = 3$, and $4\delta - 3$.

If these graphs are not isomorphic, you cannot hear the shape of these graphs. That is, two synthesizers tuned to the frequencies of the two graphs would sound the same. Nevertheless the graphs are not the same. We will have much more to say about this subject in Chapter 22. For now, some references are Cvetković, Doob, and Sachs [1980], Buser [1988 and 1992], Benson and Jacobs [1972], Baker [1966], and Fisher [1966].

Without using any representation theory, we can write down some eigenfunctions of the adjacency operators for our graphs $X_q(\delta, a)$. Here we restrict ourselves to the simplest – an analogue of the function y^s on the Poincaré upper half plane (see Terras [1985, Vol. I, Ch. 3]) and Chapter 23.

Definition. The *finite power function* $p_\chi(z)$ for $z \in H_q$ and χ a character of the multiplicative group $\mathbb{F}_q^*$ is defined by

$$p_\chi(z) = \chi(\text{Im } z).$$

*Lemma 1. **Finite Power Functions are Eigenfunctions of Adjacency Operators.*** Let A be the adjacency operator of the graph $X_q(\delta, a)$. Then the finite power function is an eigenfunction of A, that is,

$$A p_\chi = R_\chi p_\chi,$$

where the (1-dimensional) eigenvalue R_χ is given by

$$R_\chi = \sum_{w \in S_q(\delta, a)} \chi(\text{Im } w).$$

Recall that $S_q(\delta, a)$ was defined in Theorem 2.

Proof. By Theorem 2, we know that

$$A p_\chi \left\{ \begin{pmatrix} y & x \\ 0 & 1 \end{pmatrix} \sqrt{\delta} \right\} = \sum_{\begin{pmatrix} v & u \\ 0 & 1 \end{pmatrix} \in S_q(\delta, a)} \chi\left(\text{Im} \begin{pmatrix} y & x \\ 0 & 1 \end{pmatrix} \begin{pmatrix} v & u \\ 0 & 1 \end{pmatrix} \sqrt{\delta} \right)$$

$$= \sum_{\begin{pmatrix} v & u \\ 0 & 1 \end{pmatrix} \in S_q(\delta, a)} \chi(yv) = \chi(y) \sum_{\begin{pmatrix} v & u \\ 0 & 1 \end{pmatrix} \in S_q(\delta, a)} \chi(v)$$

$$= R_\chi p_\chi(z).$$

This completes the proof. ∎

We will discuss other eigenfunctions of the adjacency operators of these graphs later. First let us investigate what the representation theory of Aff(q) tells us about the spectrum of the adjacency operators of these Cayley graphs.

As a result of Theorem 1 of Chapter 17, we see that the spectrum of the adjacency operator for the graph $X_q(\delta, a)$ has two parts. Part 1 is the set of $R\chi$ (defined in Lemma 1), for χ a character of $\mathbb{F}_q^*$. Part 2 is the spectrum of the matrix M_π, where π is the $(q - 1)$-dimensional representation of Aff(q); that is,

$$M_\pi = \sum_{s \in S_q(\delta,a)} \pi(s). \tag{12}$$

We used this result to compute the eigenvalues of the adjacency operators in Angel et al. [1992], Celniker et al. [1993], and Terras [1991]. Theorem 1 of Chapter 17 makes the computation of the 1-dimensional eigenvalues easy. And it reduces the rest of the problem to finding eigenvalues of a $(q - 1) \times (q - 1)$ matrix instead of a $q(q - 1) \times q(q - 1)$ matrix.

Exercise. Check the values of the 1-dimensional eigenvalues in Lemma 1 above for small values of q. Check that they satisfy the Ramanujan bound $|R_\chi| \leq 2\sqrt{q}$, for χ not the trivial multiplicative character of $\mathbb{F}_q^*$.

Number theorists and algebraic geometers have devoted much energy to bounding sums such as the R_χ of Lemma 1. See Weil [1979, Vol. 1, pp. 387–9, 399–410] for foundational papers developing the connections between estimating the number of points on curves over $\mathbb{F}_q$, the Riemann hypothesis for the zeta function of the function field attached to the curve, and estimates of exponential sums. Other references are Ireland and Rosen [1982], Li [1996a,b], and Schmidt [1976]. Weil observes that such results go back to Gauss in special cases.

Both R. Evans and H. Stark noticed that the sum R_χ of Lemma 1 is a character sum of the type that can be bounded by elementary methods from number theory. In particular, the following theorem from Schmidt [1976, Theorem 2C, p. 43] suffices.

Theorem 3. Suppose that χ is a nontrivial multiplicative character of $\mathbb{F}_q^*$ such that d is the minimal positive exponent with $\chi^d = 1$. And suppose that the polynomial $f(x) \varepsilon \mathbb{F}_q[x]$ has exactly m distinct zeros and that $y^d - f(x)$ is absolutely irreducible, that is, irreducible over $\mathbb{F}_q$ and every algebraic extension of $\mathbb{F}_q$. Then

$$\left| \sum_{x \in \mathbb{F}_q} \chi(f(x)) \right| \leq (m - 1)\sqrt{q}.$$

Proof. See Schmidt [1976]. ∎

Corollary. Using the notation of Lemma 1, the 1-dimensional eigenvalues of the adjacency operator of $X_q(\delta, a)$ for $a \neq 0, 4\delta$, and χ a nontrivial multiplicative character of $\mathbb{F}_q^*$, satisfy the Ramanujan bound, that is,

$$|R_\chi| \leq 2\sqrt{q}.$$

Proof. (R. Evans) Let ε denote the quadratic character on $\mathbb{F}_q^*$; that is,

$$\varepsilon(x) = \begin{cases} 1, & x = u^2, \quad \text{for } u \in \mathbb{F}_q^* \\ -1, & x \text{ not a square,} \\ 0, & x = 0. \end{cases} \tag{13}$$

Then

$$R_\chi = \sum_{y \in \mathbb{F}_q} \chi(y)(\varepsilon(ay + \delta(y-1)^2) + 1).$$

Next suppose that the group of multiplicative characters of $\mathbb{F}_q^*$ is generated by ξ. If d is the minimum positive exponent such that $\chi^d = 1$, we can write $\chi = \xi^{(q-1)/d}$ and $\varepsilon = \xi^{(q-1)/2}$. So, if $f(y) = ay + \delta(y-1)^2$, we have (using the orthogonality relations for characters of $\mathbb{F}_q^*$)

$$R_\chi = \sum_{y \in \mathbb{F}_q} \xi^{(q-1)/d}(y)\xi^{(q-1)/2}(f(y))$$

$$= \sum_{y \in \mathbb{F}_q} \xi^m \left(y^{(q-1)/dm} f(y)^{(q-1)/2m} \right),$$

where

$$m = \text{g.c.d.} \left(\frac{q-1}{d}, \frac{q-1}{2} \right).$$

Then the polynomial

$$z^{(q-1)/m} - y^{(q-1)/dm} f(y)^{(q-1)/2m}.$$

is absolutely irreducible (using the following lemma of Schmidt [1976]) which we do not prove. ∎

Lemma 2. Suppose that $F(x)$ is a polynomial with coefficients in $\mathbb{F}_q$. The following are equivalent:

1) $y^d - F(x)$ is absolutely irreducible.
2) $y^d - cF(x)$ is absolutely irreducible for all $c \in \mathbb{F}_q^*$.
3) If $F(x) = a(x-x_1)^{d_1} \cdots (x-x_s)^{d_s}$ is the factorization of F over the algebraic closure of $\mathbb{F}_q$, with $x_i \neq x_j$, for $i \neq j$, then g.c.d.$(d, d_1, \ldots, d_s) = 1$.

For a proof, see Schmidt [1976, p. 11].

In our case, we have

$$F(y) = y^{(q-1)/dm}(ay + \delta(y-1)^2)^{(q-1)/2m} \quad \text{and}$$
$$1 = \text{g.c.d.} \left(\frac{q-1}{m}, \frac{q-1}{dm}, \frac{q-1}{2m} \right).$$

To see that $F(y)$ satisfies condition 3 of Lemma 2, use the fact that $f(y) = ay + \delta(y-1)^2$ has no multiple roots if $a \neq 0$ or 4δ since the discriminant is $a(a - 4\delta)$. ∎

Remarks. To find out whether the eigenvalues of the M_π defined by formula (12) above satisfy the Ramanujan bound, we will need to investigate K-Bessel and spherical functions on H_q. See the next chapter.

Lubotzky [1994] notes that the expander graphs are mainly of interest if one has a family of (n, k, c)-expanders with n going to infinity, and k, c fixed. Of course, our examples do not have this property.

Exercise.
1) Parameterize the points $z = x + y\sqrt{\delta}$ with $x \neq -1$ on the finite circle

$$x^2 - \delta y^2 = 1$$

by

$$x = \frac{1 + \delta t^2}{1 - \delta t^2}, \quad y = \frac{2t}{1 - \delta t^2}, \quad t \in \mathbb{F}_q.$$

Since t is arbitrary, this shows that the finite circle consists of $q + 1$ points.
2) Use Theorem 3 above to show that if χ is a nontrivial multiplicative character of $\mathbb{F}_q^*$ and we define $\chi(0) = 0$, as usual, then

$$\left| \sum_{Nz=1} \chi(y) \right| \leq 2\sqrt{q}.$$

Hint. We can replace $\chi(y)$ with

$$\chi(y(t)) = \chi\left(\frac{\text{linear}}{\text{quadratic}}\right).$$

Use the fact that χ has some order d and then multiply by a power of the quadratic denominator to obtain

$$\chi(y(t)) = \chi(\text{linear} \cdot \text{denominator}^{\text{power}}).$$

The number of distinct roots of the polynomial argument of χ is 3.

Question. Can we use the method of this exercise to estimate R_χ?

Chapter 20

Special Functions on $H_q - K$-Bessel and Spherical

I still dream of bears, but in the dreams I watch. I don't run. I don't speak. The bears watch me, silent, waiting. I am not afraid of bears, I tell myself. I am not afraid.

S. Simpson [1994, p. 22]

In this chapter the bears are finite analogues of the special functions known as Bessel functions and spherical harmonics. First we seek out the lairs of the Bessel functions and then move on to the spherical harmonics. The latter will require some extraordinary efforts to catch them in their lairs – for example, a detailed study of the representations of $GL(2, \mathbb{F}_q)$ to be found in the next chapter. Bounding the spherical functions will be necessary to show that the finite upper half plane graphs of the previous chapter are indeed Ramanujan. In fact, we will have to leave the details of that proof to others (Nick Katz [1993] and Winnie Li [1996a,b]).

Our Bessel and spherical functions on H_q are certain eigenfunctions of the adjacency operators A_a of $X_q(\delta, a)$ for all $a \neq 0$ or 4δ. Equivalently, we will consider eigenfunctions of all the Laplacians $\Delta = A_a - (q+1)I$ for H_q. Note that, as for the noneuclidean Laplacian on the Poincaré upper half plane, the adjacency operators are invariant under left translation by elements of the general linear group. This means that for each $g \in G = GL(2, \mathbb{F}_q)$ and for each $f : H_q \to \mathbb{C}$, defining $\tau_g f(z) = f(gz)$, for $z \in H_q$, we have

$$\tau_g A f = A \tau_g f.$$

Exercise. Prove the preceding result.
Hint. See formula (5) below.

The simplest eigenfunctions of the adjacency operators A_a for the graphs $X_q(\delta, a)$ are the *power functions* defined before Lemma 1 of the preceding chapter. That is, for any multiplicative character χ of $\mathbb{F}_q^*$ we define the power function p_χ by

$$p_\chi(z) = \chi(\operatorname{Im} z). \tag{1}$$

Lemma 1 of Chapter 19 said that p_χ is an eigenfunction of the adjacency operators for all of our graphs:

$$A p_\chi = R_\chi p_\chi, \quad \text{where } R_\chi = \sum_{w \in S_q(\delta, a)} \chi(\operatorname{Im} w).$$

Here the set $S_q(\delta, a)$ is defined in Theorem 2 of Chapter 19 to be the $z = x + y\sqrt{\delta} \in H_q$ such that $x^2 = ay + \delta(y-1)^2$. It is a symmetric set of generators of the affine group $\operatorname{Aff}(q)$, for $a \neq 0$ or 4δ. And if we include $a = 0$ and 4δ, we obtain all the K-double cosets of $G = GL(2, \mathbb{F}_q)$, where K is the isotropy subgroup fixing $\sqrt{\delta}$. We call the eigenvalues R_χ *one dimensional* because they come from one-dimensional representations of $\operatorname{Aff}(q)$. Note that if χ is the trivial representation, that is, $\chi(y) = 1$ for all $y \in \mathbb{F}_q^*$, then $R_\chi = q + 1$ and p_χ is constant.

K-Bessel Functions

We can use the power functions to build up other eigenfunctions of the adjacency operator by summing over various subgroups of $GL(2, \mathbb{F}_q)$, just as we did for the Poincaré upper half plane in Terras [1985, Vol. I, Ch. 3]. In this section we consider analogues of K-Bessel functions. We use the uppercase K to denote Kloosterman sums defined by (9) below. These are related by Proposition 1 below to the lower case animal defined by Equation (4) below. These k-Bessel functions are invariant with respect to the abelian subgroup N of G defined by

$$N = \left\{ \begin{pmatrix} 1 & x \\ 0 & 1 \end{pmatrix} \,\middle|\, x \in \mathbb{F}_q \right\}.$$

Note that the group N is isomorphic to the additive group of $\mathbb{F}_q$.

The real K-Bessel functions are well known to number theorists, since they appear in Fourier expansions of modular forms such as the Eisenstein series, although they are not always named (see Selberg [1989, pp. 367–70, 521–45] and Terras [1985, Vol. I]). There are applications to the Kronecker limit formula and the theory of complex multiplication, for example. The finite analogue or the Kloosterman sum appears in Fourier coefficients of modular forms as well. See Selberg [1989, pp. 506–20]. Recall that we encountered Klooster-man sums in Chapters 4 and 5 as eigenvalues of adjacency operators of Win-nie Li's graphs and finite euclidean graphs. We will see in formula (13) of Chapter 21 that they arise as matrix entries of representations of $GL(2, \mathbb{F}_q)$.

What do we mean by a k-Bessel function? Let's imitate the definition in Terras [1985, p. 136].

Definition. A k-*Bessel function* $f : H_q \to \mathbb{C}$ is an eigenfunction of all the adja-cency operators A_a of $X_q(\delta, a)$, such that $f(z)$ transforms by N according to the nontrivial additive character $\psi(x)$ of $\mathbb{F}_q$; that is,

$$A_a f = \lambda f, \quad \text{for all } a \in \mathbb{F}_q,$$
$$f(z + u) = \psi(u) f(z), \quad \text{for all } z \in H_q, \ u \in \mathbb{F}_q. \tag{2}$$

Here we will usually assume that when $q = p^r$, for the odd prime p, our additive character is

$$\psi(x) = \exp(2\pi i \operatorname{Tr}(x)/p), \tag{3}$$

where $\operatorname{Tr}(x)$ denotes the trace down to $\mathbb{F}_p$.

It is easy to construct such a k-Bessel function making use of the power function. Define the k-Bessel function $k(z \mid \chi, \psi)$, for $z \in H_q$, χ a multiplicative character, and ψ the additive character of $\mathbb{F}_q$ defined in formula (3), by

$$k(z \mid \chi, \psi) = \sum_{u \in \mathbb{F}_q} \chi \left[\operatorname{Im} \left(\frac{-1}{z + u} \right) \right] \psi(-u). \tag{4}$$

It is not hard to see that as a function of z, $k(z \mid \chi, \psi)$ satisfies (2) with the eigenvalue $\lambda = R_\chi$, the same eigenvalue as the power function p_χ. Note that if χ is identically 1 (i.e., the trivial character), then $k(z \mid 1, \psi) = 0$. Why? So we will exclude this case.

*Lemma 1. **Properties of k-Bessel functions on H_q.*** Under the assumptions of the definitions (3) and (4) above, we have the following properties of k-Bessel functions:

1) Let $z = x + y\sqrt{\delta}$. Then

$$k(z \mid \chi, \psi) = \chi(y) \psi(x) \sum_{u \in \mathbb{F}_q} \overline{\chi(u^2 - \delta y^2)} \, \psi(-u).$$

2) Let $f(z) = k(z \mid \chi, \psi)$. Then $f(z)$ is an eigenfunction of the adjacency operator A for $X_q(\delta, a)$, with the same eigenvalue as the power function p_χ. In particular, $f(z)$ is not identically zero, assuming that χ is a nontrivial multiplicative character.

3) For every $u \in \mathbb{F}_q$ and $z \in H_q$, we have

$$k(z + u \mid \chi, \psi) = \psi(u)k(z \mid \chi, \psi).$$

Proof.

1) See the exercise below.
2) Here you just need to use the fact that A is a left G-invariant operator; that is, A commutes with left translation. To see this, note that

$$Af(g\sqrt{\delta}) = \sum_{s \in S_q(\delta, a)} f(gs\sqrt{\delta}). \tag{5}$$

This means that it does not matter whether you replace g by hg before or after you apply the adjacency operator A.

Thus when A acts on the variable z:

$$A k(z \mid \chi, \psi) = \sum_{u \in \mathbb{F}_q} A\, p_\chi \left\{ \begin{pmatrix} 0 & -1 \\ 1 & 0 \end{pmatrix} \begin{pmatrix} 1 & u \\ 0 & 1 \end{pmatrix} z \right\} \psi(-u),$$

Since p_χ is an eigenfunction of A with eigenvalue R_χ, the same holds for $k(z \mid \chi, \psi)$, provided that it is not the zero function.

Now we must show that $k(z \mid \chi, \psi)$ isn't identically zero, assuming that χ is not trivial. To do this, take the formula in Part 1 of this lemma, multiply it by $\overline{\chi(y)\psi(x)}$, and sum it over $y \in \mathbb{F}_q^*$. This gives

$$\sum_{y \in \mathbb{F}_q^*} \overline{\chi(y)}\, k(x + y\sqrt{\delta} \mid \chi, \psi) = \sum_{\substack{y \in \mathbb{F}_q^* \\ v \in \mathbb{F}_q}} \overline{\chi(v^2 - \delta y^2)}\psi(-v)$$

$$= \sum_{w \in H_q} \overline{\chi(Nw)}\psi(-\mathrm{Tr}(w)/2)$$

$$= \Gamma_{q^2}\left(\overline{\chi \circ N}, \psi \circ \tfrac{1}{2}\mathrm{Tr} \right) - \Gamma_q(\bar{\chi}^2, \bar{\psi}),$$

where Γ_{q^2} and Γ_q are Gauss sums. We define, for a multiplicative character ω of $\mathbb{F}_q^*$ and an additive character λ of $\mathbb{F}_q$, the *Gauss sum*

$$\Gamma_q(\omega, \lambda) = \sum_{z \in \mathbb{F}_q} \omega(z)\lambda(z). \tag{6}$$

Here we define $\omega(0) = 0$. In the exercise below it is shown that $|\Gamma_q(\omega, \lambda)| = \sqrt{q}$, assuming that neither ω nor λ is the trivial character. It follows that the difference $\Gamma_{q^2} - \Gamma_q$ cannot vanish.

3) See the exercise below. ∎

Exercise.

1) Prove Part 1 of Lemma 1.
2) Prove that defining $\Gamma_q(\omega, \lambda)$ by (6), we have

$$|\Gamma_q(\omega, \lambda)| = \sqrt{q},$$

provided that neither the multiplicative character ω nor the additive character λ is trivial.
3) Prove Part 3 of Lemma 5.

Hint. 2. Note that the Gauss sum (6) is the additive Fourier transform of the multiplicative character ω. We already considered such sums in Chapter 8. However, we did not give the easiest possible proof of Part 2. It does follow from Lemma 2 of Chapter 8 when $q = p =$ an odd prime and ω is the Legendre symbol.

We are now using the notation Γ for Gauss sum to make a connection not only with Gauss but also with the gamma function. For, just as our k is a finite analogue of the Bessel function for the real upper half plane, the Gauss sum is a finite analogue of the Euler integral for the gamma function which is an integral over the positive real numbers of the product of an additive and a multiplicative character.

The Gauss sums were used by Gauss to give one of his many proofs of the quadratic reciprocity law – a proof close to that which we gave in Chapter 8. See Berndt, Evans and Williams [1998] and Ireland and Rosen [1993, pp. 147, 162] for more information about them.

It is not news that many classical special functions have counterparts over finite fields. See Evans [1986, 1993] for some other examples.

Question. When are the k-Bessel functions $k(z \mid \chi, \psi)$ and $k(z \mid \chi', \psi)$ orthogonal with respect to our usual inner product on $L^2(H_q)$? They are certainly orthogonal if they correspond to different eigenvalues $R_\chi \neq R_{\chi'}$, since the adjacency matrix is symmetric and thus eigenfunctions corresponding to different eigenvalues must be orthogonal. Evans [1994] shows that $k(z \mid \chi, \psi)$ and $k(z \mid \chi', \psi)$ are orthogonal iff $\chi' \neq \bar{\chi}$.

The next proposition says what happens when the additive characters rather than the multiplicative characters are distinct.

Proposition 1.

1) The k-Bessel functions $k(z \mid \chi, \psi)$ and $k(z \mid \chi, \psi')$ are orthogonal (with respect to the standard inner product on $L^2(H_q)$) if ψ and ψ' are distinct nontrivial additive characters of $\mathbb{F}_q$.

2) The k-Bessel function $k(z \mid \chi, \psi)$ is orthogonal to the power function $p_{\chi'}$ if ψ is any nontrivial additive character and if χ, χ' are any two multiplicative characters of $\mathbb{F}_q^*$.

Proof.

1) Note that

$$\sum_{z \in H_q} k(z \mid \chi, \psi)\overline{k(z \mid \chi, \psi')}$$

$$= \sum_{x \in \mathbb{F}_q} \psi(x)\overline{\psi'(x)} \sum_{y \in \mathbb{F}_q^*} k(y\sqrt{\delta} \mid \chi, \psi)\overline{k(y\sqrt{\delta} \mid \chi, \psi')}.$$

The sum over x is 0 as long as ψ and ψ' are different, by the orthogonality relations on the additive group $\mathbb{F}_q$.

2) The inner product we must compute is

$$\sum_{z \in H_q} k(z \mid \chi, \psi)\overline{p_{\chi'}(z)} = \sum_{x \in \mathbb{F}_q} \psi(x) \sum_{y \in \mathbb{F}_q^*} \chi'(y)k(y\sqrt{\delta} \mid \chi, \psi).$$

The sum over x vanishes assuming ψ is a nontrivial additive character of $\mathbb{F}_q$. ∎

Corollary. The eigenvalues R_χ of the adjacency matrix A for the graph $X_q(\delta, a)$ corresponding to the nontrivial one-dimensional representations of $\mathrm{Aff}(q)$ coming from multiplicative characters χ of $\mathbb{F}_q^*$ must occur with multiplicity at least q. All of these eigenvalues satisfy the Ramanujan bound

$$|R_\chi| \leq 2\sqrt{q}.$$

The rest of the eigenvalues λ of A with $\lambda \neq q + 1$ occur as eigenvalues of M_π, defined by

$$M_\pi = \sum_{s \in S_q(\delta, a)} \pi(s), \tag{7}$$

where π is the $(q-1)$-dimensional irreducible representation of $\mathrm{Aff}(q)$ induced from a nontrivial additive character ψ of $\mathbb{F}_q$ acting on the subgroup N of matrices of the form

$$\begin{pmatrix} 1 & x \\ 0 & 1 \end{pmatrix}$$

as in Proposition 2 of Chapter 16. All these eigenvalues of M_π have multiplicity $\geq q - 1$ in the adjacency operator A_a of $X_q(\delta, a)$. The only eigenvalue of A with multiplicity 1 is $q + 1$.

When looking at Table II.5 of eigenvalues (see the previous chapter), you will notice that the spectrum of the adjacency matrix for $X_q(\delta, a)$ is related to that for $X_q(\delta, 4\delta - a)$. You will also note that when $a = 2\delta$, there are $(q-1)/2$ zeros in the 1-dimensional eigenvalues as well as in the eigenvalues of M_π. These things are explained in the next proposition.

Proposition 2. (Nancy Celniker [1994]). Suppose that $a_2 = 4\delta - a_1$. For $i = 1, 2$, let $R_{\chi,i}$ denote the one-dimensional eigenvalue associated to the graph $X_q(\delta, a_i)$ and let M_i be the matrix M_π associated to the graph $X_q(\delta, a_i)$ as (7). Let W be the $(q-1) \times (q-1)$ shift matrix defined by

$$W = \begin{pmatrix} 0 & 1 & 0 & \cdot & \cdot & 0 \\ 0 & 0 & 1 & \cdot & \cdot & 0 \\ 0 & 0 & 0 & \cdot & \cdot & 0 \\ \cdot & \cdot & \cdot & \cdot & & \cdot \\ \cdot & \cdot & \cdot & & \cdot & 1 \\ 1 & 0 & 0 & \cdot & \cdot & 0 \end{pmatrix}.$$

Then we have the following conclusions.

1) The one-dimensional eigenvalues are related by $R_{\chi,2} = \varepsilon R_{\chi,1}$, where ε is ± 1. If

$$\chi(g) = \chi_j(g) = \exp\left(\frac{2\pi i j}{q-1}\right), \quad \text{for} = 0, 1, 2, \ldots, q - 2,$$

then

$$\varepsilon = \begin{cases} +1, & \text{if } j \text{ is even,} \\ -1, & \text{if } j \text{ is odd.} \end{cases}$$

Moreover, if $a_1 = 2\delta$, then $a_2 = a_1$, and $R_{\chi,1}$ must vanish whenever $\chi = \chi_j$, and j is odd.

2) The matrices associated to the $(q - 1)$-dimensional representation π of Aff(q) satisfy: $M_2 = M_1 W^r$, where $r = (q - 1)/2$. Thus the eigenvalues of M_1 and M_2 are the same up to a factor of -1. More precisely, $(q - 1)/2$ of the eigenvalues of M_2 are the same as the corresponding eigenvalues of M_1. And when $a_1 = 2\delta$, then $a_2 = a_1$, and, in this case, $(q - 1)/2$ of the eigenvalues of M_1 must vanish.

Proof.

1) See the exercise below.
2) Write $S_i = S_q(\delta, a_i)$, and let $z_i = x_i + y_i \sqrt{\delta} \in S_i$, for $i = 1, 2$. Note that $a_2 = 4\delta - a_1$ implies that

$$a_1 y_1 + \delta(y_1 - 1)^2 = -a_2 y_1 + \delta(-y_1 - 1)^2.$$

So the conjugation mapping $\gamma(z_1) = \gamma(x_1 + y_1\sqrt{\delta}) = \bar{z}_1 = x_1 - y_1\sqrt{\delta}$ takes the set S_1 one-to-one onto the set S_2.

Writing $y = g^u$, for $u \in \mathbb{Z}/(q - 1)\mathbb{Z}$, where g is a generator of the multiplicative group $\mathbb{F}_q^*$, we can say $u = \log y$. Note that this *finite analogue of the logarithm* has the usual properties, for example, it changes multiplication in $\mathbb{F}_q^*$ to addition in $\mathbb{Z}/(q - 1)\mathbb{Z}$. From the fact that $g^{(p-1)/2} = -1$, we have

$$\log(-1y) = \log(-1) + \log y = \frac{q - 1}{2} + \log y \pmod{(q - 1)}.$$

It follows that

$$M_2 = \sum_{z \in S_2} D(x)W^{\log y} = \sum_{z \in S_1} D(x)W^{\log(-y)}$$

$$= \sum_{z \in S_1} D(x)W^{\log y} W^{(q-1)/2} = M_1 W^{(q-1)/2}.$$

Thus we have proved that $M_2 = M_1 W^{(q-1)/2}$.

Note that

$$W^{(q-1)/2} = \begin{pmatrix} 0 & I \\ I & 0 \end{pmatrix}$$

and that we can write

$$M_1 = \begin{pmatrix} A & B \\ B & A \end{pmatrix},$$

where M_1 is symmetric. This is proved in the exercise below. Then

$$M_2 = \begin{pmatrix} A & B \\ B & A \end{pmatrix}\begin{pmatrix} 0 & I \\ I & 0 \end{pmatrix} = \begin{pmatrix} B & A \\ A & B \end{pmatrix}.$$

Let

$$F = \begin{pmatrix} I & I \\ -I & I \end{pmatrix}$$

and note that

$$F^{-1} = \frac{1}{2}\begin{pmatrix} I & -I \\ I & I \end{pmatrix}.$$

It follows that

$$FM_1F^{-1} = \begin{pmatrix} A+B & 0 \\ 0 & A-B \end{pmatrix} \quad \text{and}$$

$$FM_2F^{-1} = \begin{pmatrix} B+A & 0 \\ 0 & B-A \end{pmatrix}.$$

The result follows. In particular, if $a_1 = 2\delta$, then $a_1 = a_2$, and $M_1 = M_2$. So we find that

$$FM_1F^{-1} = \begin{pmatrix} 2A & 0 \\ 0 & 0 \end{pmatrix},$$

since $A = B$. ∎

Exercise.
1) Prove that

$$W^{(q-1)/2} = \begin{pmatrix} 0 & I \\ I & 0 \end{pmatrix}$$

and that we can write

$$M_1 = \begin{pmatrix} A & B \\ B & A \end{pmatrix}.$$

2) Prove Part 1 of Proposition 2 using methods similar to those which proved Part 2.
Hints. (See Celniker et al. [1993].)
2) To see that M_1 is symmetric, use the fact that S_1 is stable under inversion and the fact that π is a unitary representation. To see that M_1 has the desired

block form, use the fact that S_1 is stable under the map that sends $x + \sqrt{\delta}y$ to $-x + \sqrt{\delta}y$. Then use the formula for π given in Proposition 3 of Chapter 16.

Some Tables of Eigenvalues

For a connected regular graph X, define

$$\mu(X) = \max\{|\lambda| \mid \lambda \in \text{spectrum adjacency operator } X, \ |\lambda| \neq \text{degree}\}. \quad (8)$$

In Table II.6, we print the largest values of $\mu(X_p(\delta, a))$ which were found for $q = $ an odd prime ≤ 41. The largest μ occurs at $a = 2\delta$, except when $p = 29$. In Table II.7, we list the values of μ for the graphs $X_p(\delta, 2\delta)$ for primes p between 43 and 103. For 103, the values of $2\sqrt{p}$ and μ agree for the first 6 digits. But the computer, with a supposed knowledge of 16 digits, says the two values are actually different. Table II.8 gives the spectrum of finite upper half plane graphs over the field with 9 elements.

Just as on the real Poincaré upper half plane, there is another formula for a Bessel function. In this case, it is also called a Kloosterman sum. Recall that

Table II.6. *Largest Values of*
$\mu(X_p(\delta, a))$ *as in (8)*

p	d	a	$2\sqrt{p}$	μ
3	2	1	3.4641	2
5	2	4	4.4721	4
7	3	6	5.2915	4.8284
11	2	4	6.6332	6.4721
13	2	4	7.2111	7.2078
17	3	6	8.2462	8
19	2	4	8.7178	8.6275
23	5	10	9.5917	9.5915
29	2	2, 6	10.7703	10.5668
		1, 7		10.5429
		4		10.4721

$p = $ prime ≤ 29. Here μ does not occur as a 1-dimensional eigenvalue, except for $p = 11$ and when $p = 29$, for $a = 1$ and 7. Note that $p = 29$ is the first time that the largest μ does not occur when $a = 2\delta$. (From Celniker et al. [1993].)

Table II.7. *Values of*
$\mu(X_p(\delta, a))$, *defined by (8), for*
$43 \leq p \leq 73$. *(From Celniker*
et al. [1993])

p	d	$2\sqrt{p}$	μ
43	3	13.1149	12.9879
47	5	13.7113	13.5383
53	2	14.5602	14.5589
59	2	15.3623	15.3613
61	2	15.6205	15.6158
67	2	16.3707	16.3366
71	7	16.8523	16.8284
73	5	17.0880	17.0807

Table II.8. *Spectrum of* $X_9(1 + \sqrt{2}, \alpha)$. *(From Angel et al. [1992])*

α	1-Dimensional eigenvalues	Rest of spectrum
1	$10, (\pm 1.4142, 4), -2$	$-2, -2, -2, 4, \pm 4.4721, \pm 1.4142$
2	$10, (-2, -.8284, 4.8284), 2$	$-3.8541, -3.6180, -2, -1.3820,$ $-.8284, 2, 2.8541, 4.8284$
$1 + 2\sqrt{2}$	$10, (-4.8284, -2, .8284), 2$	$-4.8284, -3.8541, -2, .8284,$ $1.3820, 2, 2.8541, 3.6180$
$2 + 2\sqrt{2}$	$10, (-2, 0, 0), -6$	$-6, -2, 0, 0, 0, 0, 4, 4$

such sums appeared in Chapter 5 as eigenvalues of adjacency matrices of finite euclidean graphs.

Definition. Given an additive character ψ of $\mathbb{F}_q$ and a multiplicative character χ of $\mathbb{F}_q^*$, define the *Kloosterman sum* for $a, b \in \mathbb{F}_q$ by

$$K_\psi(\chi \mid a, b) = \sum_{t \in \mathbb{F}_q^*} \chi(t)\psi(at + bt^{-1}). \tag{9}$$

This should perhaps be called a generalized Kloosterman sum.

Such sums have been much studied by number theorists, since they occur in Fourier coefficients of modular forms. Thus bounds for them were wanted in connection with the Ramanujan conjectures. See Selberg [1989, pp. 506–20]. Recently there has been a certain amount of Kloostermania associated to them. See Huxley [1985] and Iwaniec [1989]. Gelfand, Graev, and Piatetski-

Shapiro [1990, p. 160] construct p-adic analogues of this function as matrix entries of continuous series representations of $SL(2, \mathbb{Q}_p)$. In a footnote, they also construct discrete series representations of $SL(2, \mathbb{F}_q)$ with a similar sort of function, which is a sum over $t \in \mathbb{F}_{q^2}$ such that $Nt = 1$ (loc. cit. p. 185). We will discuss this more in Chapter 21.

One obtains the relation between the k- and K-Bessel functions in the same way as in the continuous case (see Terras [1985, Vol. I, pp. 136–7]). For this we need Gauss sums again (defined in (6)). As we said earlier, Gauss sums can be viewed as finite analogues of the gamma function.

The formula relating the k- and K-Bessel functions for H_q is given in the following proposition.

Proposition. **Relation between k- and K-Bessel Functions.** Suppose that ψ is a nontrivial additive character of $\mathbb{F}_q$ while χ is a nontrivial multiplicative character. Given definitions (4), (6), and (9) above, we have

$$\Gamma(\chi, \psi)k(z \mid \chi, \psi) = g\chi(y)\psi(x)K(\chi' \mid -\delta y^2, -1/4), \tag{10}$$

where

$$g = \sum_{w \in \mathbb{F}_q} \psi(w^2),$$

is also a Gauss sum, and $\chi'(y) = \varepsilon(y)\chi(y)$, with

$$\varepsilon(y) = \begin{cases} 1, & \text{if } y \text{ is a square in } \mathbb{F}_q, \\ -1, & \text{otherwise.} \end{cases}$$

When $q = p$ is a prime, $\varepsilon(y)$ is the Legendre symbol from Chapter 8.

Proof. The proof proceeds as for the real analogue in Terras [1985, Vol. I, Ch. 3]. Note that

$$\Gamma(\chi, \psi)k(z \mid \chi, \psi) = \sum_{a \in \mathbb{F}_q^*} \chi(a)\psi(a)\chi(y)\psi(x) \sum_{v \in \mathbb{F}_q} \overline{\chi(v^2 - \delta y^2)}\psi(-v)$$

$$= \chi(y)\psi(x) \sum_{\substack{a \in \mathbb{F}_q^* \\ v \in \mathbb{F}_q}} \psi(-v)\chi(a(v^2 - \delta y^2)^{-1})\psi(a)$$

$$= \chi(y)\psi(x) \sum_{\substack{b \in \mathbb{F}_q^* \\ v \in \mathbb{F}_q}} \chi(b)\chi(bv^2 - b\delta y^2 - v).$$

Here we set $b = a(v^2 - \delta y^2)^{-1}$.

Complete the square next to see that

$$b(v^2 - b^{-1}v + (2b)^{-2}) - b\delta y^2 - (4b)^{-1} = b\left(v - \frac{1}{2b}\right)^2 - \frac{1}{4b} - \delta b y^2.$$

Then let $w = v - 1/2b$ and obtain

$$\Gamma(\chi, \psi)k(z \mid \chi, \psi)$$
$$= \chi(y)\psi(x) \sum_{b \in \mathbb{F}_q^*} \chi(b)\psi\left(\frac{-1}{4b} - b\delta y^2\right) \sum_{w \in \mathbb{F}_q} \psi(bw^2).$$

The proof is completed by doing the exercise below, that is, by showing that the sum over w is a Gauss sum:

$$\sum_{w \in \mathbb{F}_q} \psi(bw^2) = \lambda(b)g.$$

∎

Exercise. Prove that for $b \in \mathbb{F}_q^*$, and ψ a nontrivial additive character of $\mathbb{F}_q$, we have

$$\sum_{w \in \mathbb{F}_q} \psi(bw^2) = \varepsilon(b) \sum_{w \in \mathbb{F}_q} \psi(w^2),$$

where ε is defined as in Proposition 3.

Hint. See Chapter 8.

Spherical Functions

Next we study spherical functions and their connections with representation theory and then we finally go back to our example H_q. We model our discussion on that in Terras [1988, Vol. II, pp. 69–70]. Here we suppose that we have a finite symmetric space G/K with finite groups G and K as in Theorem 1 of Chapter 19. Again, the example of H_q appears to be a good one to keep in mind. We will be studying functions $f : G \to \mathbb{C}$ that are K-bi-invariant, that is, $f(kxh) = f(x)$ for all $k, h \in K$ and all $x \in G$. We can also think of f as a function on the symmetric space G/K itself or as a function on the double coset space $K \backslash G / K$. Define

$$L^2(K \backslash G / K) = \{f : G \to \mathbb{C} \mid f(kxh) = f(x) \,\forall k, h \in K, \forall x \in G\}. \quad (11)$$

Laplace and Legendre introduced spherical harmonics, that is, eigenfunctions of the Laplacian on the sphere, in the 1780s. See any undergraduate

Fourier series and boundary value problems course for a discussion or Terras [1985, Vol. I, Chapter 2]. The analogue for the real Poincaré upper half plane is the Legendre function (which is an integral of the power function y^s over $K = SO(2)$). The theory of spherical functions for continuous Lie groups has been carried out by many people in this century (e.g., Cartan, Weyl, Gelfand, and Harish–Chandra). Cartan noted the various characterizations of spherical functions on compact symmetric spaces G/K in 1929. He saw these functions as matrix entries of representations π of G with $\pi(K)$ having a fixed vector. Gelfand studied the convolution algebra of K-bi-invariant functions on G. The main point is that this is a commutative algebra if G/K is a symmetric space (see Theorem 1 of Chapter 19). Gelfand theory emphasizes the convolution integral operators rather than the Laplacian. Of course, in the finite case, these coincide. The analogue for p-adic groups exists also (see MacDonald [1971]). And there is an analogue for free groups (see Figá-Talamanca and Picardello [1983] and Chapter 24). Selberg made use of the theory of spherical functions in deriving the Selberg trace formula.

In the case that the symmetric space is $\mathbb{F}_2^n$, we saw in Chapter 10 that the spherical functions are Krawtchouk polynomials, which are of interest in coding theory (see Jessie MacWilliams and Sloane [1988]). Some references for the finite spherical functions are: Bannai and Ito [1984], Diaconis [1988], Lubotzky [1988], Poulos [1991], Soto-Andrade [1987], Stanton [1990], and Velasquez [1991].

We can view spherical functions through many different lenses: Fourier analysis on symmetric spaces, representation theory, the theory of permutation groups, association schemes, graph theory via the concept of collapsed adjacency matrices, or Hecke operators.

Recall that in formula (6) of Chapter 15, we defined *convolution* of functions f, g on G by

$$(f * g)(u) = \sum_{v \in G} f(v)g(v^{-1}u).$$

Theorem 1. **Three Equivalent Definitions of Spherical Functions.** A K-bi-invariant function $h : G \to \mathbb{C}$ is spherical if it satisfies any of the following equivalent criteria:

1) The function h takes the value 1 at the identity e of G and is an eigenfunction of all the convolution operators; that is, for all $f \in L^2(K\backslash G/K)$

$$(f * h)(x) = \lambda_f h(x) \quad \text{for some } \lambda_f \in \mathbb{C}.$$

2) The mapping of $f \in L^2(K\backslash G/K)$ to $(f * h)(e)$ yields an algebra homomor-phism of $L^2(K\backslash G/K)$ onto $\mathbb{C}$.

3) The nonzero function h is an eigenfunction of the mean-value operators:

$$\frac{1}{|K|} \sum_{k \in K} h(xky) = h(x)h(y), \quad \text{for all } x, y \in G. \tag{12}$$

Proof.

● **Part 1 implies Part 2.**

Assume criterion 1). That is, suppose that $h(e) = 1$ and that $f * h = \lambda_f h$, for all $f \in L^2(K\backslash G/K)$. Then $(f * h)(e) = \lambda_f$. Since convolution is associative, we see that

$$\lambda_{f*g} = ((f * g) * h)(e) = (f * (g * h))(e)$$
$$= [f * (\lambda_g h)](e) = \lambda_f \lambda_g h(e) = \lambda_f \lambda_g.$$

This is criterion 2). Why is the map onto? (*Hint*: Let $f = \delta_K$.)

● **Part 2 implies Part 3.**

Assume that

$$(h * (f * g))(e) = ((f * g) * h)(e) = (f * h)(e)(g * h)(e).$$

This implies

$$\sum_{x, y \in G} f(x^{-1})g(y^{-1})h(yx) = \sum_{s, t \in G} f(s^{-1})g(t^{-1})h(s)h(t).$$

So we can replace s and t on the right by x and y and then replace y on both sides by yk and sum over $k \in K$ to get

$$\sum_{x, y \in G} f(x^{-1})g(y^{-1}) \sum_{k \in K} h(ykx) = |K| \sum_{x, y \in G} f(x^{-1})g(y^{-1})h(x)h(y).$$

This implies 3), since f and g are arbitrary in $L^2(K\backslash G/K)$.

● **Part 3 implies Part 1.**

Suppose that

$$\frac{1}{|K|} \sum_{k \in K} h(xky) = h(x)h(y), \quad \text{for all } x, y \in G.$$

Then if f is K-bi-invariant,

$$(f * h)(x) = \sum_{v \in G} f(v) h(v^{-1} x) = \frac{1}{|K|} \sum_{\substack{v \in G \\ k \in K}} f(v^{-1}) h(vkx)$$

$$= \sum_{v \in G} f(v^{-1}) h(v) h(x) = (f * h)(e) h(x) = \lambda_f h(x),$$

where the eigenvalue $\lambda_f = (f * h)(e)$. Note that if you set $x = e$ in the formula of Part 3), you find that $h(e) = 1$, assuming h is nonzero. That is, $h(e) = 0$ iff h is identically 0. ∎

Moral of 3 in Theorem 1

eigenfunctions = eigenvalues.

When you say this to most mathematicians, their hair stands on end. Nevertheless it is true for spherical functions on symmetric spaces. And thus in our quest for good formulas for eigenvalues of finite upper half plane graphs, we now see that it suffices to find good formulas for the spherical functions.

Application. Spherical functions are both eigenfunction and eigenvalue of the adjacency operators on the finite upper half plane graphs.

Next let's work out the connection of all this with representation theory. Here we follow discussions in Diaconis [1988] and Vilenkin [1968].

*Theorem 2. **Equivalent Conditions for a Gelfand Pair.*** The following are equivalent:

1) (G, K) is a Gelfand pair.
2) No $\pi_i \in \hat{G}$ occurs more than once in the decomposition

$$\text{Ind}_K^G 1 = \pi_1 \oplus \cdots \oplus \pi_r;$$

that is, the induced representation is multiplicity-free.
3) For every $\pi_i \in \hat{G}$, with $\pi : G \to GL(V)$, occurring in $\text{Ind}_K^G 1$, there is a basis of V such that for every $f \in L^2(K \backslash G / K)$, the Fourier transform $\hat{f}(\pi)$ has the form

$$\hat{f}(\pi) = \begin{pmatrix} c & 0 \\ 0 & 0 \end{pmatrix}, \quad \text{where } c \in \mathbb{C}.$$

Proof.

• **Part 2 implies Part 3.**

Suppose that the induced representation is multiplicity-free. By the Frobenius reciprocity law, this says that the trivial representation has multiplicity one in the restriction of π_i to K. If $\pi_i : G \to GL(V_i)$, then this means that the space of K-fixed vectors in V_i is one dimensional; that is,

$$V_i^K = \{v \in V_i \mid \pi_i(k)v = v, \quad \text{for all } k \in K\} = \mathbb{C}s_i.$$

Because the induced representation of G occurs on the space $L^2(G)$, we can view s_i as a function on G. We can assume that the norm of s_i is 1.

Complete s_i to a basis for V_i so that

$$\pi_i(k) = \begin{bmatrix} 1 & 0 \\ 0 & * \end{bmatrix}, \quad \text{for all } k \in K.$$

This is just the matrix statement of the fact that

$$\operatorname{Res}_K^G \pi_i = 1 \oplus \text{(other inequivalent representations in } \hat{K}).$$

Now look at $f \in L^2(K \backslash G / K)$, using the fact that $f(xk) = f(x)$ for all $x \in G$ and $k \in K$. We see that the Fourier transform of f at π_i has the following expression:

$$\hat{f}(\pi_i) = \sum_{g \in G} f(g) \pi_i(g) = \sum_{x \in G/K} \sum_{k \in K} f(xk) \pi_i(xk)$$

$$= \sum_{x \in G/K} f(x) \pi_i(x) \sum_{k \in K} \pi_i(k).$$

The orthogonality relations for the matrix entries of representations in $\hat{K}$ imply that

$$\sum_{k \in K} \pi_i(k) = \begin{pmatrix} |K| & 0 \\ 0 & 0 \end{pmatrix}.$$

Thus

$$\hat{f}(\pi_i) = \begin{pmatrix} * & 0 \\ * & 0 \end{pmatrix}, \quad \text{where the upper left block is } 1 \times 1.$$

You can do a similar calculation using the fact that $f(kx) = f(x)$ for all $k \in K$ and $x \in G$ to see that

$$\hat{f}(\pi_i) = \begin{pmatrix} * & * \\ 0 & 0 \end{pmatrix}, \quad \text{where the upper left block is } 1 \times 1.$$

This gives the statement of Part 3.

- **Part 3 implies Part 1**

Suppose that f, g are in $L^2(K\backslash G/K)$. Then from Part 3 we see that the Fourier transforms of f and g commute; that is,

$$\hat{f}(\pi)\hat{g}(\pi) = \hat{g}(\pi)\hat{f}(\pi) \quad \text{for all } \pi \epsilon \hat{G} \text{ which occur in } \mathrm{Ind}_K^G 1.$$

By the convolution property of the Fourier transform (and the fact that it is invertible), it follows that $f * g = g * f$.

- **Part 1 implies Part 2**

Suppose that some $\pi \in \hat{G}$ occurs with multiplicity $j > 1$ in $\mathrm{Ind}_K^G 1$. Pick a basis of V, the vector space on which π acts, such that the first j coordinates span the space V^K of K-fixed vectors. The Frobenius reciprocity law says that V^K has dimension j over $\mathbb{C}$. We can take two $j \times j$ matrices A_1 and A_2 such that $A_1 A_2 \neq A_2 A_1$ and define functions $f_i \in L^2(K\backslash G/K)$ by $f_i(\rho) = 0$, for $\rho \neq \pi$, and

$$\hat{f}_i(\pi) = \begin{pmatrix} A_i & 0 \\ 0 & 0 \end{pmatrix}, \quad \text{for } i = 1, 2.$$

Then, writing $\mathscr{F}f = \hat{f}$, we have $\mathscr{F}(f_1 * f_2) \neq \mathscr{F}(f_2 * f_1)$. It follows that $f_1 * f_2 \neq f_2 * f_1$, which is a contradiction. ∎

When we are interested in Fourier analysis of functions on G/K or $K\backslash G/K$, we will need to define

$$\hat{G}^K \{\pi \in \hat{G} \mid \pi \text{ occurs in } \mathrm{Ind}_K^G 1\}. \tag{13}$$

As we have seen above, these are the irreducible unitary representations of G which have a K-fixed vector. And the space of K-fixed vectors is one dimensional.

Corollary 1. **Fourier Expansions on the Double Coset Space $K\backslash G/K$ and Spherical Functions.** Suppose (G, K) is a Gelfand pair and that the representation π is in $\hat{G}^K$ defined by (13). Then if $\pi : G \to GL(V)$, there is a one-dimensional space V^K of K-fixed vectors in V; that is,

$$V^K = \{v \in V \mid \pi(K)v = v, \text{ for all } k \in K\} = \mathbb{C}s_\pi.$$

Here we assume that s_π has norm 1. The *spherical function corresponding to* π is

$$h_\pi(x) = \langle \pi(x)s_\pi, s_\pi \rangle. \tag{14}$$

Thus the spherical function $h_\pi(x)$ can be viewed as the upper left matrix entry of π.

1) If $f \in L^2(K \backslash G / K)$, then the Fourier expansion of f from Part 3 of Theorem 2 in Chapter 15 is given by the following expansion in spherical functions:

$$f(x) = \frac{1}{|G|} \sum_{\pi \in \hat{G}^K} d_\pi \langle f, h_\pi \rangle h_\pi(x), \tag{15}$$

where $\langle f, h_\pi \rangle$ is our usual inner product on $L^2(G)$.
2) The spherical function h_π has the expression

$$h_\pi(x) = \frac{1}{|K|} \sum_{k \in K} \chi_\pi(kx), \tag{16}$$

where χ_π denotes the character of π.
3) The function h_π has all the properties listed in Theorem 1.

Proof.

1) Since (G, K) is a Gelfand pair, we know that the trivial representation of K occurs in the restriction $\text{Res}_K^G \pi$ with multiplicity one and so by Part 3 of Theorem 2, if $f \in L^2(K \backslash G / K)$,

$$\hat{f}(\pi) = \begin{pmatrix} c & 0 \\ 0 & 0 \end{pmatrix}, \quad \text{where } c \in \mathbb{C}.$$

Thus from Chapter 15, Theorem 2, Part 3 (the Fourier inversion formula on G), we have

$$f(x) = \frac{1}{|G|} \sum_{\pi \in \hat{G}^K} d_\pi \sum_{i,j=1}^{d_\pi} \langle f, \pi_{ij} \rangle \pi_{ij}(x) = \frac{1}{|G|} \sum_{\pi \in \hat{G}^K} d_\pi \langle f, \pi_{11} \rangle \pi_{11}(x).$$

The last equality happens because of vanishing of all but (perhaps) one entry of $\hat{f}(\pi)$. Since $\pi_{11}(x) = h_\pi(x)$, we are done.

2) By Part 1, we know that $h_\pi(x) = \pi_{11}(x)$ for $\pi \in \hat{G}^K$. Then using the orthogonality relations for the matrix entries of representations of K:

$$\frac{1}{|K|} \sum_{k \in K} \mathrm{Tr}(\pi(kx)) = \frac{1}{|K|} \mathrm{Tr}\left(\sum_{k \in K} \pi(k)\pi(x) \right)$$

$$= \frac{1}{|K|} \mathrm{Tr}\left(\begin{pmatrix} |K| & 0 \\ 0 & 0 \end{pmatrix} \pi(x) \right) = \pi_{11}(x) = h_\pi(x).$$

3) Prove as an exercise.

Hint. The only problem is to show that h_π is K-bi-invariant and satisfies Part 3 of Theorem 1. Use the formula in Part 2 of this corollary. To see that h_π is an eigenfunction of the mean-value operators in Part 3 of Theorem 1, you need to use the orthogonality relations for the matrix entries of representations of K again. ∎

Exercise. Show the following inner product formula for spherical functions:

$$\langle h_\pi, h_\rho \rangle = \begin{cases} 0, & \text{if } \pi \not\cong \rho, \\ |G|/d_\pi, & \text{if } \pi \cong \rho. \end{cases} \tag{17}$$

Here $\langle \,,\, \rangle$ denotes our usual inner product on $L^2(G)$.

The Spherical Transform of K-Bi-Invariant Functions on G

Fourier analysis is simpler on $L^2(K \backslash G / K)$ than on G. It involves the spherical transform, which takes K-bi-invariant functions on G to complex-valued functions on $\hat{G}^K$. It is not a matrix-valued transform!

Definition. The *spherical transform* of $f \in L^2(K \backslash G / K)$ at $\pi \in \hat{G}^K$ is

$$(\mathscr{S}f)(\pi) = \sum_{x \in G} f(x) h_\pi(x^{-1}) = \sum_{x \in G} f(x) \overline{h_\pi(x)} = \langle f, h_\pi \rangle.$$

The spherical transform has been much studied on real symmetric spaces such as the Poincaré upper half plane (see Terras [1985]). It has most of the properties of the Fourier transform on $\mathbb{R}$ or $\mathbb{Z}/n\mathbb{Z}$. For example, $\mathscr{S}f(\pi)$ is a complex number rather than a matrix, and we have already proved an inversion formula (15).

Exercise. Prove that if f and g are in $L^2(K \backslash G / K)$,

$$\mathscr{S}(f * g)(\pi) = \mathscr{S}f(\pi) \cdot \mathscr{S}g(\pi).$$

This exercise says that the spherical transform takes convolution to ordinary pointwise product. Since pointwise product is commutative, this can only hold if convolution is commutative, that is, if (G, K) is a Gelfand pair.

There are also several uncertainty principles for the spherical transform. One of them can be found in Elinor Velasquez [1998]. Another says that for nonzero $f \in L^2(K \backslash G / K)$

$$|\text{supp } f| \sum_{\pi \in \text{supp}(\mathcal{S}f)} d_\pi \geq |G|.$$

Exercise. Prove the uncertainty principle above assuming that G/K is the finite upper half plane H_q.
Hint. Imitate the proof of the corresponding result in Chapter 14. See Angel, Poulos, Terras, Trimble, and Velasquez [1994].

We will see the spherical transform again when we investigate the Selberg trace formula on H_q.

Remarks.

1) The spherical function h_π is *positive definite*, meaning that the matrix

$$(h_\pi(x^{-1}y))_{x,y \in G} \quad \text{is positive difinite.}$$

To see this, recall that as in Corollary 1, we have $h_\pi(x) = \langle s_\pi, \pi(x)s_\pi \rangle$, where s_π is a K-fixed vector of norm 1. Since we may assume that π is a unitary representation,

$$\sum_{x,y \in G} a(x)\overline{a(y)} \, h_\pi(x^{-1}y) = \sum_{x,y \in G} \langle a(x)s_\pi, a(y)\pi(x^{-1}y)s_\pi \rangle$$

$$= \sum_{x,y \in G} \langle a(x)\pi(x)s_\pi, a(y)\pi(y)s_\pi \rangle$$

$$= \langle w, w \rangle \geq 0,$$

where

$$w = \sum_{x \in G} a(x)\pi(x)s_\pi.$$

2) One can show that every positive definite spherical function h comes from some π in $\hat{G}^K$. Create the vector space V as the span of right translates of

h. Thus the eigenvalues of adjacency operators on finite upper half plane graphs correspond to representations $\pi \in \hat{G}^K$.

*Corollary 2. **Fourier Expansions on G/K and Associated Spherical Functions.*** Suppose that (G, K) is a Gelfand pair. Then any $f \in L^2(G/K)$ has a Fourier expansion of the form

$$f(x) = \frac{1}{|G|} \sum_{\pi \in \hat{G}^K} d_\pi \sum_{i=1}^{d_\pi} \langle f, \pi_{i1} \rangle \pi_{i1}(x).$$

Here, again we sum only over the $\pi \in \hat{G}$ with a K-fixed vector s_π (uniquely determined by $\|s_\pi\| = 1$) and we use only the matrix entries from the first column of π. These matrix entries in the first column of π are called *associated spherical functions*.

Proof. Vilenkin [1968]. The proof is very similar to that of Part 1 of Corollary 1. Start with Part 3 of Theorem 2 of Chapter 15 (the Fourier expansion of functions on G). Then note that $\pi \in \hat{G}$ and $f \in L^2(G/K)$ implies that

$$\hat{f}(\bar\pi) = \sum_{x \in G} f(x)\overline{\pi(x)} = \sum_{x \in G/K} f(x)\overline{\pi(x)} \sum_{k \in K} \overline{\pi(k)}.$$

Now $\sum_{k \in K} \overline{\pi(k)} = 0$ unless the restriction $\mathrm{Res}_K^G \pi$ contains the trivial representation; that is, unless $\pi \in \hat{G}^K$.

And $\pi \in \hat{G}^K$ implies

$$\sum_{k \in K} \overline{\pi(k)} = \begin{pmatrix} |K| & 0 \\ 0 & 0 \end{pmatrix}$$

by the orthogonality relations for representations of K. Thus, if

$$M = \sum_{x \in G/K} f(x)\overline{\pi(x)},$$

we have

$$\hat{f}(\bar\pi) = M \begin{pmatrix} |K| & 0 \\ 0 & 0 \end{pmatrix} = (*\,0).$$

Therefore only the first column of matrix entries of π contributes to the Fourier expansion of $f \in L^2(G/K)$. ■

Now we return to our favorite example.

Example. Spherical Functions on Finite Upper Half Planes

A *spherical function* $f : H_q \to \mathbb{C}$ is a K-invariant eigenfunction of all the Laplacians for the graphs $X_q(\delta, a)$ that we have attached to H_q such that $f(\sqrt{\delta}) = 1$, where K is the isotropy subgroup of $\sqrt{\delta}$ in $G = GL(2, \mathbb{F}_q)$.

Recall that we can identify the finite upper half plane H_q with G/K and that (G, K) is a Gelfand pair. In the discussion that follows we shall identify functions on H_q with functions on G by identifying $f(g)$ with $f(g\sqrt{\delta})$ for $g \in G = GL(2, \mathbb{F}_q)$. We will also often identify functions on G with functions on G/K or $K \backslash G/K$.

We write A_a for the adjacency operator associated to the generating set $S_a = S_q(\delta, a)$, for $a \in \mathbb{F}_q$, defined in Theorem 2 of Chapter 19. For what follows, it is important to remember Part 5 of that Theorem 2, which says that the S_a represent the K-orbits in H_q or the K-double cosets of G.

We also need to recall that the S_a are invariant under inversion, that is, that $g \in S_a$ implies that $g^{-1} \in S_a$.

Definition. The *adjacency operator* A_a of the graph $X_q(\delta, a)$ is defined by

$$A_a f(z) = \sum_{d(z,w) = a} f(w).$$

Note that $A_0 = I$ and $A_{4\delta} f(z) = f(\bar{z})$. The other A_a are adjacency operators of $(q + 1)$-regular graphs. Identify a function $f(z)$ on H_q with a function on the group $G = GL(2, \mathbb{F}_q)$ by writing $f(g) = f(g\sqrt{\delta})$, for $g \in G$.

It follows that the adjacency operator A_a is actually convolution with the delta function for the set S_a; that is, $A_a f = f * \delta_{S_a}$, where $\delta_S(u) = 1$ for $u \in S$ and 0 otherwise.

From Chapter 19, we know that H_q is a symmetric space G/K and thus the adjacency operators A_i, $i \in \mathbb{F}_q$, form a commutative set of self-adjoint operators. Thus there is a common basis of orthogonal eigenfunctions for the adjacency operators A_i, $i \in \mathbb{F}_q$. They are orthogonal with respect to the standard inner product

$$\langle f, g \rangle = \sum_{t \in G} f(t\sqrt{\delta}) \overline{g(t\sqrt{\delta})}.$$

This basis is a basis of spherical functions, once renormalized to have value 1 at the origin.

In particular, if $f : G/K \to \mathbb{C}$ is a K-invariant eigenfunction of all the adjacency operators A_a, then it is also an eigenfunction of the *mean value operators* from Part 3 of Theorem 1:

$$M_y f(x) = |K|^{-1} \sum_{k \in K} f(xky), \quad \text{for each } x, y \in G.$$

The reason is that M_y is a constant multiple of the adjacency operator for the graph $X_q(\delta, a)$ if y is in the generating set $S_q(\delta, a)$ defined in Theorem 2 of Chapter 19.

Suppose then that $M_y f = \lambda_y f$ for some $\lambda_y \in \mathbb{C}$. Then $f(I) = 1$ implies that the eigenvalue $\lambda_y = f(y)$. To see this, write out what it means to say $M_y f(x) = \lambda_y f(x)$ at $x = $ the identity, assuming that f is K-invariant. Thus we see that the eigenvalues of the mean value operators are the eigenfunctions (as noted after Theorem 1). As a corollary we see that our spherical functions are automatically real-valued in this case.

The adjacency operators could be studied using algebraic combinatorics as in Bannai and Ito (1). The adjacency relations form a symmetric association scheme. But we do not have a P- or Q-polynomial association scheme. Our space is not two-point homogeneous. Nevertheless the space has many of the properties obtained by Bannai and Ito [1984] and Stanton [1981, 1984, 1990].

See also the permutation groups version of things in Neumann [1977] where the ring of adjacency operators is connected to the centralizer ring of the permutation group G acting on G/K.

We want to imitate the theory of the Poincaré upper half plane as described, for example, in Terras [1985, Vol. I]. In that case one has $G = SL(2, \mathbb{R}) = KAK$, where $K = SO(2)$ and $A = $ the diagonal matrices in G. This does not happen for $GL(2, \mathbb{F}_q)$, since the sets S_a need not be represented by diagonal matrices. And in the Poincaré upper half plane, one has Harish-Chandra's integral formula for spherical functions, which writes *any* spherical function as an integral over K of a power function (formula (4) of Chapter 23 or Terras [1988; Vol. II, p. 70]). This result does produce some spherical functions in the finite case – but *only* those corresponding to 1-dimensional eigenvalues and not arbitrary spherical functions. See formula (18) below. We know from Corollary 1 to Theorem 2 that we can get all the spherical functions on H_q by averaging characters of representations of $GL(2, \mathbb{F}_q)$ that have nonzero K-fixed vectors. Soto-Andrade [1987] has worked out the details of this result. See also Evans [1994]. And see formula (19) below as well as Theorem 4 in Chapter 21.

Poulos [1991] shows that our graphs $X_q(\delta, a)$, for $a \neq 0, 4\delta$, are not only regular, but also highly regular as defined in formula (18) of Chapter 10. This allows us to collapse the adjacency matrix to a $q \times q$ matrix with the same minimal polynomial.

Definition. A connected graph $X(V, E)$ is *highly regular* with collapsed adjacency matrix $C = (c_{ij})$ iff for every vertex $v \in V$, there is a partition of V into sets V_i, $i = 1, \ldots, n$, with $V_1 = \{v\}$, such that each vertex $y \in V_i$ is adjacent to exactly c_{ij} vertices in V_j.

Theorem 3. (Poulos [1991]). **Finite Upper Half Plane Graphs are Highly Regular.**

1) For $a \neq 0, 4\delta$, the graph $X_q(\delta, a)$ is highly regular.
2) We can partition H_q as a union of sets $S_i = S_q(\delta, i)$, $i \in \mathbb{F}_q$. Then the collapsed adjacency operator $C = (c_{ij})$ associated to the graph $X_q(\delta, a)$ has i, j entry c_{ij} equal to the number of vertices $y \in S_j$ to which each vertex $v \in S_i$ is adjacent in $X_q(\delta, a)$. This number is independent of the choice of v in S_i.
3) When $(i, j) \neq (0, a)$ or $(4\delta, 4\delta - a)$, we have $c_{ij} \leq 2$.

Proof. See Poulos [1991] and Angel [1993] for the proof of this result.

Example 1. H_3. In this case, $\delta = -1$ and we write $i = \sqrt{-1}$. There is only one 4-regular graph, the octahedron of Figure II.2. The K-orbits in H_3 are

$$S_0 = \{i\}, \quad S_1 = \{\pm i \pm 1\}, \quad S_{-1} = \{-i\}.$$

The spherical functions are determined by their values on the S_i. And the collapsed adjacency matrix is easily seen to be

$$\begin{pmatrix} 0 & 4 & 0 \\ 1 & 2 & 1 \\ 0 & 4 & 0 \end{pmatrix}.$$

The eigenvalues are 0, 4, and -2. The eigenvalues 0 and 4 are one-dimensional, but -2 is not. The eigenvectors are $(1, 0, -1)$, $(1, 1, 1)$, and $(1, -1/2, 1)$. These eigenfunctions give the spherical functions for the octahedron H_3.

The collapsed adjacency matrix can also be interpreted as the adjacency matrix of a multigraph, that is, a collection of vertices connected by 0, 1, or more directed edges.

Since the spherical function f is constant on S_i, we write $f(S_i) = f(i)$. Then $Af = \lambda f$ iff

$$4f(1) = \lambda f(0),$$
$$f(0) + 2f(1) + f(-1) = \lambda f(1),$$
$$4f(1) = \lambda f(-1).$$

In this case, we are looking at a three-term recursion for the spherical functions. However, as soon as q is larger than 3, this fails (except possibly for the $a = 2\delta$ graph), since Poulos [1991] proves that $z \in H_q$ with $z \neq \pm\sqrt{\delta}$ is adjacent to at most two-elements of any S_i. This means that if you arrange the collapsed adjacency matrix so that the first and last rows correspond to $\sqrt{\delta}$ and $-\sqrt{\delta}$, respectively, then the middle rows have entries that are ≤ 2.

Example 2. H_5. In what follows we take $\delta = 2$ and list the sets S_i in the following order to take advantage of symmetry:

$$S_0, \; S_{-4\delta\,=\,2}, \; S_{-8\delta\,=\,4}, \; S_{-12\delta\,=\,1}, \; S_{-16\delta\,=\,3\,=\,4\delta}.$$

The collapsed adjacency matrix for $X_5(2, 1)$ is

$$\begin{pmatrix} 0 & 0 & 0 & 6 & 0 \\ 0 & 1 & 2 & 2 & 1 \\ 0 & 2 & 2 & 2 & 0 \\ 1 & 2 & 2 & 1 & 0 \\ 0 & 6 & 0 & 0 & 0 \end{pmatrix}.$$

The eigenvalues are $6, -2, -3, 2$, and 1.
The collapsed adjacency matrix for $X_5(2, 2)$ is

$$\begin{pmatrix} 0 & 6 & 0 & 0 & 0 \\ 1 & 2 & 2 & 1 & 0 \\ 0 & 2 & 2 & 2 & 0 \\ 0 & 1 & 2 & 2 & 1 \\ 0 & 0 & 0 & 6 & 0 \end{pmatrix}.$$

The eigenvalues are $6, 3, 1, -2$, and -2.
The collapsed adjacency matrix for $X_5(2, 4)$ is

$$\begin{pmatrix} 0 & 0 & 6 & 0 & 0 \\ 0 & 2 & 2 & 2 & 0 \\ 1 & 2 & 0 & 2 & 1 \\ 0 & 2 & 2 & 2 & 0 \\ 0 & 0 & 6 & 0 & 0 \end{pmatrix}.$$

The eigenvalues are $6, -4, 2, 0$, and 0.

The common eigenfunctions for these three collapsed adjacency matrices are

$$\begin{pmatrix} 1 \\ 1 \\ 1 \\ 1 \\ 1 \end{pmatrix}, \begin{pmatrix} 6 \\ 1 \\ -4 \\ 1 \\ 6 \end{pmatrix}, \begin{pmatrix} 6 \\ -2 \\ 2 \\ -2 \\ 6 \end{pmatrix}, \begin{pmatrix} 6 \\ 3 \\ 0 \\ -3 \\ -6 \end{pmatrix}, \begin{pmatrix} 1 \\ -2 \\ 0 \\ 2 \\ -1 \end{pmatrix}.$$

The middle three vectors should be multiplied by $1/6$ to make them correspond to spherical functions.

Harish-Chandra's integral formula for spherical functions on the real Poincaré upper half plane (formula (4) of Chapter 23) has an analogue in the finite case for χ a character of the multiplicative group $\mathbb{F}_q^*$:

$$h_\chi(z) = |K|^{-1} \sum_{k \in K} p_\chi(kz). \tag{18}$$

Unfortunately this gives only the spherical functions corresponding to the one-dimensional eigenvalues of Aff(q). These can also be viewed as the spherical functions corresponding to what in the real case would be called continuous series representations of G. Since we already know, by the corollary to Theorem 3 of Chapter 19, that the one-dimensional eigenvalues satisfy the Ramanujan bound, we are not too excited by (18).

What are the non-one-dimensional spherical functions and how can we bound them or, equivalently, their eigenvalues? We have in formula (16) a similar result to (18) with p_χ replaced by the character of a higher dimensional representation of G found by decomposing $\mathrm{Ind}_K^G 1$. Soto-Andrade [1987] has worked out the details of this and found the following formula for the rest of the spherical functions on H_q. We will give the details of the proof in our discussion of Theorem 4 in the next chapter (using the method of Evans [1994]).

Suppose that ω is a nontrivial representation of the subgroup U of $\mathbb{F}_q(\sqrt{\delta})^*$ consisting of elements w such that $Nw = w^{1+q} = 1$. Let ε denote the character of the multiplicative group $\mathbb{F}_q^*$ that is 1 on squares, -1 on nonsquares, and 0 on 0. The *Soto–Andrade formula for the spherical function* associated to ω is

$$(q+1)h_\omega(z)$$

$$= \sum_{\substack{w = u + v\sqrt{\delta} \\ Nw = 1}} \varepsilon\left(2u + \frac{d(z, \sqrt{\delta})}{\delta} - 2\right)\omega(w), \quad \text{if } d(z, \sqrt{\delta}) \neq 0, 4\delta. \tag{19}$$

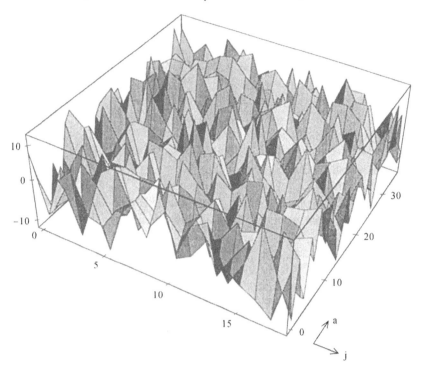

Figure II.5. Soto–Andrade spherical functions for $p = 37$. $(q + 1)h_{\omega_0^j}(S_a), a \neq 0, 4\delta$.

This is a spherical function associated to a discrete series representation of $GL(2, \mathbb{F}_q)$. We will discuss such representations in the next chapter. One reference is Piatetski–Shapiro [1983]. Here we must assume that a is not 0 or 4δ. And we must assume that ω is nontrivial on U. If ω_0 generates $\hat{U}$, then you get a complete list of the h_ω, from

$$\omega \in \left\{ \omega_0^j \mid j = 1, 2, \ldots \frac{q-1}{2} \right\}.$$

We find that half of the eigenvalues of M_π from the $(q - 1)$-dimensional representation of $\text{Aff}(q)$ come from formula (19) and the rest from formula (18).

Figure II.5 shows all of the eigenvalues (19) for $p = 37$ as a surface plot in Mathematica. It looks like pretty rough terrain.

It might seem surprising that the discrete series of $GL(2, \mathbb{F}_q)$ occurs in the left regular representation of G on $L^2(H_q)$. This did not happen in the real case. Harish-Chandra's integral formula gives all the spherical functions for the Poincaré upper half plane and it involves only the characters of principal series representations. One may conjecture that the difference here is due to the fact that our groups are finite and thus more like the compact group $O(3)$, all of whose representations occur in L^2 of the sphere.

In order to estimate these exponential sums, one needs Weil's 1948 proof of the Riemann hypothesis for zeta functions associated to curves over finite fields (see Weil [1979]). Nick Katz [1993] used this plus l-adic étale cohomology to show that the sum on the right side of (19) is less than or equal to $2\sqrt{q}$ in absolute value. Winnie Li [1996a,b] found another proof using class field theory. Both proofs require Weil's proof of the Riemann hypothesis for zeta functions of curves. This proves that all the graphs $X_q(\delta, a)$ are Ramanujan. We should note, however, that in the case that $q = 9$ already, equality can occur; that is, one sees eigenvalues of absolute value $2\sqrt{q}$ when q is p^2. See the earlier tables. Let's state the result despite the fact that we have not given the details of the proofs of (19) or the estimates of (19). See the next chapter for the discussion of (19). We will not cover the estimation of (19) here. See Katz [1993] and Li [1996a,b] for the details on the estimation of (19).

Theorem 4. *Finite Upper Half Plane Graphs are Ramanujan.*
The finite upper half plane graphs $X_q(\delta, a)$, for $a \neq 0, 4\delta$, are Ramanujan graphs.

Other sorts of special functions on H_q can be considered (e.g., analogues of modular forms); see Angel et al. [1992] and Harish-Chandra [1984].

Exercise.
a) Check that formula (19) does produce all the non-one-dimensional eigenvalues of M_π from the $(p-1)$-dimensional representation of Aff(p) for $p = 3$, 5, 7, 11.

b) Can you graph the spectrum of M_π as $p \to \in \infty$?

We can consider the *three questions about spectra of graphs* that we posed in Chapter 5.

Question 1. Are the graphs $X_q(\delta, a)$ Ramanujan?

Question 2. What can you say about the distribution of the eigenvalues of the adjacency operators of the graphs $X_q(\delta, a)$?

Question 3. What can you say about the "level curves" of the eigenfunctions of the adjacency operator of the graphs $X_q(\delta, a)$?

We have seen (without proof) that the answer to Question 1 is "Yes."

Next we proceed to Question 2. The question is: Do the eigenvalues have the semicircle or Sato–Tate distribution? That is, does

$$\frac{1}{q-1}\#\left\{\lambda \,\middle|\, \frac{\lambda}{\sqrt{q}} \in E\right\} \sim \frac{1}{2\pi}\int_E \sqrt{4 - x^2}\,dx, \quad \text{as } q \to \infty?$$

Again we neglect multiplicities and look only at the $q - 1$ eigenvalues given in formulas (18) and (19). This question is still open. There is some evidence for an answer of "Yes." See Kuang [1997] for a proof that the first and second moments of the eigenvalues asymptotically match those of the semicircle distribution. McKay [1981] shows that, under certain hypotheses, the semicircle distribution is the distribution of the eigenvalues of a large regular graph. But McKay's result assumes that the degree of the graph is fixed and our graphs have degrees going to infinity with q. See Winnie Li [in press] where it is shown that finite upper half plane graphs are quotients of graphs of Morgenstern, which are analogues of the Lubotzky, Phillips, and Sarnak graphs for function fields.

The histograms shown in Figures II.6 and II.7 give some evidence for the conjecture.

You might ask what happens if you replace $\mathbb{F}_q$ by $\mathbb{Z}/q\mathbb{Z}$? That is, we are replacing the field with q elements with the ring with q elements. In this case the graphs fail to be Ramanujan for all $q = p^2$, with prime $p > 3$ and the eigenvalue distribution looks quite different. See Angel et al. [1995].

Finally, we consider Question 3 for finite upper half plane graphs. So we look at level curves of finite spherical functions h on H_q. For an odd prime p,

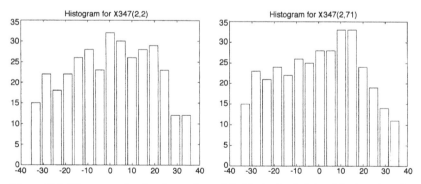

Figure II.6. Histograms for the eigenvalues of $X_q(\delta, a)$. The height of the bar gives the number of normalized eigenvalues in a subinterval. Figures made by Bernadette Shook using Matlab.

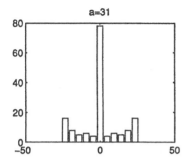
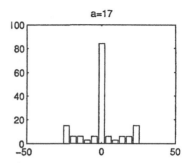

Figure II.7. Histogram of eigenvalues for finite upper half planes over the rings $\mathbb{Z}/13^2\mathbb{Z}$ with $\delta = 2$. Figures made by B. Shook using Matlab.

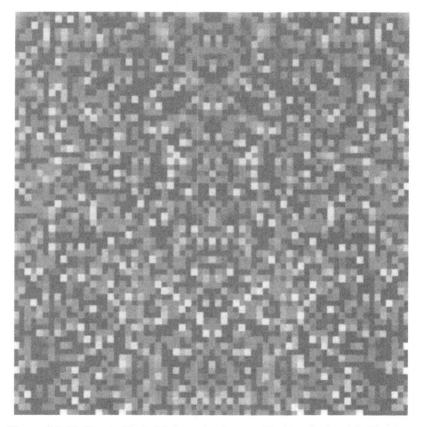

Figure II.8. ListDensityPlot in Mathematica for $p = 67$ with point (x, y) in H_{67} given a color or shade of grey determined by the value of the finite Poincaré distance from the origin.

Figure II.9. ListDensityPlot in Mathematica for $p = 127$ with point (x, y) in H_{127} given a color or shade of grey determined by the value of the finite Poincaré distance from the origin.

H_p is a rectangular grid

$$-\frac{p-1}{2} \le x \le \frac{p-1}{2}, \quad 1 \le j \le p - 1.$$

We color the square at (x, j) according to the value of the distance

$$d(x + y\sqrt{\delta}, \sqrt{\delta}) = \frac{x^2 - \delta(y-1)^2}{y}, \quad y = \delta^j.$$

A level "curve" is obtained by connecting the dots of the same color. The result for various primes can be found in the Figures II.8 and II.9. These look much more chaotic than the analogous figures for the euclidean graphs (see Figures I.22 and I.23 in Chapter 5).

Table II.9. *Comparisons of finite euclidean and noneuclidean graphs*

	Euclidean (Chapter 5)	Noneuclidean
space $X \cong G/K$	$\mathbb{F}_q^n$	$H_q \subset \mathbb{F}_q(\sqrt{\delta})$
distance	$(x_1 - y_1)^2 + \cdots + (x_n - y_n)^2$	$N(z - w)/(\mathrm{Im}z\ \mathrm{Im}w)$
group G	$\mathbb{F}_q^n \cdot O(n, \mathbb{F}_q)$	$GL(2, \mathbb{F}_q)$
subgroup K	$O(n, \mathbb{F}_q)$	all matrices $\begin{pmatrix} a & b\delta \\ b & a \end{pmatrix}$
graphs	$E_q(n, a)$, for $a \in \mathbb{F}_q$	$X_q(n, a)$, for $a \in \mathbb{F}_q$
degree	$q^{n-1} + \text{error}$	$q + 1$
# conn. graphs	≤ 3	$\leq q - 2$
edge set	$\{x \in \mathbb{F}_q^n \mid d(x, 0) = a\}$	$\{z \in \mathbb{F}_q \mid d(z, \sqrt{\delta}) = a\}$
eigenvalues/eigen-		power, Kloosterman
functions of	Kloosterman sums	and Soto–Andrade sums,
adjacency operator		spherical functions
Is graph Ramanujan?	asymptotically as $q \to \infty$	Yes
Is spectrum	not semicircle if n odd	?
semicircle?	semicircle if n even	
level curves	finite Fresnel patterns	look chaotic[†]
of eigenfunctions		
Is 0 in the spectrum?	Yes, for n odd	known to be yes
	No, for n even	if $a = 2\delta$, else unknown

[†]One of the referees of Terras [1996] asked the interesting questions: "Are the non-Euclidean graphs chaotic also from the point of view of 'non-Euclidean eyes'?" And "What is the minimum g such that $X_q(\delta, a)$ can be drawn on a surface of genus g? Does it look less chaotic there?"

The exact Mathematica command used in Figures II.8 and II.9 is (with $g = \delta = $ generator of multiplicative group $\mathbb{F}_p^*$):

```
ListDensityPlot[Table[Mod[(i^2-g*(PowerMod[g,j,p]-1)^2)*
    PowerMod[g,-j,p],p], {j,1,p-1}, {i,-(p-1)/2,(p-1)/2}],
    Mesh->False,Frame->False].
```

Here $g = \delta$. You can insert "ColorFunction –> Hue" if you want a color plot.

We close this chapter with a table summarizing the comparisons between finite euclidean and noneuclidean graphs. See Table II.9.

Chapter 21

The General Linear Group $GL(2, \mathbb{F}_q)$

How could they have been so blind? Now that it had been
pointed out to them, it was all too perfectly obvious. The
structure of C_{60} was not only the most wonderfully
symmetrical molecular structure they had ever contemplated,
it was also absurdly commonplace. A modern soccer ball is 20
white leather hexagons and 12 black leather pentagons stitched
together, with each pentagon surrounded by five hexagons. It
has 60 vertices; 60 points where the corners of the pentagons
and the hexagons meet along the seams. How many times had
each of them looked at a soccer ball without really registering
these simple facts?

Baggott [1996, p. 70]

We have found many reasons to study the representations of $GL(2, \mathbb{F}_q)$. In
Chapter 13, for example, we discussed the molecule buckminsterfullerene or
C_{60} and noted that an understanding of the spectral lines of this molecule
requires a knowledge of the representations of $A_5 \cong PSL(2, \mathbb{F}_5)$. In the
last chapter we found that an understanding of the representations of
$GL(2, \mathbb{F}_q)$ seems necessary to bound the spherical functions on the finite upper half plane. Also the Ramanujan graphs of Lubotzky, Phillips, and Sarnak
[1988] are Cayley graphs for either $PGL(2, \mathbb{F}_q)$ or $PSL(2, \mathbb{F}_q)$, with prime
q, using generating sets with $p + 1$ elements (p denoting a different prime).
See also Lubotzky [1994, pp. 96 ff] and Sarnak [1990, pp. 73 ff]. Finally,
the representations of $SL(2, \mathbb{F}_q)$ are needed in the paper of Lafferty and
Rockmore [1992] where spectra of degree four Cayley graphs of $SL(2, \mathbb{F}_q)$
are studied.

With this motivation we dive into the subject, which turns out to be more difficult than the representation theory of the affine and solvable groups discussed
earlier.

Exercise. Prove that we have the following group isomorphisms:

$$SL(2, \mathbb{F}_2) \cong S_3; \quad PSL(2, \mathbb{F}_3) \cong A_4; \quad PSL(2, \mathbb{F}_5) \cong A_5.$$

Exercise. Let G be the group of symmetries of a cube. Let H be the subgroup fixing a point. Show that $|G| = 48$ and that (G, H) is a Gelfand pair. Find the eigenvalues of the adjacency matrix of the cube viewed as a graph.

Hint. See Chapter 10.

One can show that except for small values of q, $PSL(n, \mathbb{F}_q)$ is simple for $n \geq 2$. The group $GL(n, \mathbb{F}_q)$ is often called reductive. See Digne and Michel [1991]. Group theorists often write $L_n(q) = PSL(n, q)$. See Conway and Sloane [1993, p. 266] and Iyanaga and Kawada [1, Vol. I, p. 536].

Exercise.
a) Check the character table for $PSL(2, \mathbb{F}_5)$ given in Table II.10.
b) Show that $PSL(2, \mathbb{F}_5)$ is simple.

Hint.
a) See James and Liebeck [1993].
b) One method to show that a group is simple makes use of the character table. One can show (see Burrow [1993, pp. 91–3] or James and Liebeck [1993, p. 172]) that a group G is simple iff for each $g \in G$ and for each character χ_π, for $\pi \in \hat{G}$, $\pi \neq 1$, we have $\chi_\pi(g) \neq \chi_\pi(e)$ if $g \neq e$. Here e is the identity in G.

Note. The group $PSL(2, \mathbb{F}_5)$ is the icosahedral group – the group of symmetries of the icosahedron.

Exercise. Show that $|GL(2, \mathbb{F}_q)| = (q^2 - 1)(q^2 - q)$.

Table II.10. *Character table for the icosahedral group $A_5 \cong PSL(2, \mathbb{F}_5)$*

	C_1	C_2	C_3	C_4	C_5
χ_1	1	1	1	1	1
χ_2	3	−1	0	$(1 + \sqrt{5})/2$	$(1 - \sqrt{5})/2$
χ_3	3	−1	0	$(1 - \sqrt{5})/2$	$(1 + \sqrt{5})/2$
χ_4	4	0	1	−1	−1
χ_5	5	1	−1	0	0

Research Project. Make use of the character table of A_5 to explain the spectral lines of the buckyball. The infrared spectrum has four lines. The Raman spectrum has ten lines. For this project you will need to read the papers of Chung and Sternberg [1992, 1993]. See also Chung [1996] and Sternberg [1994].

Next we proceed to find the character table for $GL(2, \mathbb{F}_q)$, q *odd*. It is possible to find the character table more speedily without writing down all the representations. See Fulton and Harris [1991] and Steinberg [1994]. G. Frobenius (1896), I. Schur (1907), and H. Jordan (1907) gave the characters of $PSL(2, \mathbb{F}_q)$, $SL(2, \mathbb{F}_q)$, and $GL(2, \mathbb{F}_q)$. H. W. Brinkmann (1921) gave the characters of $PSL(3, q)$. Steinberg [1994] has the exact references and he determines the characters of $GL(n, \mathbb{F}_q)$ and $PGL(n, \mathbb{F}_q)$, for $n = 2, 3$, as well as some of the characters of $GL(4, \mathbb{F}_q)$. Silberger [1969] shows how to use the characters to determine the matrix entries of the representations for $SL(2, \mathbb{F}_q)$ and $GL(2, \mathbb{F}_q)$. Silberger notes that in 1946 H. D. Kloosterman [1946] was the first to describe all the irreducible representations of $SL(2, \mathbb{Z}/q\mathbb{Z})$ using theta functions. Note that this is not the special linear group over the field $\mathbb{F}_q$ unless $q = $ prime. In 1964 A. Weil [1979, Vol. III, pp. 1–69] gave an alternative version of this construction. See also S. Tanaka [1966]. Here we will instead follow the discussion in Piatetski–Shapiro [1983]. See also Naimark and Stern [1982, pp. 118–31] as well as Gelfand, Graev, and Piatetski–Shapiro [1990, pp. 158 and 185].

The characters of $GL(n, \mathbb{F}_q)$ were found by J. A. Green [1955]. Generalizations to other finite groups of Lie type or Chevalley groups due to Lusztig and Deligne are discussed by Bhama Srinivasan [1979] as well as Digne and Michel [1991]. It requires a classification of maximal tori and construction of a family of virtual representations for each maximal torus using l-adic cohomology. Here a *torus* is defined to be a subgroup isomorphic to a product of multiplicative groups of finite fields. In the case $G = GL(2, \mathbb{F}_q)$, q odd, we found two maximal tori:

$D = $ the diagonal matrices $\cong \mathbb{F}_q^* \times \mathbb{F}_q^*$ and
the subgroup K defined to be the group elements fixing $\sqrt{\delta}$ in the finite upper half plane H_q.

The Type I representations of $GL(2, \mathbb{F}_q)$ below are connected to characters α, β of the multiplicative group $\mathbb{F}_q^*$ and the Type II representations of $GL(2, \mathbb{F}_q)$ below are connected to characters ν of K, which is isomorphic to $\mathbb{F}_{q^2}^*$.

Before constructing the representations of $GL(2, \mathbb{F}_q)$, we need to find the conjugacy classes in the group. To do this, you must remember the Jordan forms of 2×2 matrices over $\mathbb{F}_q$. You also need to note that the number of elements in a conjugacy class

$$\{g\} = \{xgx^{-1} \mid x \in G\}$$

is

$$|\{g\}| = |G|/|G_g|, \tag{1}$$

where G_g = centralizer of g in G, that is,

$$G_g = \{x \in G \mid xg = gx\}.$$

Exercise. Prove formula (1).
Hint. Consider the onto map $f : G \rightarrow \{g\}$ defined by $f(x) = xgx^{-1}$. That is, G acts on itself by conjugation. An orbit is a conjugacy class. The isotropy subgroup of G fixing g is the centralizer G_g.

We chose the terminology central, parabolic, hyperbolic, and elliptic in analogy to the terminology used for fractional linear transformations from $SL(2, \mathbb{Z})$. See Terras [1985]. Note that the elliptic conjugacy class corresponds to elements of the subgroup K defined by formula (3) of Chapter 19. The eigenvalues of the matrix

$$\begin{pmatrix} x & y\delta \\ y & x \end{pmatrix}$$

are the numbers $z = x + y\sqrt{\delta}$ and $z^q = \bar{z} = x - y\sqrt{\delta}$ in $\mathbb{F}_q(\sqrt{\delta})$, $x, y \in \mathbb{F}_q$. Over the field $\mathbb{F}_q(\sqrt{\delta})$, the matrix is similar to a matrix of the form

$$\begin{pmatrix} z & 0 \\ 0 & \bar{z} \end{pmatrix}.$$

We can also take the matrix representatives for the elliptic conjugacy classes to have the form

$$\begin{pmatrix} 0 & -Nz \\ 1 & Trz \end{pmatrix}.$$

Here $Nz = z\bar{z}$ and $Trz = z + \bar{z}$.

Exercise. Check Table II.11.
Hint. Compare Fulton and Harris [1991] and Piatetski–Shapiro [1983].

Table II.11. *Conjugacy classes for $GL(2, \mathbb{F}_q)$, q odd, for δ a nonsquare in $\mathbb{F}_q$*

Representative	# Elements in class	# Classes
$\begin{pmatrix} r & 0 \\ 0 & r \end{pmatrix}$, central	1	$q - 1$
$\begin{pmatrix} r & 1 \\ 0 & r \end{pmatrix}$, parabolic	$q^2 - 1$	$q - 1$
$\begin{pmatrix} r & 0 \\ 0 & s \end{pmatrix}$, $r \neq s$, hyperbolic	$q^2 + q$	$\frac{(q-1)(q-2)}{2}$
$\begin{pmatrix} r & s\delta \\ s & r \end{pmatrix}$, $s \neq 0$, elliptic	$q^2 - q$	$\frac{q(q-1)}{2}$

Next let's find the irreducible representations of $G = GL(2, \mathbb{F}_q)$. Since Table II.11 says that there are $q^2 - q$ conjugacy classes, we must find this same number of inequivalent, irreducible, unitary representations.

The Irreducible Representations of $GL(2, \mathbb{F}_q)$

Type I. Representations Associated to Characters of $\mathbb{F}_q^*$

One-Dimensional Representations

For each multiplicative character α of $\mathbb{F}_q^*$ there is a *one-dimensional representation* given by

$$\alpha(g) = \alpha(\det g), \quad \text{for } g \in G. \tag{2}$$

Principal Series Representations

Next consider the *Borel subgroup B* of $G = GL(2, \mathbb{F}_q)$ defined by

$$B = \left\{ \begin{pmatrix} a & b \\ 0 & d \end{pmatrix} \in G \right\}. \tag{3}$$

Then G/B can be identified with the projective line $\mathbb{P}^1(\mathbb{F}_q)$, which is defined to be the quotient of nonzero 2-vectors modulo multiplication by nonzero scalars, that is,

$$\mathbb{P}^1(\mathbb{F}_q) = \left(\mathbb{F}_q^2 - \{0\} \right)/\mathbb{F}_q^*.$$

The induced representation $\pi = \text{Ind}_B^G 1$ is not irreducible because it contains a copy of the trivial representation. But if we take the complement of the trivial representation in π, we do get an irreducible representation π_0. This representation π_0 acts on the space V_0 defined below by right shift

$$V_0 = \{f : G \to \mathbb{C} \mid f(bx) = f(x) \text{ for all } x \in G, b \in B \text{ and } \langle f, 1 \rangle = 0\}$$
$$[\pi_0(g)f](x) = f(xg), \quad \text{for } f \in V_0, x, g \in G. \tag{4}$$

We can multiply (tensor) the representation π_0 by the one-dimensional representation α to obtain an irreducible representation

$$\pi_\alpha = \pi_0 \otimes \alpha. \tag{5}$$

Exercise. Compute the character of π_α and show that it is irreducible. Then show that $\pi_\alpha \cong \pi_\beta$ iff $\alpha = \beta$.

Next suppose we have two multiplicative characters α, β of $\mathbb{F}_q^*$ and define a one-dimensional representation of the Borel subgroup B by

$$\mu_{\alpha,\beta} \begin{pmatrix} a & b \\ 0 & d \end{pmatrix} = \alpha(a)\beta(b). \tag{6}$$

Then let

$$\rho_{\alpha,\beta} = \text{Ind}_B^G \mu_{\alpha,\beta}. \tag{7}$$

This is irreducible if $\alpha \neq \beta$. But $\rho_{\alpha,\alpha} \cong \alpha \oplus \pi_\alpha$.

Characters of $\rho_{\alpha,\beta}$

To compute the characters of $\rho_{\alpha,\beta}$ defined by (7), we use the Frobenius formula from the corollary to Proposition 1 in Chapter 16:

$$\chi_{\rho_{\alpha,\beta}}(g) = \frac{1}{|B|} \sum_{x \in G} \tilde{\mu}_{\alpha,\beta}(xgx^{-1}), \quad \text{with } \tilde{\mu}_{\alpha,\beta}(x) = \begin{cases} \mu_{\alpha,\beta}(x), & x \in B, \\ 0, & x \notin B. \end{cases} \tag{8}$$

Center. Formula (8) implies that

$$\chi_{\rho_{\alpha,\beta}} \begin{pmatrix} r & 0 \\ 0 & r \end{pmatrix} = \frac{|G|}{|B|} \alpha(r)\beta(r) = (q+1)\alpha(r)\beta(r).$$

Parabolic. From (8), we have

$$\chi_{\rho_{\alpha,\beta}} \begin{pmatrix} r & 1 \\ 0 & r \end{pmatrix} = \frac{1}{|B|} \sum_{x \in G} \tilde{\mu}_{\alpha\beta} \left(x \begin{pmatrix} r & 1 \\ 0 & r \end{pmatrix} x^{-1} \right).$$

Now if

$$x = \begin{pmatrix} a & b \\ c & d \end{pmatrix},$$

we have

$$x \begin{pmatrix} r & 1 \\ 0 & r \end{pmatrix} x^{-1} = (ad - bc)^{-1} \begin{pmatrix} * & * \\ -c^2 & * \end{pmatrix} \in B \quad \text{iff } c = 0 \text{ and } x \in B.$$

If $x \in B$, one finds

$$x \begin{pmatrix} r & 1 \\ 0 & r \end{pmatrix} x^{-1} = \begin{pmatrix} r & a/d \\ 0 & r \end{pmatrix}.$$

Thus

$$\chi_{\rho_{\alpha,\beta}} \begin{pmatrix} r & 1 \\ 0 & r \end{pmatrix} = \alpha(r)\beta(r).$$

Hyperbolic. This time we need to compute for $x = \begin{pmatrix} a & b \\ c & d \end{pmatrix}, r \neq s$:

$$x \begin{pmatrix} r & 0 \\ 0 & s \end{pmatrix} x^{-1} = (ad - bc)^{-1} \begin{pmatrix} * & * \\ cd(r - s) & * \end{pmatrix} \in B \quad \text{iff } c = 0 \text{ or } d = 0.$$

So we have two cases: $c = 0$ and $d = 0$.

Case 1.

$$c = 0 \Rightarrow x \in B \quad \text{and} \quad x \begin{pmatrix} r & 0 \\ 0 & s \end{pmatrix} x^{-1} = \begin{pmatrix} r & b(s - r)/d \\ 0 & s \end{pmatrix}.$$

Case 2.

$$d = 0 \Rightarrow x \begin{pmatrix} r & 0 \\ 0 & s \end{pmatrix} x^{-1} = \begin{pmatrix} s & a(r - s)/c \\ 0 & r \end{pmatrix}.$$

Note that the number of x contributing a nonzero element to the sum in Case 1 is the same as that in Case 2, namely $|B|$.

It follows that

$$\chi_{\rho_{\alpha,\beta}} \begin{pmatrix} r & 0 \\ 0 & s \end{pmatrix} = \alpha(r)\beta(s) + \alpha(s)\beta(r).$$

Elliptic. Now

$$x \begin{pmatrix} r & \delta s \\ s & r \end{pmatrix} x^{-1}$$

is never in B. Thus

$$\chi_{\rho_{\alpha,\beta}} \begin{pmatrix} r & \delta s \\ s & r \end{pmatrix} = 0.$$

Exercise.
a) Show that $\rho_{\alpha,\beta}$ is irreducible iff $\alpha \neq \beta$.
b) Show that $\rho_{\alpha,\beta} \cong \rho_{\beta,\alpha}$.

It follows that we get $(q-1)(q-2)/2$ inequivalent irreducible representations $\rho_{\alpha,\beta}$, for $\alpha \neq \beta$.

We have explained how to find all but the last row of Table II.12 – the discrete series.

Type1 II. Representations Associated to Characters of $\mathbb{F}_{q^2}^*$

These representations are associated to the maximal torus

$$K = \left\{ \begin{pmatrix} a & b\delta \\ b & a \end{pmatrix} \in GL(2, \mathbb{F}_q) \right\}, \tag{9}$$

which is isomorphic to $\mathbb{F}_{q^2}^*$.

Table II.12. *Character table for* $GL(2, \mathbb{F}_q)$

	$\begin{pmatrix} r & 0 \\ 0 & r \end{pmatrix}$	$\begin{pmatrix} r & 1 \\ 0 & r \end{pmatrix}$	$\begin{pmatrix} r & 0 \\ 0 & s \end{pmatrix}, r \neq s$	$\begin{pmatrix} x & y\delta \\ y & x \end{pmatrix}, y \neq 0$
Type	Central	Parabolic	Hyperbolic	Elliptic
#{g}	1	$q^2 - 1$	$q^2 + q$	$q^2 - q$
# Classes	$q - 1$	$q - 1$	$(q-1)(q-2)/2$	$q(q-1)/2$
α	$\alpha(r)^2$	$\alpha(r)^2$	$\alpha(rs)$	$\alpha(Nz)$
π_α	$q\alpha(r)^2$	0	$\alpha(rs)$	$-\alpha(Nz)$
$\rho_{\alpha,\beta}, \alpha \neq \beta$	$(q+1)\alpha(r)\beta(r)$	$\alpha(r)\beta(r)$	$\alpha(r)\beta(s) + \alpha(s)\beta(r)$	0
$\sigma_\nu, \nu \neq \nu^q$	$(q-1)\nu(r)$	$-\nu(r)$	0	$-\nu(z) - \nu(\bar{z})$

Here δ is a nonsquare in $\mathbb{F}_q$, q odd; $z = x + y\sqrt{\delta}$; α is a multiplicative character of $\mathbb{F}_q^*$, ν is a multiplicative character of $\mathbb{F}_{q^2}^* = \mathbb{F}_q(\sqrt{\delta})^*$, and $Nz = z\bar{z} = x^2 - \delta y^2$.

Discrete Series Representations

Sometimes the discrete series representations are called cuspidal representations. We take our description from Piatetski-Shapiro [1983]. Suppose ν
is a multiplicative character of $\mathbb{F}_{q^2}^* = \mathbb{F}_q(\sqrt{\delta})^*$, where δ is a nonsquare in
$\mathbb{F}_q$ (q odd). We will assume

$$\nu \neq \chi \circ N,$$

for any multiplicative character χ of $\mathbb{F}_q^*$, where N is the norm from $\mathbb{F}_{q^2}^*$ down to
$\mathbb{F}_q^*$. Such characters are called *nondecomposable*. Equivalently, one has Part 1
of the following lemma.

Lemma 1.
1) A character ν of $\mathbb{F}_{q^2}^*$ is nondecomposable iff $\nu \neq \bar{\nu}$.
2) If ν is a nondecomposable character of $\mathbb{F}_{q^2}^*$, then

$$\sum_{Nx = a} \nu(x) = 0 \quad \text{for every } a \in \mathbb{F}_q^*.$$

Proof. Do it as an exercise. ∎
Hint. See Piatetski–Shapiro [1983, pp. 33–4].

Since the subgroup K defined by formula (9) is isomorphic to the multiplicative group $\mathbb{F}_{q^2}^*$, we know that ν defines a 1-dimensional representation of
K. One might hope that $\text{Ind}_K^G \nu$ would be irreducible. But this is false and it is
impossible to get the discrete series *directly* out of this representation. However,
see Silberger [1969].

Exercise.
a) Verify that $\text{Ind}_K^G \nu$ is reducible.
b) Show that the virtual character (meaning it is a $\mathbb{Z}$-linear combination of
   characters)

$$\chi_{\pi_1 \otimes \rho_{\alpha,1}} - \chi_{\rho_{\alpha,1}} - \chi_{\text{Ind}_K^G \nu}$$

has the same values on conjugacy classes of $GL(2, \mathbb{F}_q)$ as the character of
σ_ν in the last row of character Table II.12.

The discrete series representation σ_ν will be defined as in Piatetski–Shapiro
[1983] using generators and relations for $GL(2, \mathbb{F}_q)$.

Theorem 1. The group $GL(2, \mathbb{F}_q)$ is generated by the *Borel subgroup B* and the *Weyl group element*

$$w = \begin{pmatrix} 0 & 1 \\ -1 & 0 \end{pmatrix}$$

with defining relations obtained by setting

$$t = \begin{pmatrix} 1 & 1 \\ 0 & 1 \end{pmatrix}, \quad u = wt.$$

(1) $w \begin{pmatrix} r & 0 \\ 0 & s \end{pmatrix} w^{-1} = \begin{pmatrix} s & 0 \\ 0 & r \end{pmatrix};$ (2) $w^2 = -I;$ (3) $u^3 = I.$

Proof. See Piatetski–Shapiro [1983, pp. 31–2]. The proof makes use of the *Bruhat decomposition* which writes

$$G = B \cup BwA \text{ (disjoint)}, \quad \text{where } A = \left\{ \begin{pmatrix} 1 & x \\ 0 & 1 \end{pmatrix} \middle| x \in \mathbb{F}_q \right\}. \tag{10}$$

■

By Theorem 1, it suffices to define the representation σ_ν on $B \cup \{w\}$. We will take $\sigma_\nu(g)$ to be an operator acting on the vector space

$$V = \{ f : \mathbb{F}_q^* \to \mathbb{C} \}.$$

Also, choose ψ to be a fixed nontrivial additive character of $\mathbb{F}_q$; for example,

$$\psi(x) = \exp\left(\frac{2\pi i \operatorname{Tr}(x)}{p} \right), \quad \text{if } q = p^r, p = \text{ prime.}$$

Then we make the following definition.

Definition of Discrete Series

For an element of the Borel subgroup B, define

$$\left[\sigma_\nu \begin{pmatrix} a & b \\ 0 & d \end{pmatrix} f \right](x) = \nu(d)\psi(bd^{-1}x)f(ad^{-1}x), \tag{11}$$

and for the Weyl group element w, define

$$[\sigma_\nu(w)f](y) = -\sum_{x \in \mathbb{F}_q^*} \nu(x^{-1})j(yx)f(x). \tag{12}$$

Here $j(u)$ is defined to be a *generalized Kloosterman sum*:

$$j(u) = \frac{1}{q} \sum_{\substack{t \in \mathbb{F}_q(\sqrt{\delta})^* \\ Nt = u}} \psi(t + \bar{t})v(t). \tag{13}$$

Note that Piatetski–Shapiro [1983, p. 35] does not include the minus sign in formula (12), but it is necessary for the correct character values.

Exercise.
i) Prove that if

$$g = \begin{pmatrix} a & b \\ c & d \end{pmatrix} \in GL(2, \mathbb{F}_q)$$

and $c \neq 0$ then

$$\begin{pmatrix} a & b \\ c & d \end{pmatrix} = \begin{pmatrix} (bc - ad)c^{-1} & -a \\ 0 & -c \end{pmatrix} \begin{pmatrix} 0 & 1 \\ -1 & 0 \end{pmatrix} \begin{pmatrix} 1 & c^{-1}d \\ 0 & 1 \end{pmatrix}. \tag{14}$$

ii) Show that if g is as in Part i) we have

$$[\sigma_v(g)f](y) = \sum_{x \in \mathbb{F}_q^*} k(y, x; g)f(x), \tag{15}$$

where

$$k(y, x; g) = \frac{-1}{q} \psi\left(\frac{ay + dx}{c}\right) \sum_{\substack{t \in \mathbb{F}_q(\sqrt{\delta})^* \\ Nt = yx^{-1}\det(g)}} \psi\left(\frac{-x(t + \bar{t})}{c}\right)v(t). \tag{16}$$

The verification that we get a representation from definitions (11)–(13) requires Theorem 1 and the verification that the definitions preserve the relations given in the theorem. See Piatetski–Shapiro [1983, pp. 35–40] for the details, which require some properties of the Kloosterman sum given in the following lemma.

Lemma 2.

1)
$$\sum_{v \in \mathbb{F}_q^*} j(uv)j(v)v(v^{-1}) = \begin{cases} v(-1), & \text{if } u = 1, \\ 0, & \text{if } u \neq 1. \end{cases}$$

2) $$\sum_{v \in \mathbb{F}_q^*} j(xv)j(yv)v(v^{-1})\psi(v) = v(-1)\psi(-x-y)j(xy).$$

Proof. The proof is left as an exercise.

Hint. See Piatetski–Shapiro [1983, pp. 35–8]. ■

Next let's check that the representation defined in (11)–(13) has the character values given in the last row of Table II.12.

Character of the Discrete Series Representation

Center. The matrix of

$$\sigma_v \begin{pmatrix} r & 0 \\ 0 & r \end{pmatrix}$$

is $v(r)I_{q-1}$. Thus the trace is $v(r)(q-1)$.

Parabolic. Let g be a generator of the multiplicative group $\mathbb{F}_q^*$ and take the basis of V to be the delta functions δ_u, $u = g^j$, $j = 0, 1, 2, \ldots, q-2$. Then the matrix of

$$\sigma_v \begin{pmatrix} r & 1 \\ 0 & r \end{pmatrix}$$

is diagonal with jth diagonal entry $\psi(r^{-1}g^j)v(r)$. So the trace is

$$v(r) \sum_{x \in \mathbb{F}_q^*} \psi(r^{-1}x) = -v(r).$$

Hyperbolic. The matrix of

$$\sigma_v \begin{pmatrix} r & 0 \\ 0 & s \end{pmatrix}, \quad r \neq s,$$

is a shift matrix and thus its trace is 0.

Elliptic. Let

$$g = \begin{pmatrix} a & b\delta \\ b & a \end{pmatrix},$$

with $b \neq 0$ and $z = a + b\sqrt{\delta}$. Then using Equations (15) and (16), we see that

$$\chi_{\sigma_v}(g) = \sum_{x \in \mathbb{F}_q^*} k(x, x; g)$$

$$= \frac{-1}{q} \sum_{x \in \mathbb{F}_q^*} \psi\left(\frac{2ax}{b}\right) \sum_{\substack{Nt = Nz \\ t \in \mathbb{F}_{q^2}^*}} \psi\left(\frac{-x(t + \bar{t})}{b}\right) v(t)$$

$$= \frac{-1}{q} \sum_{\substack{Nt = Nz \\ t \in \mathbb{F}_{q^2}^*}} v(t) \sum_{x \in \mathbb{F}_q^*} \psi\left(\frac{x(2a - t - \bar{t})}{b}\right).$$

Here $Nz = \text{norm } z = z\bar{z} = z^{1+q}$.

By Lemma 1, the inner sum over x is $q - 1$ if $2a = \text{Tr} z = \text{Tr} t$ and -1 otherwise. Here $\text{Tr} z = \text{trace } z = z + \bar{z}$. Note that there are only two values of t with $\text{Tr} t = \text{Tr} z$, namely, $t = z$ and $t = \bar{z}$. Thus

$$\chi_{\sigma_v}(g) = \frac{-1}{q} \left\{ (q - 1)(v(z) + v(\bar{z})) - \sum_{\substack{Nt = Nz \\ \text{Tr}t \neq \text{Tr}z \\ t \in \mathbb{F}_{q^2}^*}} v(t) \right\}$$

$$= \frac{-1}{q} \left\{ q(v(z) + v(\bar{z})) + \sum_{\substack{t \in \mathbb{F}_{q^2}^* \\ Nt = Nz}} v(t) \right\} = -v(z) - v(\bar{z}).$$

In the last equality we used Part 2 of Lemma 1.

Exercise.

i) Show that the representation σ_v is irreducible.

ii) Show that $\sigma_v \cong \sigma_\mu$ iff $v = \mu$.

Finally, we have a description of all the inequivalent, irreducible, unitary representations of $GL(2, \mathbb{F}_q)$, q odd.

Theorem 2. The Type I inequivalent, irreducible, unitary representations of $GL(2, \mathbb{F}_q)$ associated to characters α, β of the multiplicative group $\mathbb{F}_q^*$ are:

$(q - 1)$ one-dimensional representations α defined by (2),

$(q - 1)$ representations π_α having degree q defined by (5),

$(q - 1)(q - 2)/2$ principal series representations $\rho_{\alpha, \beta}$, of degree $q + 1$ defined by (7), and the Type II inequivalent, irreducible, unitary representations of $GL(2, \mathbb{F}_q)$ associated to nondecomposable multiplicative characters ν of $\mathbb{F}_{q^2}^*$ are:

$(q-1)/2$ discrete series representations σ_ν of degree $q-1$
defined by (11)–(13).

Note. The number of the four types of representations of $G = GL(2, \mathbb{F}_q)$ correspond exactly to the numbers of the four types of conjugacy classes of G.

Exercise. Finish the proof of Theorem 2.

Soto–Andrade's Formula for Spherical Functions

Our next problem is to obtain Soto-Andrade's formula (19) of Chapter 20 for spherical functions corresponding to the discrete series representations σ_ν. We will follow the method of Evans [1994]. First one must compute which representations from Theorem 2 occur in $\mathrm{Ind}_K^G 1$, where K is the subgroup of matrices

$$\begin{pmatrix} a & b\delta \\ b & a \end{pmatrix},$$

that is, those matrices fixing $\sqrt{\delta}$ in the finite upper half plane H_q. That is, we must compute $\hat{G}^K$ as defined in formula (13) of Chapter 20. We know from Theorem 2 of Chapter 20 that since (G, K) is a Gelfand pair, any $\pi \in \hat{G}$ occurring in $\mathrm{Ind}_K^G 1$ must occur with multiplicity 1.

Theorem 3. As usual q is odd and δ a nonsquare in $\mathbb{F}_q$. If $G = GL(2, \mathbb{F}_q)$ and K is the subgroup of matrices

$$\begin{pmatrix} a & b\delta \\ b & a \end{pmatrix},$$

then

$$\hat{G}^K = \{\pi \in G \mid \pi \text{ occurs in } \mathrm{Ind}_K^G 1\}$$

consists of the following representations:

Type I. $(q+1)/2$ representations coming from multiplicative characters α of $\mathbb{F}_q^*$:
1-dimensional: trivial,

q-dimensional: π_ε, $\varepsilon^2 = 1$ on $\mathbb{F}_q^*$, ε nontrivial (i.e., ε the character that is 1 on squares and -1 on nonsquares),

$(q+1)$-dimensional: $\rho_{\alpha,\alpha^{-1}}$, where α^2 nontrivial; $\alpha = \xi^j$, $j = 1, \ldots,$ $(q-3)/2$, if ξ is a character of order $q - 1$ of $\mathbb{F}_q^*$.

Type II. $(q-1)/2$ discrete series representations coming from multiplicative characters ν of $\mathbb{F}_{q^2}^*$:

$(q-1)$-dimensional: σ_ν with $\nu \neq \nu^q$ and ν is trivial on $\mathbb{F}_q^*$, where $\nu = \omega^{j(q-1)}$, $j = 1, \ldots, (q-1)/2$, assuming that ω is a character of $\mathbb{F}_{q^2}^*$ of order $q^2 - 1$.

Proof. Note first that the degrees do add up correctly to $q(q-1)$, since

$$q(q-1) = 1 + q + (q+1)(q-3)/2 + (q-1)(q-1)/2.$$

To prove Theorem 3, we need to compute the character of $\tau = \mathrm{Ind}_K^G 1$. By the Frobenius formula for the character of an induced representation given in the corollary to Proposition 1 of Chapter 16, we have

$$\chi_\tau(g) = \sum_{z \in H_q} \tilde{\chi}_K(a_z g a_z^{-1}), \quad \text{where } a_z = \begin{pmatrix} y & x \\ 0 & 1 \end{pmatrix}, \quad \text{if } z = x + y\sqrt{\delta}, \quad (17)$$

and

$$\tilde{\chi}_K(g) = \begin{cases} 1, & g \in K \\ 0, & \text{otherwise.} \end{cases}$$

Let

$$g = \begin{pmatrix} a & b \\ c & d \end{pmatrix}.$$

Then

$$a_z g a_z^{-1} = \begin{pmatrix} a + y^{-1}xc & x(d-a) - y^{-1}x^2c + yb \\ y^{-1}c & d - y^{-1}xc \end{pmatrix} \in K$$

if and only if

$$a + y^{-1}xc = d - y^{-1}xc \quad \text{and} \quad x(d-a) - y^{-1}x^2c + yb = y^{-1}c\delta^{-1}.$$

This is equivalent to

$$d - a = 2y^{-1}xc \quad \text{and} \quad c\{\delta^{-1} + x^2\} = y^2 b. \quad (18)$$

So $c = 0$ implies that $b = 0$ and $a = d$. This means that

$$\chi_\tau \begin{pmatrix} a & 0 \\ 0 & a \end{pmatrix} = q(q-1) \quad \text{and} \quad \chi_\tau \begin{pmatrix} a & b \\ 0 & d \end{pmatrix} = 0, \quad \text{if } a \neq d \text{ or } b \neq 0.$$

That takes care of the first three types of conjugacy classes in Table II.11. Only the elliptic class, with

$$g = \begin{pmatrix} a & b \\ b\delta^{-1} & a \end{pmatrix},$$

remains. Use Equation (18) to see that if $d = a$ and $c = b\delta^{-1}$, $b \neq 0$, we have $x = 0$ and $\delta^{-2} = y^2$. Thus there are only two nonzero terms in the sum (17). It follows that

$$\chi_\tau \begin{pmatrix} a & b \\ b\delta^{-1} & a \end{pmatrix} = 2.$$

The list of character values of the induced character τ is given in Table II.13.

To figure out which representations $\pi \in \hat{G}$ of Table II.12 will occur in $\tau = \text{Ind}_K^G 1$, we need to compute which inner products $\langle \chi_\tau, \chi_\pi \rangle = |G|$. There are four types of representations of G classified according to their dimensions.

1-Dimensional.

$$\langle \chi_\alpha, \chi_\tau \rangle = \sum_{r \in \mathbb{F}_q^*} q(q-1)\alpha(r^2) + \sum_{z \in H_q} q(q-1)\alpha(\text{N}z)2/2$$

$$= q(q-1) \sum_{r \in \mathbb{F}_{q^2}^*} \alpha(\text{N}z) = q(q-1) \begin{cases} q^2 - 1, & \text{if } \alpha \circ N = 1 \\ 0, & \text{otherwise} \end{cases}.$$

Next note that α composed with the norm from $\mathbb{F}_{q^2}^*$ down to $\mathbb{F}_q^*$ is trivial iff α is trivial on $\mathbb{F}_q^*$, since the norm maps onto.

Table II.13. *Character values for* $\tau = \text{Ind}_K^G 1$

	$\begin{pmatrix} r & 0 \\ 0 & r \end{pmatrix}$	$\begin{pmatrix} r & 1 \\ 0 & r \end{pmatrix}$	$\begin{pmatrix} r & 0 \\ 0 & s \end{pmatrix}, r \neq s$	$\begin{pmatrix} x & y\delta \\ y & x \end{pmatrix}, y \neq 0$
Type	Central	Parabolic	Hyperbolic	Elliptic
τ	$q(q-1)$	0	0	2

Here $G = G(2, \mathbb{F}_q)$ and K is the subgroup fixing $\sqrt{\delta}$ in H_q.

q-Dimensional.

$$\langle \chi_{\pi_\alpha}, \chi_\tau \rangle = \sum_{r \in \mathbb{F}_q^*} q(q-1)q\alpha(r^2) - \sum_{z \in H_q} q(q-1)\alpha(Nz)$$

$$= q(q-1) \left\{ (q+1) \sum_{r \in \mathbb{F}_q^*} \alpha(r^2) - \sum_{z \in \mathbb{F}_{q^2}^*} \alpha(Nz) \right\}.$$

The first sum in braces vanishes unless $\alpha^2 = 1$ on $\mathbb{F}_q^*$ and the second sum in braces vanishes unless $\alpha \circ N = 1$ on $\mathbb{F}_{q^2}^*$. And, since the norm is onto, this last condition means that $\alpha = 1$ on $\mathbb{F}_q^*$. Note that if $\alpha = 1$ on $\mathbb{F}_q^*$, the two sums cancel. So the only nonzero case occurs when α is the quadratic character that is 1 on squares and -1 on nonsquares in $\mathbb{F}_q^*$.

(q + 1)-Dimensional.

$$\langle \chi_{\rho_{\alpha,\beta}}, \chi_\tau \rangle = \sum_{r \in \mathbb{F}_q^*} (q+1)q(q-1)\alpha(r)\beta(r).$$

This vanishes unless $\beta = \alpha^{-1}$. Here we need $\alpha \neq \beta$ and thus $\alpha^2 \neq 1$. Note that also $\rho_{\alpha,\beta} \cong \rho_{\beta,\alpha}$ and thus we need only half the remaining $\rho_{\alpha,\alpha^{-1}}$. This is why $\alpha = \xi^j$, with j only summed from 1 to $(q-3)/2$. Of course, $\xi^{(q-1)/2} = \varepsilon$, the quadratic character with $\varepsilon^2 = 1$.

(q − 1)-Dimensional.

$$\langle \chi_{\rho_\nu}, \chi_\tau \rangle = \sum_{r \in \mathbb{F}_q^*} q(q-1)^2\nu(r) - \sum_{z \in H_q} q(q-1)\{\nu(z) + \nu(\bar{z})\}$$

$$= q(q-1) \left\{ \sum_{r \in \mathbb{F}_q^*} \{q-1+2\}\nu(r) - \sum_{z \in \mathbb{F}_{q^2}^*} \{\nu(z) + \nu(\bar{z})\} \right\}$$

Since $\nu \neq \nu^q$ implies ν is nontrivial, the last sum in braces is zero. And the first sum is $q^2 - 1$, provided that ν restricted to $\mathbb{F}_q^*$ is trivial. These are exactly the $\nu = \xi^{(q-1)j}$, $j = 1, \ldots, q-1$. We need only half of these values of j, since $\sigma_\nu \cong \sigma_{\nu^q}$. ∎

Next we need to show that the spherical functions corresponding to the discrete series representations σ_ν are indeed given by Soto–Andrade's formula (19) of Chapter 20.

Theorem 4 *Soto–Andrade's Formula for the Spherical Functions on Finite Upper Half Planes Corresponding to Discrete Series Representations.*

Assume that $z \in H_q$ with $d(z, \sqrt{\delta}) = a$, $a \neq 0, 4\delta$. Then, for $\nu \neq \bar{\nu}$ a character of $\mathbb{F}_q(\sqrt{\delta})^*$,

$$(q+1)h_{\sigma_\nu}(z) = \sum_{\substack{w = u + v\sqrt{\delta} \\ Nw = 1}} \varepsilon\left(2u - 2 + \frac{a}{\delta}\right)\nu(w).$$

Proof. (Evans [1994].) Start with formula (16) in Chapter 20:

$$h_{\sigma_\nu}(g) = \frac{1}{|K|} \sum_{k \in K} \chi_{\sigma_\nu}(kg). \tag{19}$$

Then note that since the spherical function is K-bi-invariant, we can replace the $g \in GL(2, \mathbb{F}_q)$ with $z \in H_q$, or, equivalently,

$$g = g_z = \begin{pmatrix} y & x \\ 0 & 1 \end{pmatrix}$$

satisfying

$$x = ay + \delta(y-1)^2. \tag{20}$$

An element $k \in K$ has the form

$$k = k_{c,d} = \begin{pmatrix} c & d\delta \\ d & c \end{pmatrix},$$

and thus

$$kg = \begin{pmatrix} c & d\delta \\ d & c \end{pmatrix}\begin{pmatrix} y & x \\ 0 & 1 \end{pmatrix} = \begin{pmatrix} cy & cx + d\delta \\ dy & dx + c \end{pmatrix}.$$

Let N be the determinant of kg and T be the trace of kg. Let w be an eigenvalue of $k_{c,d}g_z$. We can rewrite (19) using Table II.12 as a sum over $w \in \mathbb{F}_{q^2}^*$ provided that we keep track of the number R_w of $c, d \in \mathbb{F}_q^*$ such that $k_{c,d}g_z$ has w as an eigenvalue. One sees that

$$N = y(c^2 - \delta d^2),$$
$$T = c(y+1) + dx.$$

So

$$Nx^2 y^{-1} = (cx)^2 - \delta(T - c(y+1))^2.$$

This means c satisfies the quadratic equation

$$c^2(x^2 - \delta(y+1)^2) + 2\delta(y+1)Tc - \delta T^2 - x^2Ny^{-1} = 0.$$

Use (20) to see that the discriminant of this quadratic equation for c is

$$4x^2\delta(T^2 - (4 - b\delta^{-1})N).$$

This implies that

$$R_w = 1 + \varepsilon(\delta T^2 + (b - 4\delta)N).$$

Thus we can rewrite (19) as

$$h_{\sigma_v}(z) = \frac{1}{|K|} \sum_{T,N \in \mathbb{F}_q} v\left(\frac{T + \sqrt{T^2 - 4N}}{2}\right)$$
$$\times \{\varepsilon(T^2 - 4N) - 1\}\{1 + \varepsilon(\delta T^2 + (b - 4\delta)N)\}.$$

Next change variables, writing

$$w = \frac{T + V\sqrt{\delta}}{2}, \qquad \delta V^2 = T^2 - 4N,$$

and see

$$h_{\sigma_v}(z) = \frac{-1}{|K|} \sum_{w \in \mathbb{F}_{q^2}} v(w)\varepsilon\left(\delta T_w^2 + (b - 4\delta)N_w\right),$$

where N_w and T_w denote the norm and trace, respectively of w down to $\mathbb{F}_q$. Note that

$$4\delta N_w(w^{q-1} + w^{1-q} - 2 + b\delta^{-1}) = 4\delta T_w^2 + 4N_w(b - 4\delta).$$

Now write $v = \xi^{j(q-1)}$, where ξ is a character on $\mathbb{F}_{q^2}^*$ of order $q^2 - 1$. Since

$$\varepsilon(N_w) = \xi^{(q-1)/2}(w^{q+1}) = \xi^{(q+1)/2}(w^{q-1})$$

and $\varepsilon(\delta) = -1$, we have

$$h_{\sigma_v}(z) = \frac{1}{|K|} \sum_{w \in \mathbb{F}_{q^2}} \xi^j(w^{q-1})\xi^{(q+1)/2}(w^{q-1})\varepsilon(w^{q-1} + w^{1-q} - 2 + b\delta^{-1}).$$

For any $t \in \mathbb{F}_{q^2}$ of norm 1, there are $q - 1$ elements w of $\mathbb{F}_{q^2}$ such that $w^{q-1} = t$. Therefore

$$h_{\sigma_v}(z) = \frac{1}{q+1} \sum_{\substack{t \in \mathbb{F}_{q^2} \\ Nt = 1}} \xi^{j-(q+1)/2}(t)(t + t^{-1} - 2 + b\delta^{-1}).$$

■

Note. Evans [1995] evaluates the spherical functions in characteristic 2. The formula is slightly different.

Character Table for *GL*(3, $\mathbb{F}_q$)

We close this chapter with the character table of $GL(3, \mathbb{F}_q)$ from Steinberg [1951]. We will see that this time there are three maximal tori – corresponding to

$$\left(\mathbb{F}_q^*\right)^3, \quad \mathbb{F}_q^* \times \mathbb{F}_{q^2}^*, \quad \mathbb{F}_{q^3}^*.$$

But first we need the conjugacy classes. See Table II.14.

Note that here in the last two rows of the table one must use a similarity transform outside of the base field $\mathbb{F}_q$ to produce matrices in the listed form. Thus in the case of classes of the form B_1, the actual matrix in $GL(3, \mathbb{F}_q)$, for q odd, would be similar to a matrix of the form

$$\begin{pmatrix} r & 0 & 0 \\ 0 & s & \delta t \\ 0 & t & s \end{pmatrix}, \quad \text{where } \delta \text{ is a nonsquare in } \mathbb{F}_q \text{ and } t \neq 0.$$

The 2×2 lower right-hand corner matrix corresponds to $z = s + t\sqrt{\delta} \in \mathbb{F}_{q^2}^*$.

For classes of type C_1, let us restrict ourselves to the special case that there is some noncube in $\mathbb{F}_q^*$, which implies $q \equiv 1 \pmod 3$. Why? (Answer this as an exercise.)

Suppose that ε is a noncube in $\mathbb{F}_q^*$. Then the actual matrix in $GL(3, \mathbb{F}_q)$ representing a conjugacy class of type C_1 will be an element of the group K_3 defined by

$$K_3 = \left\{ k = \begin{pmatrix} a & c\varepsilon & b\varepsilon \\ b & a & c\varepsilon \\ c & b & a \end{pmatrix} \middle| \det k \neq 0 \right\}, \quad \varepsilon \text{ a noncube in } \mathbb{F}_q^*. \qquad (21)$$

Table II.14. *Conjugacy classes in $GL(3, \mathbb{F}_q)$ from Steinberg [1951]*

Element	Matrix Rep.	# Classes	# Elements in class
A_1	$\begin{pmatrix} r & 0 & 0 \\ 0 & r & 0 \\ 0 & 0 & r \end{pmatrix}$	$q-1$	1
A_2	$\begin{pmatrix} r & 1 & 0 \\ 0 & r & 0 \\ 0 & 0 & r \end{pmatrix}$	$q-1$	$(q-1)(q+1)$ $\times (q^2 + q + 1)$
A_3	$\begin{pmatrix} r & 1 & 0 \\ 0 & r & 1 \\ 0 & 0 & r \end{pmatrix}$	$q-1$	$q(q-1)^2(q+1)$ $\times (q^2 + q + 1)$
A_4	$\begin{pmatrix} r & 0 & 0 \\ 0 & r & 0 \\ 0 & 0 & s \end{pmatrix}, r \neq s$	$(q-1)(q-2)$	$q^2(q^2 + q + 1)$
A_5	$\begin{pmatrix} r & 1 & 0 \\ 0 & r & 0 \\ 0 & 0 & s \end{pmatrix}, r \neq s$	$(q-1)(q-2)$	$q^2(q-1)(q+1)$ $\times (q^2 + q + 1)$
A_6	$\begin{pmatrix} r & 0 & 0 \\ 0 & s & 0 \\ 0 & 0 & t \end{pmatrix} \begin{smallmatrix} r \neq s \\ r \neq t \\ s \neq t \end{smallmatrix}$	$(q-1)(q-2)(q-3)/6$	$q^3(q+1)$ $\times (q^2 + q + 1)$
B_1	$\begin{pmatrix} r & 0 & 0 \\ 0 & z & 0 \\ 0 & 0 & z^q \end{pmatrix}, \begin{smallmatrix} z \in \mathbb{F}_{q^2}^* \\ z \notin \mathbb{F}_q^* \end{smallmatrix}$	$q(q-1)^2/2$	$q^3(q-1)$ $\times (q^2 + q + 1)$
C_1	$\begin{pmatrix} w & 0 & 0 \\ 0 & w^q & 0 \\ 0 & 0 & w^{q^2} \end{pmatrix}, \begin{smallmatrix} w \in \mathbb{F}_{q^3}^* \\ w \notin \mathbb{F}_q^* \end{smallmatrix}$	$q(q-1)(q+1)/3$	$q^3(q-1)^2(q+1)$

The number of conjugacy classes is $q(q-1)(q+1)$.

Exercise. Assume $q \equiv 1 \pmod 3$ and that ε is a noncube in $\mathbb{F}_q^*$.
a) Show that the group K_3 defined by (21) is isomorphic to the multiplicative group $\mathbb{F}_{q^3}^* = \mathbb{F}_q^*(\varepsilon^{1/3})$.
b) Show that

$$\det \begin{pmatrix} a & c\varepsilon & b\varepsilon \\ b & a & c\varepsilon \\ c & b & a \end{pmatrix} = \text{norm}(a + b\theta + c\theta^2),$$

where $\theta = \varepsilon^{1/3}$ and the norm denotes the norm from $\mathbb{F}_q^*(\varepsilon^{1/3})$ down to $\mathbb{F}_q^*$.

Thus we may view K_3 as an analogue of the subgroup K of $GL(2, \mathbb{F}_q)$ consisting of matrices of the form

$$\begin{pmatrix} a & b\delta \\ b & a \end{pmatrix},$$

with nonzero determinant. Here δ is a nonsquare in $\mathbb{F}_q$. Note that we can embed K into $GL(3, \mathbb{F}_q)$ as the subgroup

$$K_2 = \left\{ k = \begin{pmatrix} a & b\delta & 0 \\ b & a & 0 \\ 0 & 0 & 1 \end{pmatrix} \,\middle|\, \det k \neq 0 \right\}. \tag{22}$$

Thus the finite group analogue of the orthogonal group $O(2, \mathbb{R})$ is not contained in the finite analogue of $O(3, \mathbb{R})$ in $GL(3, \mathbb{F}_q)$. This is very different from the situation for $GL(3, \mathbb{R})$.

As we said earlier, the groups K_i are often called *tori* in $GL(3, \mathbb{F}_q)$. Of course, the diagonal matrices in $GL(3, \mathbb{F}_q)$ form another torus. A torus is a subgroup isomorphic to a product of multiplicative groups of fields. Maximal tori are important for the Deligne–Lusztig theory of the representations of finite groups of Lie type. See Digne and Michel [1991, Chapter 15] for the example of $GL(2, \mathbb{F}_q)$ worked out in this framework.

Now we give the character table for $GL(3, \mathbb{F}_q)$, which we take more or less from Steinberg [1951]. It will require two tables to do this: Tables II.15 and II.16.

Exercise. Check Tables II.15 and II.16. Note that the number of representations of each type is exactly the same as the number of conjugacy classes of the corresponding type (listed in the same order).

Table II.15. *Character table for $GL(3, \mathbb{F}_q)$, Part I*

	A_1	A_2	A_3	A_4
α	$\alpha(r^3)$	$\alpha(r^3)$	$\alpha(r^3)$	$\alpha(r^2 s)$
π_α	$(q^2 + q)\alpha(r^3)$	$q\alpha(r^3)$	0	$(q + 1)\alpha(r^2 s)$
π'_α	$q^3\alpha(r^3)$	0	0	$q\alpha(r^2 s)$
$\pi_{\alpha,\beta}$	$(q^2 + q + 1)$ $\cdot \alpha(r^2)\beta(r)$	$(q + 1)\alpha(r^2)\beta(r)$	$\alpha(r^2)\beta(r)$	$(q + 1)\alpha(rs)\beta(r)$ $+ \alpha(r^2)\beta(s)$
$\pi'_{\alpha,\beta}$	$q(q^2 + q + 1)$ $\cdot \alpha(r^2)\beta(r)$	$q\alpha(r^2)\beta(r)$	0	$(q + 1)\alpha(rs)\beta(r)$ $+ q\alpha(r^2)\beta(s)$
$\pi_{\alpha,\beta,\gamma}$	$(q + 1)(q^2 + q + 1)$ $\cdot \alpha(r)\beta(r)\gamma(r)$	$(2q + 1)$ $\cdot \alpha(r)\beta(r)\gamma(r)$	$\alpha(r)\beta(r)\gamma(r)$	$(q + 1)\{\alpha(s)\beta(r)\gamma(r)$ $+ \beta(s)\alpha(r)\gamma(r)$ $+ \gamma(s)\alpha(r)\beta(r)\}$
$\rho_{\alpha,\nu}$	$(q - 1)(q^2 + q + 1)$ $\cdot \alpha(r)\nu(r)$	$-\alpha(r)\nu(r)$	$-\alpha(r)\nu(r)$	$(q - 1)\alpha(s)\nu(r)$
σ_μ	$(q - 1)^2(q + 1)$ $\cdot \mu(r)$	$-(q - 1)\mu(r)$	$\mu(r)$	0

Here α, β, γ are (distinct) characters of $\mathbb{F}_q^*$; ν is a character of $\mathbb{F}_{q^2}^*$ with $\nu \neq \nu^q$; and μ is a character of $\mathbb{F}_{q^3}^*$, with $\mu \neq \mu^q$.

Table II.16. *Character table for $GL(3, \mathbb{F}_q)$, Part II*

	A_5	A_6	B_1	C_1
α	$\alpha(r^2 s)$	$\alpha(rst)$	$\alpha(rNz)$	$\alpha(Nw)$
π_α	$\alpha(r^2 s)$	$2\alpha(rst)$	0	$-\alpha(Nw)$
π'_α	0	$\alpha(rst)$	$-\alpha(rNz)$	$\alpha(Nw)$
$\pi_{\alpha,\beta}$	$\alpha(rs)\beta(r)$ $+\alpha(r^2)\beta(s)$	$\beta(r)\alpha(st)$ $+\beta(s)\alpha(rt)$ $+\beta(t)\alpha(rs)$	$\alpha(Nz)\beta(r)$	0
$\pi'_{\alpha,\beta}$ $\alpha \neq \beta$	$\alpha(rs)\beta(r)$	$\beta(r)\alpha(st)$ $+\beta(s)\alpha(rt)$ $+\beta(t)\alpha(rs)$	$-\alpha(Nz)\beta(r)$	0
$\pi_{\alpha,\beta,\gamma}$ α, β, γ distinct	$\alpha(s)\beta(r)\gamma(r)$ $+\beta(s)\alpha(r)\gamma(r)$ $+\gamma(s)\alpha(r)\beta(r)$	$\alpha(r)\beta(s)\gamma(t)$ $+$ all permutations	0	0
$\rho_{\alpha,\nu}$ $\nu \neq \nu^q$	$-\alpha(s)\nu(r)$	0	$-\alpha(r)(\nu(z) + \nu(z^q))$	0
σ_ν	0	0	0	$\mu(w) + \mu(w^q)$ $+ \mu(w^{q^2})$

There has been progress in generalizing the work on finite upper half planes to $GL(3, \mathbb{F}_q)$. See the Ph.D. theses of Nancy Allen [1996] and Maria Martinez [1998].

See Anna Helversen–Pasotto [1997] for connections between discrete series representations of $GL(n, \mathbb{F}_q)$ and identities involving Gauss sums.

Chapter 22

Selberg's Trace Formula and Isospectral Non-isomorphic Graphs

> Selberg's theory is one of the biggest guns.
> T. Tamagawa [1960]

> Some readers may feel that Hejhal has created a "monster."
> D. Hejhal [1992]

> All of this looks to be elaborate and extremely difficult, as indeed it is; so it is very helpful to understand it for simple examples . . . .
> R. Langlands (in Aubert et al. [1989, p. 152])

Selberg's trace formula was formulated in 1956 by Selberg [1989, Vol. I, pp. 423–63] for arithmetic groups Γ like $SL(2, \mathbb{Z})$ acting on the Poincaré upper half plane H defined in (2) of Chapter 23. Despite the slightly fearsome quotes above, it has proved to be extremely useful. Applications include:

- a derivation of Weyl's law on the distribution of eigenvalues of the Laplacian on $\Gamma \backslash H$,
- an answer to the question: Can you hear the shape of $\Gamma \backslash H$?,
- the discussion of the analytic properties of Selberg's zeta function (which are essentially equivalent to the trace formula).

We will summarize some of this in Chapter 23. Our main goal here is to consider finite analogues of some of these results. At this point the reader might wish to review the discussion "beginnings of a trace formula" at the end of Chapter 3 as well as Chapter 12 (Poisson's summation formula).

Finite versions of the trace formula do not seem to have been much studied. We first began to think about these things after reading Arthur's derivation of

385

the Frobenius reciprocity law from the trace formula in Aubert et al. [1989, pp. 11–27].

In this chapter we consider an application of the pre-trace formula to graph theory, which ultimately led to a solution of an old problem of Mark Kac [1966]: "Can you hear the shape of a drum?" That is, if two plane drums have the same (Dirichlet) spectra for the eigenvalues of the Laplacian, must the drums be obtained from each other by euclidean motion? To see that this reformulation is equivalent to the original question of Kac one must note that the eigenvalues of the Laplacian determine the fundamental frequencies and modes of vibration of the drum (via separation of variables for the wave equation). See Chapter 13 for the connection between eigenvalues of adjacency operators and frequencies of vibration for systems of masses attached by springs as well as molecules. Another reference is Cvetković et al. [1980].

See Courant and Hilbert [1953, pp. 300–2] for a discussion of a square drum. The eigenvalues of the Laplacian for a unit square drum are $\lambda_{n,m} = \pi(n^2 + m^2)$, $n, m = 1, 2, 3, \ldots$. The corresponding eigenfunctions are

$$u_{n,m}(x, y) = \sin(nx)\sin(my).$$

It is interesting to let Mathematica or Matlab do list density plots of sums of eigenfunctions such as $v = u_{1,4}$ or the more interesting sum $v = u_{1,4} + (2/3)^{1/2}u_{4,1}$. The picture might well be compared with Figures I.22 and I.23 of Chapter 5 as well as Figures II.8 and II.9 in Chapter 20. See Figure I.20 in Chapter 5 or D. L. Powers [1987, cover and Chapter 5] for a circular drum where the eigenfunctions of the Laplacian in polar coordinates involve the J-Bessel function. Incidentally Kac attributes the original drum question to S. Bochner.

Here we will consider the graph theoretical analogue of the Kac–Bochner question. That is, we will seek nonisomorphic isospectral graphs, that is, graphs whose adjacency matrices have the same eigenvalues with the same multiplicities but which are not isomorphic.

Examples of such graphs have been known for a long time. See Cvetković, Doob, and Sachs [1980, pp. 24 and 256]. The first examples are due to Collatz and Singowitz in 1957. See Figure II.10.

Definition. Given a subgroup Γ of a finite group G and a generating subset S of G, which we will call the edge set, the *Schreier graph* $X = X(\Gamma \backslash G, S)$ has as vertices the cosets Γg, $g \in G$. Two vertices Γg and Γgs are joined by an oriented edge for all $s \in S$.

Note that when $\Gamma = \{e\}$, Schreier graphs are Cayley graphs. As with Cayley graphs, we should really attach a color to each element of S and color the edges

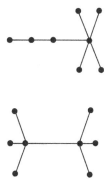

Figure II.10. Isospectral nonisomorphic graphs found by Collatz and Singowitz in 1957.

with the colors. The Schreier graphs are directed with multiple edges and loops, in general.

Example. (Buser [1988, 1992].) Let $G = SL(3, \mathbb{F}_2) = GL(3, \mathbb{F}_2) \cong PSL(2, \mathbb{F}_7) =$ the simple group of order 168. Set

$$\Gamma_1 = \left\{ \begin{pmatrix} 1 & * & * \\ 0 & * & * \\ 0 & * & * \end{pmatrix} \right\}, \quad \Gamma_2 = {}^t\Gamma_1,$$

$$S = \left\{ A = \begin{pmatrix} 0 & 1 & 1 \\ 0 & 1 & 0 \\ 1 & 0 & 0 \end{pmatrix}, B = \begin{pmatrix} 1 & 0 & 0 \\ 0 & 0 & 1 \\ 0 & 1 & 1 \end{pmatrix} \right\}.$$

See Figure II.11 for the two Schreier graphs.

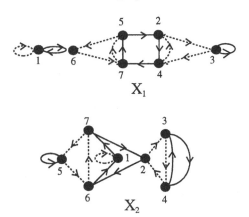

Figure II.11. Two Schreier graphs, which are isospectral but not isomorphic. (From Buser [1988, 1992].)

To find the first graph X_1, you identify G/Γ_1 as the space of column vectors in $\mathbb{F}_2^3 - 0$, with the elements of S acting on the left. This means that

$$A \begin{pmatrix} a \\ b \\ c \end{pmatrix} = \begin{pmatrix} b+c \\ b \\ a \end{pmatrix} \quad \text{and} \quad B \begin{pmatrix} a \\ b \\ c \end{pmatrix} = \begin{pmatrix} a \\ c \\ b+c \end{pmatrix}.$$

Connect

$$\begin{pmatrix} a \\ b \\ c \end{pmatrix} \quad \text{to vector} \quad \begin{pmatrix} b+c \\ b \\ a \end{pmatrix}$$

with a solid arrow and connect

$$\begin{pmatrix} a \\ b \\ c \end{pmatrix} \quad \text{to} \quad \begin{pmatrix} a \\ c \\ b+c \end{pmatrix}$$

with a dotted arrow.

To find the second graph X_2, you identify $\Gamma_2 \backslash G$ with the space of row vectors in $\mathbb{F}_2^3 - 0$, with the elements of S acting on the right.

You can see that the graphs in Figure II.11 are not isomorphic. We will use the pre-trace formula to see that the graphs are nevertheless isospectral; that is, their adjacency operators have the same spectra.

You might ask whether there isn't an easier method to see that these drums are isospectral, making use of the group automorphism $\mu(x) = {}^t x^{-1}$, $x \in G$, which sends Γ_1 to Γ_2. However, μ does not preserve the set S. So you cannot use it to transplant eigenfunctions of the adjacency operator on X_1 to an eigenfunction on X_2.

Exercise. Check that Figure II.11 is correct and that the two graphs in the figure are not isomorphic.

Hint. Look at vertex 1 on X_1. It is only connected to one other vertex.

The graphs in Figure II.11 led to an answer of "no" for the question of Kac–Bochner. There are planar drums whose shape can't be heard. Gordon, Webb, and Wolpert [1992] found that the shape of the drums in Figure II.12 can't be heard. These figures were found by fattening up the graphs in Figure II.11 to the Riemannian manifolds in Figure II.13. Then use is made of symmetry to flatten it out. This answers the question of Kac for non-convex planar drums. It is still open for convex drums. Gordon and Webb [1996] explain the method of transplantation for proving that certain drums are isospectral without appealing

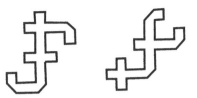

Figure II.12. Two isospectral drums whose shape cannot be heard (Gordon, Webb, and Wolpert [1992].)

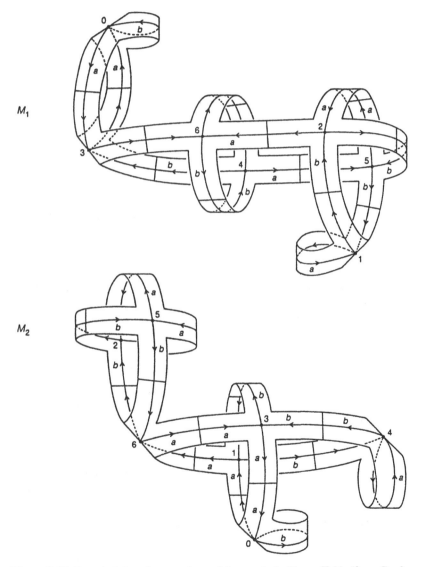

Figure II.13. Buser's fattened up versions of the graphs in Figure II.11. (from Gordon, Webb, and Wolpert [1992].)

to group theory. See Peter Doyle's web site for a large number of examples of drums that cannot be heard. Another interesting reference is the earlier article of Gordon [1989] which comes with a recording of drum music by Dennis DeTurck.

It turns out that the isospectral property of the graphs is related to an algebraic property of the subgroups Γ_i, $i = 1, 2$, called almost conjugacy.

Definition. Two subgroups Γ_i, $i = 1, 2$, of G are called *almost conjugate* if for every conjugacy class

$$\{g\} = \{xgx^{-1} \mid x \in G\}, \quad \text{we have } \#(\{g\} \cap \Gamma_1) = \#(\{g\} \cap \Gamma_2). \tag{1}$$

Exercise. Show that the subgroups Γ_1 and Γ_2 of the preceding example are indeed almost conjugate.

Hint. See Buser [1992, p. 289].

Some History

The examples of almost conjugate subgroups were used by number theorists (e.g., Gassmann in 1925) to show that, in some sense, you can't use the Dedekind zeta function to hear the shape of an algebraic number field. See Perlis [1977]. There are algebraic number fields $K_i = \mathbb{Q}(\theta_i)$, $i = 1, 2$, with common Galois closure N such that the Galois groups are $\Gamma_i = \text{Gal}(N/K_i)$, for $i = 1, 2$. Here we can take θ_1 to be a root of $x^7 - 7x + 3 = 0$ and θ_2 to be a root of $x^7 + 14x^4 - 42x^2 - 21x + 9 = 0$. Then define the Dedekind zeta function of a number field K as follows:

$$\zeta_K(s) = \sum_a N \, a^{-s} = \prod_{\not p} (1 - N \not p^{-s})^{-1}, \quad \text{if Re } s > 1.$$

The sum is over all ideals a in the ring of integers o_K of K and the norm Na is the order of the quotient ring o_K/a . The product is over all prime ideals $\not p$ of o_K.

The Dedekind zeta function shares the usual properties of the Riemann zeta function such as the Euler product above, the analytic continuation to $s \in \mathbb{C}$ with a pole at $s = 1$, a functional equation relating the value at s and the value at $1 - s$. It actually is the Riemann zeta function when $K = \mathbb{Q}$. This function is used to study the distribution of prime ideals in the ring o_K, unique factorization or the lack thereof in o_K, and the group of units of o_K.

It turns out that, for the fields above, $\zeta_{K_1} = \zeta_{K_2}$, even though the number fields K_1 and K_2 are not isomorphic. The proof that the two zeta functions coincide is essentially the same as that which we shall give to see that the graphs in Figure II.11 are isospectral (making use of Artin L-functions). Some references for Dedekind zeta functions are Cassels and Fröhlich [1967], Stark [1992], and Terras [1985].

Next we must recall some facts from earlier chapters in Part II. We will need to know that the character determines the representation up to equivalence. And we will need to consider the induced representation $\rho = \mathrm{Ind}_\Gamma^G \sigma$. Recall that if Γ is a subgroup of the finite group G and $\sigma : \Gamma \to GL(W)$ is a representation, then $\rho : G \to GL(V_\sigma)$, where

$$V_\sigma = \{f : G \to W \mid f(\gamma x) = \sigma(\gamma) f(x) \quad \text{for all } \gamma \in \Gamma, \ x \in G\}. \tag{2}$$

If $f \in V_\sigma$, $[\rho(g)f](x) = f(xg)$, for $x \in G$. That is, $\rho(g)$ acts by right translation on f. The adjacency operator of a Schreier graph is a sum of such $\rho(g)$s with $\sigma = 1$. See formula (10) below.

Next we need to think about Fourier transforms on G. Let $f : G \to \mathbb{C}$ and suppose that ρ is any representation of G. The *Fourier transform* of f at ρ is

$$\hat{f}(\rho) = \sum_{y \in G} f(y)\rho(y). \tag{3}$$

One usually takes ρ to be irreducible in formula (3), but here we'll also need to take $\rho = \mathrm{Ind}_\Gamma^G \sigma$.

The Pre-Trace Formula

The pre-trace formula is just a rewrite of $\mathrm{Tr}(\hat{f}(\rho))$, where $\rho = \mathrm{Ind}_\Gamma^G 1$. Then if

$$\begin{aligned}
\phi \in V_1 &= \{f : G \to \mathbb{C} \mid f(\gamma x) = f(x), \forall \, \gamma \in \Gamma, \ x \in G\} \\
&= L^2(\Gamma \backslash G),
\end{aligned}$$

we have

$$\begin{aligned}
[\hat{f}(\rho)\phi](x) &= \sum_{y \in G} f(y)[\rho(y)\phi](x) = \sum_{y \in G} f(y)\phi(xy) \tag{4} \\
&= \sum_{u \in G} f(x^{-1}u)\phi(u).
\end{aligned}$$

For the last equality, set $u = xy$.

Now ϕ is Γ-invariant and this implies that

$$[\hat{f}(\rho)\phi](x) = \sum_{y \in \Gamma \backslash G} \sum_{\gamma \in \Gamma} f(x^{-1}\gamma y)\phi(y). \tag{5}$$

Take a basis $\{\delta_x \mid x \in \Gamma \backslash G\}$ of V_1, with $\delta_x(y) = 0$ when $x \neq y$ and 1 if $x = y$. Then the matrix of $\hat{f}(\rho)$ with respect to this basis has as its entry corresponding to $x, y \in \Gamma \backslash G$ the sum:

$$\sum_{\gamma \in \Gamma} f(x^{-1}\gamma y).$$

It follows that the trace of $\hat{f}(\rho)$ is given by the *pre-trace formula I*:

$$\mathrm{Tr}\hat{f}(\rho) = \sum_{x \in \Gamma \backslash G} \sum_{\gamma \in \Gamma} f(x^{-1}\gamma x), \quad \text{if } \rho = \mathrm{Ind}_\Gamma^G 1. \tag{6}$$

It is also possible to rewrite the left-hand side of this formula as a sum over the *dual space* $\hat{G}$ – a complete set of inequivalent, irreducible, unitary representations π of G. This is possible since Proposition 2 of Chapter 15 says that we can decompose ρ as a direct sum of copies of the π:

$$\rho = \sum_{\pi \in \hat{G}}^{\oplus} m_G(\pi, \rho)\pi,$$

where $m_G(\pi, \rho) = m(\pi, \rho)$ is a nonnegative integer called the *multiplicity* of π in ρ.

The exact result for $\rho = \mathrm{Ind}_\Gamma^G 1$ will be called the *pre-trace formula II*:

$$\sum_{\pi \in \hat{G}} \mathrm{Tr}(\hat{f}(\pi))m(\pi, \rho) = \sum_{x \in \Gamma \backslash G} \sum_{\gamma \in \Gamma} f(x^{-1}\gamma x), \tag{7}$$

where $m(\pi, \rho)$ is the multiplicity of π in $\rho = \mathrm{Ind}_\Gamma^G 1$.

Exercise. Use formula (7) to prove the *Frobenius reciprocity law*, which says

$$m_G(\kappa, \mathrm{Ind}_\Gamma^G 1) = m_\Gamma(1, \kappa \mid_\Gamma),$$

where $\kappa \mid_\Gamma$ is the restriction of $\kappa \in \hat{G}$ to Γ.
Hint. Plug $f = \mathrm{Tr}(\kappa)$ into (7). See Arthur's paper in Aubert et al. [1989, p. 11].

Next we want to apply the pre-trace formula I to show that Schreier graphs for almost conjugate subgroups $\Gamma_i, i = 1, 2$ of G are isospectral. To see this, we set $f = \delta_{\{g\}}$ in the pre-trace formula I. That is, f is the function which is 1 on the conjugacy class $\{g\}$ and 0 off this conjugacy class. We obtain

$$\mathrm{Tr}[\hat{\delta}_{\{g\}})(\rho)] = \sum_{x \in G} \delta_{\{g\}}(x) \mathrm{Tr}(\rho(x)) = |\{g\}| \mathrm{Tr}\rho(g) = \frac{|G|}{|G_g|}\chi_\rho(g), \tag{8}$$

where $G_g = \{x \in G \mid xg = gx\} = $ *centralizer* of g in G.

So

$$\frac{|G|}{|G_g|}\chi_\rho(g) = \sum_{x\in\Gamma\backslash G}\sum_{\gamma\in\Gamma}\delta_{\{g\}}(x\gamma x^{-1}) = \frac{|G|}{|\Gamma|}\#(\{g\}\cap\Gamma). \tag{9}$$

Therefore, if Γ_1 and Γ_2 are almost conjugate, the two representations $\rho_i = \mathrm{Ind}_{\Gamma_i}^G 1$ have the same character for $i = 1, 2$ and thus are equivalent representations. This means that there is a 1-1 linear map $T : L^2(\Gamma_1\backslash G) \to L^2(\Gamma_2\backslash G)$ such that $T\rho_1(g)T^{-1} = \rho_2(g)$, for all $g \in G$.

Now the adjacency operator A_i on the Schreier graph $X(\Gamma_i\backslash G, S)$ is given for $\phi \in L^2(\Gamma_i\backslash G) = V_1$ by

$$(A_i\phi)(x) = \sum_{s\in S}\phi(xs) = \sum_{s\in S}[\rho_i(s)\phi](x), \quad \text{where } \rho_i = \mathrm{Ind}_{\Gamma_i}^G 1. \tag{10}$$

So $T\rho_1(g)T^{-1} = \rho_2(g)$ implies that $T A_1 T^{-1} = A_2$. It follows that A_1 and A_2 have the same spectrum. That is, the two Schreier graphs are isospectral. We have proved the following theorem.

Theorem 1. Suppose that Γ_1 and Γ_2 are almost conjugate subgroups of a finite group G. Then the Schreier graphs $X(\Gamma_i\backslash G, S)$ are isospectral for any edge set S in G.

Exercise. Are the adjacency operators A_i for the graphs in Figure II.11 actually diagonalizable by unitary operators? That is: are the A_i normal (meaning that $^tA_i A_i = A_i\,^tA_i$)?

Exercise. Show that we have another pair of isospectral nonisomorphic Schreier graphs $X(G/H_i, S)$ with

$$G = \mathrm{Aff}(\mathbb{Z}/8\mathbb{Z})$$
$$= \left\{(y, x) = \begin{pmatrix} y & x \\ 0 & 1 \end{pmatrix} \,\middle|\, x, y \in \mathbb{Z}/8\mathbb{Z},\ y \in (\mathbb{Z}/8\mathbb{Z})^* \right\},$$

$$H_1 = \{(1, 0),\ (3, 0),\ (5, 0),\ (7, 0)\} \quad \text{and}$$
$$H_2 = \{(1, 0),\ (3, 4),\ (5, 4),\ (7, 0)\}.$$

Hint. See Buser [1988, 1992].

Note. Other references for isospectral graphs are Lubotzky [1995], Brooks [1997], and Brooks, Gornet, and Gustafson [in press].

Chapter 23

The Trace Formula on Finite Upper Half Planes

When you are searching for something or someone, go off the beaten path.

Gabrielle Daniels [1994, p. 87]

The Original Selberg Trace Formula

Our aim in this chapter is to find a formula which is close to that of Selberg's original paper [1989, Vol. I, pp. 423–63] – except that we replace the field $\mathbb{R}$ of real numbers with the finite field $\mathbb{F}_q$. First we recall briefly the original continuous Selberg trace formula for the *special linear group* $G = SL(2, \mathbb{R})$ of 2×2 real matrices with determinant one. Longer surveys can be found in Elstrodt [1981], Hejhal [1976a], and Terras [1982]. Still more information can be found in Buser [1992], Chavel [1984], and Terras [1985]. And the monster formulation can be found in Hejhal [1976b].

With $G = SL(2, \mathbb{R})$, we take $K = SO(2)$, the *orthogonal group* of real proper rotations; that is,

$$K = \left\{ \begin{pmatrix} a & -b \\ b & a \end{pmatrix} \,\middle|\, a^2 + b^2 = 1 \right\}. \tag{1}$$

So we may set $a = \cos(u)$, $b = \sin(u)$, $u \in \mathbb{R}$. We can identify the quotient G/K with the *Poincaré upper half plane*

$$H = \{x + iy \mid x, y \in \mathbb{R}, y > 0\}, \tag{2}$$

for

$$g = \begin{pmatrix} a & b \\ c & d \end{pmatrix}$$

394

acts transitively on $z \in H$ via fractional linear transformation $gz = (az + b)/(cz + d)$. And K is the subgroup of G fixing the point i. Moreover, we have the G-invariant *Poincaré distance element*

$$ds^2 = \frac{dx^2 + dy^2}{y^2} \quad \text{with } Laplacian \ \Delta = y^2 \left(\frac{\partial^2}{\partial x^2} + \frac{\partial^2}{\partial y^2} \right). \tag{3}$$

The simplest eigenfunction of this Laplacian is the *power function* $p_s(z) = (\text{Im } z)^s$, for any $s \in \mathbb{C}$. And we can easily obtain a K-invariant eigenfunction of Δ or (zonal) *spherical function* by averaging the power function over K. Thus we obtain Harish–Chandra's integral formula for spherical functions:

$$h_s(z) = \int_{k \in K} p_s(kz) \, dk, \quad \text{where } dk = \text{Haar measure on } K. \tag{4}$$

This function can be seen to be a Legendre function using geodesic polar co-ordinates. See Terras [1985, Vol. I, p. 141]. The fact that all (zonal) spherical functions have this form is a result of Harish–Chandra for a general real symmetric space.

If f is a K-invariant function on H, that is, $f(kz) = f(z)$, for all $k \in K, z \in H$, the *spherical transform* is

$$\mathscr{S}f(s) = \int_{z \in H} f(z)h_s(z)\frac{dx \, dy}{y^2}. \tag{5}$$

This was called "the Helgason transform" in Terras [1985, Vol. I, p. 146, formula (3.24)]. It is invertible and takes convolution on G to ordinary pointwise product of spherical transforms.

In the Selberg trace formula often the K-invariant function $f(z)$ is replaced by a function of two variables called a *point-pair invariant*:

$$f(z, w) = \Phi\left(\frac{|z - w|^2}{\text{Im } z \, \text{Im } w} \right) = f(gz, gw) \quad \text{for all } g \in G, \ z, w \in H.$$

This makes sense because (see Hejhal [1976b, Vol. I, p. 6] and Terras [1985, Vol. I, pp. 154, 266])

$$\cosh d(z, w) = 1 + \frac{|z - w|^2}{2 \, \text{Im}(z) \, \text{Im}(w)}.$$

Now the trace formula involves a discrete subgroup Γ of G such as $\Gamma = SL(2, \mathbb{Z})$, the *modular group* of 2×2 matrices with integer entries and determinant one. Next we need to classify the conjugacy classes $\{\gamma\}$ of Γ according to the various possible Jordan forms of the matrix γ. We define a conjugacy class $\{\gamma\}$ in Γ to be *central* if γ commutes with every element of Γ.

The conjugacy class $\{\gamma\}$ is said to be *hyperbolic* if γ has Jordan form

$$\begin{pmatrix} t & 0 \\ 0 & t^{-1} \end{pmatrix}, \quad \text{for } t \in \mathbb{R}, \quad t \neq 0, \pm 1.$$

Then define the *norm* $\gamma = N\gamma = t^2$, which can be chosen to be >1 without loss of generality. The centralizer Γ_γ of a hyperbolic γ is cyclic:

$$\Gamma_\gamma = \langle \gamma_0 \rangle = \{ \gamma_0^j \mid j \in \mathbb{Z} \}.$$

We say the generator γ_0 is *primitive hyperbolic*.

The conjugacy class $\{\gamma\}$ is *parabolic* if γ has Jordan form

$$\pm \begin{pmatrix} 1 & a \\ 0 & 1 \end{pmatrix} \quad \text{for } a \neq 0,$$

and $\{\gamma\}$ is *elliptic* if γ has two distinct nonreal eigenvalues.

The *geodesics* in H are semicircles and half-lines orthogonal to the real axis. They are the "straight lines" in a noneuclidean geometry (since Euclid's 5th postulate fails). The *horocycles* are orthogonal to the geodesics (e.g., horizontal lines in H). The hyperbolic conjugacy classes in Γ correspond to closed geodesics in the fundamental domain $\Gamma \backslash H$.

The *horocycle transform* (also known as the Harish or the Abel transform) of a function $f : H \to \mathbb{C}$ is

$$F(y) = y^{-1/2} \int_{x \in \mathbb{R}} f(x + iy) \, dx. \tag{6}$$

It is also invertible since if we Mellin transform it with respect to y, we get the spherical transform.

The Selberg trace formula involves a sum over the discrete spectrum of the noneuclidean Laplacian Δ on $L^2(\Gamma \backslash H)$, assuming as usual that Γ is a discrete subgroup of G such that $\Gamma \backslash H$ has finite volume with respect to the Poincaré volume element $y^{-2} \, dx \, dy$. Writing $\lambda_n = s_n(1 - s_n)$ for the *discrete spectrum*, we have

$$\Delta \phi_n = \lambda_n \phi_n, \quad n \geq 0 \text{ (with } \phi_0 = \text{constant)}, \quad \phi_n \in L^2(\Gamma \backslash H).$$

Finally, after all these preliminaries, the *Selberg trace formula* says that for a K-invariant function f on H with compact support (or some weaker convergence condition) we have

$$\sum_{n \geq 0} \mathscr{S}f(s_n) = \operatorname{area}(\Gamma \backslash H)f(i) + \sum_{\substack{\{\gamma\} \\ \text{hyperbolic} \\ \Gamma_\gamma = \langle \gamma_0 \rangle}} \frac{\log N\gamma_0}{N\gamma^{1/2} - N\gamma^{-1/2}} F(N\gamma)$$

$$+ \text{parabolic and elliptic terms.} \qquad (7)$$

We will not give the formulas for the parabolic and elliptic terms here. The parabolic terms are not present when $\Gamma \backslash H$ is compact. The elliptic terms are not present if Γ acts without fixed points on H. However, the modular group $SL(2, \mathbb{Z})$ has both elliptic and parabolic conjugacy classes. Most of the references rewrite the central term $f(i)$ using the inversion formula for the spherical transform. Of course, the horocycle transform F is an inverse Mellin transform of the spherical transform $\mathscr{S}f$.

On the left in (7) is a sum over the discrete spectrum of Δ on $L^2(\Gamma \backslash H)$. On the right, the hyperbolic term is a sum over closed geodesics in $\Gamma \backslash H$. So the trace formula gives rise to a duality between the spectrum of the Laplacian on $L^2(\Gamma \backslash H)$ and the length spectrum of the closed geodesics in $\Gamma \backslash H$. This is analogous to the duality between the zeros of the Riemann zeta functions and the primes in $\mathbb{Z}$. The Selberg zeta function defined below makes the analogy even more precise.

Let us say a bit more about the length spectrum. See also Buser [1992], Hejhal [1976 a, b], and Terras [1985, Vol. 1, pp. 277–81]. If z, w are two points on the real axis (or ∞), let $C(z, w)$ denote a geodesic line or circle in H connecting these points. And let $\overline{C(z, w)}$ be the image of this geodesic in a fundamental domain $D \subset H$ for $\Gamma \backslash H$. $\overline{C(z, w)}$ is a closed geodesic iff z and w are the fixed points of a hyperbolic element γ of Γ. If a point q lies on $C(z, w)$, then so does γq, and the Poincaré distance between q and γq is $\log N\gamma$, where $N\gamma$ is the norm of γ defined above. These numbers are the lengths of the closed geodesics in $\Gamma \backslash H$.

Applications

Weyl Law

Assume $\Gamma = SL(2, \mathbb{Z})$, for example. Plug $\mathscr{S}f(s) = \exp(s(s-1)t)$, for $t > 0$, into formula (7). This spherical transform corresponds to $f =$ the heat kernel. Then one obtains

$$\sum_{n \geq 0} \exp(\lambda_n t) \sim \frac{\operatorname{area}(\Gamma \backslash H)}{4\pi t}, \quad t \to 0+.$$

A Tauberian theorem erases the Laplace transform to obtain

$$\#\{\lambda_n \mid |\lambda_n| \leq x\} \sim \frac{\text{area}(\Gamma \backslash H)x}{4\pi}, \quad x \to \infty.$$

This sort of result has not been proved for nonarithmetic Γ such that $\Gamma \backslash H$ is noncompact. Sarnak [1996] conjectures that the discrete spectrum may be finite for such Γ.

Isospectral but not Isometric Riemann Surfaces

Marie-France Vignéras [1980] found Γ_i, $i = 1, 2$, such that the Laplacian has the same spectrum on $L^2(\Gamma_i \backslash H)$, but these Riemann surfaces $\Gamma_i \backslash H$ are not isometric. See also Sunada [1988] and Buser [1988, 1992].

Analytic Properties of the Selberg Zeta Function

See Elstrodt [1981], Hejhal [1976a], and Vignéras [1979]. Here one substitutes

$$\mathscr{S}f\left(\frac{1}{2} + ir\right) = \frac{1}{r^2 + \left(\alpha - \frac{1}{2}\right)^2}$$
$$-\frac{1}{r^2 + \left(\beta - \frac{1}{2}\right)^2}, \quad \text{for Re } \alpha, \text{ Re } \beta > 1.$$

Then

$$F(e^u) = \frac{\exp\left(\left(\alpha - \frac{1}{2}\right)|u|\right)}{2\alpha - 1} - \frac{\exp\left(\left(\beta - \frac{1}{2}\right)|u|\right)}{2\beta - 1}.$$

The difference is necessary to obtain convergence. Then the hyperbolic term is

$$\frac{Z'(\alpha)}{(2\alpha - 1)Z(\alpha)} - \frac{Z'(\beta)}{(2\beta - 1)Z(\beta)},$$

where $Z(s)$ is the Selberg zeta function, which is defined as follows. Set

$$\mathscr{P}_\Gamma = \{\text{primitive hyperbolic conjugacy in } \Gamma\}.$$

Define the *Selberg zeta function* by viewing the primitive hyperbolic conjugacy classes as pseudoprimes. That is,

$$Z(s) = \prod_{\{\gamma_0\} \in \mathscr{P}_\Gamma} \prod_{\nu \geq 0} \left(1 - N\gamma_0^{-s-\nu}\right). \tag{8}$$

Compare this with the Euler product for the Riemann zeta function, which writes

$$\zeta(s) = \prod_{p \text{ prime}} (1 - p^{-s})^{-1}, \quad \text{for Re } s > 1.$$

The absence of the power -1 in (8) is a little curious.

The Selberg zeta function has many properties analogous to the Riemann zeta function, such as a functional equation. The nontrivial zeros of $Z(s)$ correspond to the discrete spectrum of the Laplacian on $L^2(\Gamma \backslash H)$. Thus the Selberg zeta function satisfies the Riemann hypothesis.

Sarnak [1996] has said that for nonarithmetic Γ with noncompact fundamental domain he doubted that one should think of $Z(s)$ as a zeta function. The reason is that Sarnak conjectures that in this case the discrete spectrum of the Laplacian on $L^2(\Gamma \backslash H)$ may be finite.

As we shall see in the last chapter, the Ihara zeta function for finite regular graphs shares many properties of the Selberg zeta function. Other references on the Selberg and other zeta functions are Kurokawa and Sunada [1992] and Ruelle [1976, 1994].

Other Applications

The original paper of Selberg gave an application to eigenvalues of Hecke operators T_p acting on $f \in L^2(\Gamma \backslash H)$ via

$$T_p f(z) = f(pz) + \sum_{0 \le j < p} f\left(\frac{z+j}{p}\right).$$

Since these operators commute with Δ, one can assume that the eigenfunctions of Δ in the left-hand side of the trace formula are also eigenfunctions of the Hecke operators. See Hejhal [1976b] for more information as well as Selberg [1989, Vol. I, pp. 423–63].

Gutzwiller [1990] shows that the Selberg trace formula is of great importance for the study of quantum chaos.

Higher Rank Groups such as GL(3, ℝ)

See Dorothy Wallace [1994] for a version of the trace formula for $GL(3, \mathbb{Z})$. Compare with adelic versions in the papers of Arthur and Langlands in Aubert et al. [1989]. Langlands says (Aubert et al. [1989, p. 135]): "In contrast to its initial purpose, which was apparently to analyze the spectrum of the Laplace–Beltrami operator on the quotient of the upper half plane by a Fuchsian group, the arithmetical applications of the trace formula usually involve a comparison of two or more trace formulas or of a trace formula with a Lefschetz formula ....." And Langlands [loc. cit., p. 138] says: "In contrast to the trace formula of Selberg, which was, as its name implies, a formula for a trace, but which could only be proven for groups of rank one, the formula developed by Arthur begins with an equality between two functions that are then integrated separately over [the adelic quotient] $G(\mathbb{Q})\backslash G^1(\mathbb{A})$ and the integrals calculated in completely different manner."

Return to the Finite Trace Analogue of Selberg's Trace Formula

We need to work on the right-hand side of pre-trace formula II, which is formula (7) of Chapter 22 in the special case of the general linear group $GL(2, \mathbb{F}_q)$, q odd. Recall that the pretrace formula II says

$$\sum_{\pi \in \hat{G}} \mathrm{Tr}(\hat{f}(\pi)) m(\pi, \rho) = \sum_{x \in \Gamma\backslash G} \sum_{\gamma \in \Gamma} f(x^{-1}\gamma x), \quad \text{if } \rho = \mathrm{Ind}_\Gamma^G 1. \qquad (9)$$

Here $m(\pi, \rho)$ denotes the multiplicity of π in ρ.

So the finite field $\mathbb{F}_q$ is our analogue of the real number field $\mathbb{R}$. Let δ be a nonsquare in $\mathbb{F}_q$. Then $\mathbb{F}_q(\sqrt{\delta}) = \mathbb{F}_{q^2}$ is our analogue of the complex number field $\mathbb{C}$. As in Chapter 19, the subgroup K analogous to $O(2, \mathbb{R})$ of G is

$$K = \left\{ \begin{pmatrix} a & b\delta \\ b & a \end{pmatrix} \middle| a, b \in \mathbb{F}_q, \, a^2 - b^2\delta \neq 0 \right\}. \qquad (10)$$

Then (see Chapter 19), we can identify G/K with the *finite upper half plane*

$$H_q = \{ z = x + y\sqrt{\delta} \mid x, y \in \mathbb{F}_q, \, y \neq 0 \}. \qquad (11)$$

The general linear group G acts on H_q by fractional linear transformation as over $\mathbb{R}$, and the finite analogue of the Poincaré distance (3) is

$$d(z, w) = \frac{N(z - w)}{\mathrm{Im}(z)\,\mathrm{Im}(w)}, \quad \text{where } z, w \in H_q, \qquad (12)$$

$$z = x + y\sqrt{\delta}, \quad \bar{z} = x - y\sqrt{\delta} = z^q, \quad Nz = z\bar{z}.$$

Again $d(gz, gw) = d(z, w)$ for all $g \in G$, $z, w \in H_q$.

If we take the sets S to be K-orbits in G/K, we find that (except for two orbits) we obtain Schreier graphs $X(G/K, S_q(\delta, a))$, which we have called *finite upper half plane graphs.* The sets $S_q(\delta, a)$ can be viewed as subsets of H_q of the form

$$S_q(\delta, a) = \{z \in H_q \mid d(z, \sqrt{\delta}) = a\}, \quad \text{for } a \in \mathbb{F}_q. \tag{13}$$

For $a \neq 0, 4\delta$, the graphs $X(G/K, S_q(\delta, a))$ are connected and $(q + 1)$-regular. We have seen in previous sections that these graphs provide examples of *Ramanujan graphs*.

Now take f in (9) to be a K-bi-invariant function on G; that is, $f(kxh) = f(x)$ for all $k, h \in K$, $x \in G$. Then the sum on the left in (9) is over $\pi \in \hat{G}^K$, which means the set of all $\pi \in \hat{G}$ which occur in $L^2(G/K)$. We have shown in Theorem 2 of Chapter 20 that $\pi \in \hat{G}^K$ is equivalent to the existence of a nonzero K-fixed vector in the representation space of π. Because the convolution algebra of functions on $K \backslash G / K$ is commutative, we can call (G, K) a Gelfand pair and G/K a *symmetric space*. The space of K-fixed vectors for π is a 1-dimensional vector space and can be identified with scalar multiples of a fixed vector s_π of norm 1. The *spherical function* h_π is defined as the top left matrix entry of π, that is, $h_\pi(x) = \langle \pi(x)s_\pi, s_\pi \rangle$, for $x \in G$. Formula (16) of Chapter 20 says that

$$h_\pi(x) = \frac{1}{|K|} \sum_{k \in K} \chi_\pi(kx), \quad \text{for } x \in G. \tag{14}$$

Here χ_π is the character of π as in Chapter 15.

These finite spherical functions are also analogues of the real spherical functions defined by formula (4). However, for $GL(2, \mathbb{F}_q)$, because discrete as well as principal series representations of G appear in $\text{Ind}_K^G 1$, we need to use (14) for the two types of representations from Theorem 3 of Chapter 21. Ultimately we get formulas (18) and (19) of Chapter 20 for the spherical functions corresponding to the two types of representations.

The spherical functions are of interest for $G = GL(2, \mathbb{F}_q)$ and K as in (10) because they are the eigenfunctions of adjacency operators of the finite upper half plane graphs $X(G/K, S_q(\delta, a))$ for $S_q(\delta, a)$ defined by (13). Moreover, Equation (12) of Chapter 20 says that in this case the eigenfunctions are the eigenvalues. We showed in preceding chapters that the explicit formulas for these functions derived from (14) by Soto–Andrade [1987] ultimately lead to bounds for these functions which prove that the finite upper half plane graphs are Ramanujan in the sense of formula (9) of Chapter 3.

To return to our discussion of the trace formula, note that

$$\text{Tr}(\hat{f}(\pi)) = \mathscr{S}f(\pi) = \sum_{x \in G} f(x)h_\pi(x) = \langle f, \overline{h_\pi} \rangle,$$
$$\text{if } f \in L^2(K \backslash G / K), \tag{15}$$

which is the *spherical transform* of f at $\pi \in \hat{G}^K$. Here $\langle f, g \rangle$ denotes the standard inner product on G:

$$\langle f, g \rangle = \sum_{x \in G} f(x)\overline{g(x)}.$$

Exercise. Prove formula (15).

Recall that we showed in Chapter 20 that the spherical transform has many properties analogous to the ordinary Fourier transform of functions on the real line as well as the spherical transform on the Poincaré upper half plane defined by (5). For example, there is an *inversion formula*:

$$f(x) = \frac{1}{|G|} \sum_{\pi \in \hat{G}^K} d_\pi \, \mathscr{S}f(\pi) h_\pi(x^{-1}).$$

And there is a *convolution property*. For $f, g \in L^2(K\backslash G/K)$ define the convolution by

$$(f * g)(x) = \sum_{x \in G} g(y) f(xy^{-1}).$$

Then

$$\mathscr{S}(f * g)(\pi) = \mathscr{S}f(\pi)\mathscr{S}g(\pi), \quad \text{for } \pi \in \hat{G}^K.$$

Now, just as Selberg noticed in the real case, given a subgroup Γ of G, one can rewrite the right-hand side of the pre-trace formula (9) as a sum over conjugacy classes. Define

$$\{\gamma\} = \{x^{-1}\gamma x \mid x \in \Gamma\} = \text{the } \textit{conjugacy class} \text{ of } \gamma \text{ in } \Gamma; \tag{16}$$
$$\Gamma_\gamma = \{x \in \Gamma \mid x^{-1}\gamma x = \gamma\} = \text{the } \textit{centralizer} \text{ of } \gamma \text{ in } \Gamma; \tag{17}$$
$$\Xi_\Gamma = \{\kappa \mid \kappa \text{ is a conjugacy class of } \Gamma\}. \tag{18}$$

Since the map $T : \Gamma_\gamma\backslash\Gamma \to \{\gamma\}$ defined by $T(\Gamma_\gamma x) = x^{-1}\gamma x$ is 1-1 and onto, we have

$$\sum_{x \in \Gamma\backslash G} \sum_{\gamma \in \Gamma} f(x^{-1}\gamma x) = \sum_{x \in \Gamma\backslash G} \sum_{\{\gamma\} \in \Xi_\Gamma} \sum_{u \in \Gamma_\gamma\backslash\Gamma} f(x^{-1}u^{-1}\gamma u x)$$

$$= \sum_{y \in \Gamma_\gamma\backslash G} \sum_{\{\gamma\} \in \Xi_\Gamma} f(y^{-1}\gamma y)$$

$$= \sum_{\{\gamma\} \in \Xi_\Gamma} \frac{|G_\gamma|}{|\Gamma_\gamma|} \sum_{y \in G_\gamma\backslash G} f(y^{-1}\gamma y).$$

Next define the *orbital sum* of f at γ to be

$$I_G(f, \gamma) = \sum_{y \in G_\gamma \backslash G} f(y^{-1} \gamma y). \tag{19}$$

Combining what we have just found gives the following theorem.

*Theorem 1. **The Selberg Trace Formula for Finite Symmetric Spaces G/K.***
Suppose that G/K is a finite symmetric space and that Γ is a subgroup of G.
If f is a K-bi-invariant function on G and $\rho = \text{Ind}_\Gamma^G 1$, then

$$\sum_{\pi \in \hat{G}^K} m(\pi, \rho) \mathscr{S} f(\pi) = \sum_{\{\gamma\} \in \Xi_\Gamma} \frac{|G_\gamma|}{|\Gamma_\gamma|} I_G(f, \gamma),$$

where $m(\pi, \rho)$ is the multiplicity of the irreducible representation $\pi \in \hat{G}$ in
$\rho = \text{Ind}_\Gamma^G 1$, assuming π occurs in $\text{Ind}_K^G 1$. $\mathscr{S} f$ is the spherical transform
defined by (15), $I_G(f, \gamma)$ is the orbital transform defined by (19), and we use
the definitions (16)–(18).

Finite Trace Formula Examples

Now we apply Theorem 1 to our example with $G = GL(2, \mathbb{F}_q)$, $q = p^r$, p an
odd prime, $r \geq 2$, K as in (10), $\Gamma = GL(2, \mathbb{F}_p)$. We need a table of conjugacy
classes in Γ. This is given in Table II.11 in Chapter 21.

We can imitate the computations of the *orbital transforms* for $G = SL(2, \mathbb{R})$
and $\Gamma = SL(2, \mathbb{Z})$ to be found in Terras [1985, Chapter 3]. We find the following terms.

Central Terms

If γ is in the center Z_Γ of Γ, then

$$\gamma = z_a = \begin{pmatrix} a & 0 \\ 0 & a \end{pmatrix}, \quad \text{for some } a \in \mathbb{F}_p^*$$

and

$$I_G(f, z_a) = \sum_{x \in G \backslash G} f(x) = f(\sqrt{\delta}).$$

So we find that the *central term* is

$$\frac{|G|}{|\Gamma|} f(\sqrt{\delta}) \sum_{h \in Z_\Gamma} 1 = \frac{|G| f(\sqrt{\delta}) |Z_\Gamma|}{|\Gamma|} = \frac{q(q-1)(q^2-1) f(\sqrt{\delta})}{p(p^2-1)}. \tag{20}$$

Hyperbolic Terms

These come from hyperbolic conjugacy classes with

$$\gamma = h_{a,b} = \begin{pmatrix} a & 0 \\ 0 & b \end{pmatrix}, \quad a \neq b.$$

The centralizer G_γ in this case consists of all diagonal matrices in G. So a fundamental domain for G_γ acting on the finite upper half plane H_q is

$$G_\gamma \backslash H_q \cong \{x + \sqrt{\delta} \mid x \in \mathbb{F}_q\}.$$

And our orbital transform, identifying G/K with H_q, is

$$I_G(f, h_{a,b}) = (q+1) \sum_{u \in \mathbb{F}_q} f\left(u + \frac{a}{b}\sqrt{\delta}\right).$$

The factor $(q+1)$ comes from $|K/Z|$, if Z denotes the center of K. This is an analogue of the horocycle transform in formula (6) above. For the real Poincaré upper half plane, the horocycle transform (6) was invertible.

We define the *horocycle transform* of $f \in L^2(K\backslash G/K)$ for $y \in \mathbb{F}_q^*$ by

$$F(y) = \sum_{u \in \mathbb{F}_q} f(u + y\sqrt{\delta}). \tag{21}$$

Question. Suppose f is in $L^2(K\backslash G/K)$. Is this transform invertible?
Answer. No, unless we put an additional condition on f, since the dimension of $L^2(K\backslash G/K)$ is q, while the dimension of $L^2(\mathbb{F}_q^*)$ is $q-1$.

Finally, our *hyperbolic term* is

$$\frac{(q+1)(q-1)^2}{2(p-1)} \sum_{\substack{c \in \mathbb{F}_p^* \\ c \neq 1}} F(c). \tag{22}$$

Once more, we can define $h_{a,b}$ to be *primitive hyperbolic* if a/b generates the multiplicative group $\mathbb{F}_q^*$.

Parabolic Terms

If

$$\gamma = p_a = \begin{pmatrix} a & 1 \\ 0 & a \end{pmatrix},$$

the centralizer G_γ consists of all matrices of the form

$$\begin{pmatrix} x & y \\ 0 & x \end{pmatrix}, \quad \text{for } x \neq 0.$$

And $G_\gamma \backslash H_q$ can be identified with the set $\{v\sqrt{\delta} \mid v \in \mathbb{F}_q^*\}$. Thus the argument of f in the orbital sum is, in this case, $g_u^{-1} p_a g_u \sqrt{\delta} = (au)^{-1} + \sqrt{\delta}$, where

$$g_u = \begin{pmatrix} u & 0 \\ 0 & 1 \end{pmatrix}, \quad u \in \mathbb{F}_q^*,$$

represents $G_\gamma \backslash H_q$.

So the *parabolic terms* add up to

$$\frac{q(q^2 - 1)}{p}(F(1) - f(\sqrt{\delta})), \tag{23}$$

where F is the horocycle transform (21).

There is a big difference here with the real case – there are no convergence problems!

Elliptic Terms

Here

$$\gamma = e_{a,b} = \begin{pmatrix} a & b\xi \\ b & a \end{pmatrix},$$

where ξ is a nonsquare in $\mathbb{F}_p$. These terms turn out to be rather confusing. There are two cases depending on whether, for $q = p^r$, the exponent r is odd or even. If r is odd, then we may assume that the nonsquare ξ in $\mathbb{F}_p$ is also a nonsquare in $\mathbb{F}_q$. So in that case we can write $\xi = \delta$. If r is even, however, then $\xi = \eta^2$ for some η in $\mathbb{F}_q$. We will need to treat each case differently (something we failed to do in the paper Angel, Poulos, Terras, Trimble, and Velasquez [1994]). *Case I. $q = p^r$, r odd.* Here

$$\gamma = e_{a,b} = \begin{pmatrix} a & b\delta \\ b & a \end{pmatrix},$$

where δ is a nonsquare in $\mathbb{F}_p$ as well as $\mathbb{F}_q$. Then the centralizer G_γ is K. So we can identify $G_\gamma \backslash G$ with H_q. For $z \in H_q$, let

$$m_z = \begin{pmatrix} y & x \\ 0 & 1 \end{pmatrix}.$$

Then the *elliptic term for odd r* is

$$\frac{1}{2} \sum_{\substack{a,b\in\mathbb{F}_p \\ b\neq 0}} \frac{q^2-1}{p^2-1} \sum_{z\in H_q} f\left(m_z^{-1} e_{a,b}\, m_z \sqrt{\delta}\right).$$

The big difference with the real case is that there seem to be lots more elliptic terms to deal with.

To simplify the elliptic term in this case that r is odd, note that for $\phi : \mathbb{F}_q \to \mathbb{C}$, we have

$$f\left(m_z^{-1} e_{a,b}\, m_z \sqrt{\delta}\right) = \phi(d(e_{a,b}z, z)).$$

Now compute

$$d(e_{a,b}z, z) = \frac{N\left(\frac{az+b\delta}{bz+a} - z\right)}{\operatorname{Im}\left(\frac{az+b\delta}{bz+a}\right)y} = \frac{N(\delta - z^2)}{((a/b)^2 - \delta)y^2},$$

where the norm N is the field norm from $\mathbb{F}_q(\sqrt{\delta})$ down to $\mathbb{F}_q$. Thus we find, writing $f(z) = \phi(d(z, \sqrt{\delta}))$, that *for r odd, the elliptic terms* sum to

$$\frac{(q^2-1)}{2(p+1)} \sum_{u\in\mathbb{F}_p} \sum_{z\in H_q} \phi\left(\frac{N(\delta - z^2)}{(u^2 - \delta)y^2}\right). \tag{24}$$

Case II. $q = p^r$, r even. In this case, the nonsquare $\xi \in \mathbb{F}_p$ becomes a square $\xi = \eta^2$, $\eta \in \mathbb{F}_{p^2} \subset \mathbb{F}_q$, $q = p^r$. This means that the centralizer $G_{a,b}$ is no longer K. Instead, since our elliptic element is similar in G to

$$\begin{pmatrix} a+\eta b & 0 \\ 0 & a-\eta b \end{pmatrix},$$

the centralizer G_y is similar to the set of diagonal matrices in G. So $G_y \backslash G$ is exactly the same as in the hyperbolic y case. However, the centralizer Γ_y is the analogue of K for Γ. Thus, for even r, the elliptic term corresponding to a, b looks like a constant times a sum over $u \in \mathbb{F}_q$ of terms

$$f\left(t_u^{-1}\begin{pmatrix} a+\eta b & 0 \\ 0 & a-\eta b \end{pmatrix} t_u\right) \quad \text{where } t_u = \begin{pmatrix} 1 & u \\ 0 & 1 \end{pmatrix}.$$

So the *elliptic term for even r* is

$$\frac{q+1}{2}\frac{(q-1)^2}{p^2-1} \sum_{\substack{a,b\in\mathbb{F}_p \\ b\neq 0}} F\left(\frac{a+\eta b}{a-\eta b}\right), \tag{25}$$

where again F denotes the horocycle transform (21). Note that

$$w = \frac{a + \eta b}{a - \eta b} \in \mathbb{F}_{p^2} = \mathbb{F}_p(\eta),$$

where $\eta^2 = \xi$. And these w are exactly the elements of $\mathbb{F}_{p^2}$ having norm 1 with $w \neq 1$.

The trace formula is obtained by putting together (20) through (25) to get the right-hand side of the formula of Theorem 1. In order to make use of the trace formula for specific functions f, we need to make a study of the horocycle transform appearing in the hyperbolic and parabolic terms.

Example 1. Take f to be the function that is identically 1 on G. You find that for $G = GL(2, \mathbb{F}_q)$ and $\Gamma = GL(2, \mathbb{F}_p)$, with p an odd prime, as above,

$$\sum_{\pi \in \hat{G}^K} \frac{\delta_1(\pi)|G|}{d_\pi} m\left(\pi, \text{Ind}_\Gamma^G 1\right) = |G|.$$

This says that $m_G(1, \text{Ind}_\Gamma^G 1) = m_\Gamma(1, 1|_\Gamma) = 1$, a formula which we already believe by the Frobenius reciprocity law.

Example 2. Plug $f(z) = \delta_0(d(z, \sqrt{\delta}))$ into the trace formula. Then the spherical transform is $\mathscr{S}f(\pi) = |K|$, for all $\pi \in \hat{G}^K$.
Case 1. r is odd.

$$\sum_{\pi \in \hat{G}^K} m\left(\pi, \text{Ind}_\Gamma^G 1\right) = \frac{q(q-1)}{p(p^2-1)} + \frac{p}{p+1}.$$

Case 2. r is even.

$$\sum_{\pi \in \hat{G}^K} m\left(\pi, \text{Ind}_\Gamma^G 1\right) = \frac{q(q-1)}{p(p^2-1)}.$$

Exercise. Try plugging $f(z) = \psi(d(z, \sqrt{\delta}))$, for ψ an additive character of $\mathbb{F}_q$, into the trace formula.

Project. Find analogues of the applications of Selberg's original formula. Plug in an analogue of the heat kernel as in Elinor Velasquez, [preprint].

Another reference for trace formulas on finite groups is Marie-France Vignéras [1992].

Chapter 24

Trace Formula for a Tree and Ihara's Zeta Function

I want to stand under the twisted magic of that tree and hear
only the wind.

Gabriel Daniels [1994, p. 83]

Generalities on the Spectrum of the Adjacency Operator on the k-Regular Tree

We will be sketchier in this chapter as it is not our main focus. More details can be found in the references: Brooks [1986, 1991], Cartier [1973], Figà-Talamanca and Nebbia [1991], Figà-Talamanca and Picardello [1983], Hashimoto [1989, 1990, 1992], Quenell [1994], Serre [1980], Stark and Terras [1996], Sunada [1986, 1988], and Venkov and Nikitin [1994]. It might help the reader to consult a basic topology book such as Massey [1967] for some of the basic concepts in this section (e.g., fundamental group, universal covering space).

Recall that a *tree* is a connected graph without cycles or circuits. Let X be a finite k-regular graph and let $\tilde{X}$ be the k-regular tree (i.e., the universal covering space of X). That is, $\tilde{X}$ is a connected k-regular infinite graph without cycles (i.e., circuits or closed paths). We first consider the spectral theory of the adjacency operator on $\tilde{X}$ and then connect it with that of $X = \Gamma \backslash \tilde{X}$, where Γ is the fundamental group of X; that is, $\Gamma = \pi_1(X, o)$, where o is the origin of the tree. Note that the case $k = 2$ is degenerate. Compare Figures II.14 and II.15. We can view $\tilde{X}$ as a Cayley graph $X(G, S)$, where G is a free group and S a set of generators.

A doubly infinite path in $\tilde{X}$ is a *geodesic*. It can be viewed as connecting two boundary points of $\tilde{X}$.

Example. $k = 2$. In this case the tree can be identified with the Cayley graph $X(\mathbb{Z}, \{\pm 1\})$ as in Figure II.14. The adjacency operator A acts on functions

Figure II.14. Part of the 2-regular tree $X(\mathbb{Z}, \{-1, 1\})$.

$f : \mathbb{Z} \to \mathbb{C}$ by $Af(x) = f(x + 1) + f(x - 1)$. Its eigenfunctions are the exponentials $e_a(n) = \exp(2\pi i an)$, for $a \in \mathbb{C}$, $n \in \mathbb{Z}$;

$$Ae_a = 2\cos(2\pi a), \quad a \in \mathbb{C}.$$

To obtain the spectral decomposition of A, we only need the exponentials e_a, with $a \in [0, 1)$, for, by the theory of Fourier series, we know that any function f in $L^2(\mathbb{Z})$ has an expression

$$f(n) = \int_{a=0}^{1} \langle f, e_a \rangle e_a(n) \, da, \quad \text{where } \langle f, e_a \rangle = \hat{f}(a) = \sum_{n \in \mathbb{Z}} f(n)\overline{e_a(n)}.$$

This is the dual of the usual Fourier series representation. Since A is self-adjoint, the eigenvalues of A must be real and thus $a \in \mathbb{R}$.

Next consider the heat kernel on $X(\mathbb{Z}, \{\pm 1\})$ as in our discussion of the finite heat kernel #2 in Chapter 7. See also Chung and Yau [1997]. The Laplacian is $\mathscr{L} = I - (1/2)A$. The eigenfunctions and eigenvalues are

$$\phi_u(a) = \exp(2\pi i au), \quad a \in \mathbb{Z}, \ u \in \mathbb{R}/\mathbb{Z};$$

$$\lambda_u = 1 - 2\cos(2\pi u) = 2\sin^2(\pi u).$$

The *heat kernel* is

$$H_t(a, b) = \int_0^1 \exp\left(-2t\sin^2(\pi u)\right) \exp(2\pi i u(a - b)) \, du$$

$$\sim \frac{1}{\sqrt{\pi t}}, \quad \text{for } t \text{ large.}$$

One finds that $H_t(a, a)$ Compare with the finite analogue in Chapter 7.

Notes on Spectral Theory

Let us say a bit about spectral theory of operators on a Hilbert space. Other examples and references can be found in Terras [1985]. If D is a subspace of a Hilbert space V (usually D dense in V), and $A : D \to V$ linear, we say A is *symmetric* iff $\langle Au, v \rangle = \langle u, Av \rangle$ for all u, v in D. A densely defined operator is *self-adjoint* iff for all u in D, $\langle Au, v \rangle = \langle u, v^* \rangle$ implies $v \in D$, $v^* = Av$.

The spectrum of $A = \sigma(A)$ consists of the complex numbers λ such that $(A - \lambda I)^{-1}$ does not exist as a bounded operator on V. Assume that A is self-adjoint. Then $\sigma(A)$ is a subset of $\mathbb{R}$ and is a union of the point spectrum $\sigma_p(A)$ and the continuous spectrum $\sigma_c(A)$. The point spectrum consists of eigenvalues λ with $Av = \lambda v$ for some nonzero vector $v \in V$. The continuous spectrum consists of λ such that $(A - \lambda I)^{-1}$ exists but is unbounded. For a compact operator $\sigma(A) = \sigma_p(A)$. So our adjacency operator is self-adjoint, but not compact. From the case $k = 2$, we expect the spectrum of the adjacency operator on the k-regular (infinite) tree to be continuous. In any case the spectral theorem says that f in V has a generalized Fourier expansion as a sum over the point spectrum and an integral over the continuous spectrum with respect to the spectral measure. Figà-Talamanca and Nebbia [1991, pp. 60–61] compute the spectral measure for the adjacency operator on the tree, as does Cartier [1973].

Proposition 1. **Properties of the Adjacency Operator on the k-Regular Tree.** Suppose that $f, g \in L^2(\tilde{X})$. The inner product is

$$\langle f, g \rangle = \sum_{x \in \tilde{X}} f(x)\overline{g(x)}.$$

If A is the adjacency operator on the k-regular tree $\tilde{X}$, we have the following facts:

1. **Self Adjointness.** $\langle f, Ag \rangle = \langle Af, g \rangle$.
2. **Boundedness.** $|\langle Af, f \rangle| \leq 2\sqrt{k-1}\langle f, f \rangle$.

Proof.

1. This is a standard fact about the adjacency operator of an undirected graph.
2. See Sunada [1988, p. 252]. Orient the edges of $\tilde{X}$ so that every vertex $x \in \tilde{X}$ has a unique oriented edge e with x as its origin. Write $x \sim y$ if 2 vertices $x, y \in \tilde{X}$ are connected by an edge (i.e, are adjacent vertices). Let E be the set of edges of X (assumed directed as described). For a directed edge e, write $o(e) = $ origin of edge e, and $\ell(e) = $ terminus of edge e. We

have

$$|\langle Af, f \rangle| = \left| \sum_{x \in \tilde{X}} \sum_{\substack{y \in \tilde{X} \\ y \sim x}} f(y)\overline{f(x)} \right|$$

$$= \left| \sum_{e \in E} \{ f(\ell(e))\overline{f(o(e))} + f(o(e))\overline{f(\ell(e))} \} \right|$$

$$\leq 2 \left| \sum_{e \in E} |f(o(e))| \, |f(\ell(e))| \right|.$$

Now use the Cauchy–Schwarz inequality to see that

$$|\langle Af, f \rangle| \leq 2 \left(\sum_{x \in \tilde{X}} |f(x)|^2 \right)^{1/2} \left((k-1) \sum_{x \in \tilde{X}} |f(x)|^2 \right)^{1/2}.$$

Here we have used the fact that the map $e \mapsto o(e)$ is 1-1 while the map $e \mapsto \ell(e)$ is $(k-1)$ to 1. The inequality in 2 follows. ∎

Remark. For prime p, we can view the $(p+1)$-regular tree as a p-adic symmetric space for $GL(2, \mathbb{Q}_p)$, where $\mathbb{Q}_p$ denotes the field of p-adic numbers. See Figà-Talamanca and Nebbia [1991] and Serre [1980]. There is also a p-adic upper half plane defined by Stark [1987]. See Trimble [1993] for the connection with $GL(2, \mathbb{Q}_p)$. We will not make any use of p-adic groups and symmetric spaces here, although it would allow us to stay closer to the preceding chapter.

Spherical Functions on the k-Regular Tree

Next we need to think about spherical functions on $(q+1)$-regular infinite trees. From now on we will assume $q > 1$. Again let o denote the origin of our tree. The 3-regular tree is pictured in Figure II.15.

Definition. We say that a function $h : \tilde{X} \to \mathbb{C}$ is a *spherical function* if

(i) $h(kx) = h(x)$ for every automorphism k of $\tilde{X}$ fixing o;
(ii) $Ah = \lambda h$, for some $\lambda \in \mathbb{C}$ (where A is the adjacency operator of $\tilde{X}$);
(iii) $h(o) = 1$.

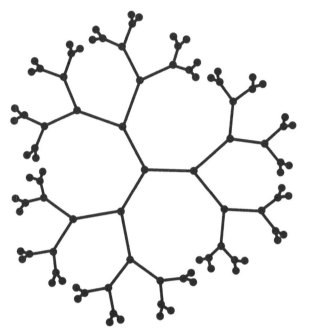

Figure II.15. Part of the 3-regular tree.

The spherical function h is *uniquely determined* by the eigenvalue λ and we proceed to do so. Note that $h(x) = h(d(x, o))$, where $d(x, o)$ is the distance of x from the origin (i.e., the number of edges in the path from o to x).

Note also that we can almost construct an eigenfunction by a method analogous to that which we used in formula (4) of Chapter 23.

Notation. $q = k - 1$, where k is the degree of the tree.

Definition. The *power function* for $s \in \mathbb{C}$ is $p_s(x) = q^{-sd}$, where $d = d(x, o)$.

So if $d \neq 0$, we have

$$Ap_s(d) = q^{-s(d-1)} + qq^{-s(d+1)} = (q^s + q^{1-s})q^{-sd}.$$

When $d = 0$, we have

$$Ap_s(0) = (q + 1)q^{-s}.$$

So p_s just misses being an eigenfunction of A. If $(q + 1)q^{-s} = (q^s + q^{1-s})$, there would be no problem, that is, if $q^{-s} = q^s$ or $q^{2s} = 1$. But we know that the spectrum is continuous and we need fairly arbitrary values of s (e.g., s such that Re $s = 1/2$).

So we concoct a function

$$h_s(d) = c(s)q^{-sd} + c(1-s)q^{-(1-s)d},$$

$$Ah_s(d) = \begin{cases} (q+1)\big(c(s)q^{-s} + c(1-s)q^{-(1-s)}\big), & \text{for } d = 0, \\ (q^s + q^{1-s})h_s(d), & \text{for } d \neq 0. \end{cases}$$

We want $h(0) = 1 = c(s) + c(1-s)$ and

$$q^s + q^{1-s} = (q+1)\big(c(s)q^{-s} + c(1-s)q^{-(1-s)}\big).$$

So we have two equations in two unknowns. A solution is

$$c(s) = \frac{q^{s-1} - q^{1-s}}{(q+1)(q^{s-1} - q^{-s})}.$$

This is all right provided that the denominator does not vanish. If the denominator does vanish (i.e., if $q^{2s-1} = 1$), then you must take a limit.

Setting $z = q^{s-1/2}$, we obtain

$$c(s) = \frac{z^2 - q}{(q+1)(z^2 - 1)}.$$

The map sending s to $1-s$ sends z to $1/z$.

So we have

$$h_s(d) = c(s)q^{-sd} + c(1-s)q^{-(1-s)d} = \frac{q(z^{2+d} - z^{-d}) - (z^d - z^{2-d})}{q^{d/2}(q+1)(z^2-1)}. \quad (1)$$

Now let $w = z^2$ approach 1, and see that $h_s(d)$ approaches the following limit:

$$h_s(d) = q^{1-d/2}\left(1 + \frac{q-1}{q+1}d\right), \quad \text{if } q^{2s-1} = 1. \quad (2)$$

Question. Are our eigenfunctions h_s in $L^p(\tilde{X})$, for $p = 1$ or 2?
It is impossible for both p_s and p_{1-s} to be in L^p, for we can easily see that

1) p_s is in $L^1(\tilde{X})$ if Re $s > 1$,
2) p_s is in $L^2(\tilde{X})$ if Re $s > 1/2$.

We know that the spectrum of the adjacency operator A lies on Re $s = 1/2$, which is the dividing line between Re $s > 1/2$ and Re $s < 1/2$. This line is also the fixed line of the map from s to $1 - s$.

The Pre-Trace Formula on the k-Regular Tree

Suppose that $f : \tilde{X} \to \mathbb{C}$ is of finite support (or "rapidly decreasing") and K-invariant, where K is the group of automorphisms of $\tilde{X}$ fixing the origin o. Then $f(x) = f(d(x, o))$, where, as usual, d is the distance (= number of edges) between points on the tree. We could also form the point-pair invariant $f(x, y) = f(d(x, y))$, for $x, y \in \tilde{X}$.

If $X = \Gamma \backslash \tilde{X}$ is a finite k-regular graph, where Γ is the fundamental group $\Gamma = \pi_1(X, o)$, let f act on functions $\phi \in L^2(X)$ via

$$(L_f \phi)(x) = \sum_{y \in X} \sum_{\gamma \in \Gamma} f(d(x, \gamma y)) \, \phi(y). \tag{3}$$

Compare with formula (5) in Chapter 22.

By reasoning analogous to that in Chapter 22, the trace of L_f on $L^2(X)$ has two expressions:

$$\sum_{\substack{\pi \in \hat{G}^K \\ \pi \in \mathrm{Ind}_\Gamma^G 1}} \mathrm{Tr}(\hat{f}(\pi)) = \sum_{x \in X} \sum_{\gamma \in \Gamma} f(d(x, \gamma x)). \tag{4}$$

Moreover, from Brooks [1991] we can identify the *left-hand side* of the trace formula:

$$\sum_{i=1}^{|X|} \langle f, h_{s_i} \rangle \tag{5}$$

where $A h_{s_i} = \lambda_i h_{s_i}$, and the $\lambda_i = q^{s_i} + q^{1-s_i}$ run through the spectrum of the adjacency operator on X. We are assuming that X is a finite $(q + 1)$-regular graph.

As usual, we rewrite the right-hand side of (4) as a sum over conjugacy classes $\{\gamma\}$ in Γ. Here Γ_γ denotes the centralizer of γ in Γ. So (4) becomes the *pre-trace formula*:

$$\sum_{\substack{s_i \text{ from} \\ \text{Spec } A \text{ on } X}} \langle f, h_{s_i} \rangle = \sum_{\{\gamma\}} \sum_{\Gamma_\gamma \backslash \tilde{X}} f(d(y, \gamma y)). \tag{6}$$

The details of this calculation go as follows:

$$\sum_{x \in X} \sum_{\gamma \in \Gamma} f(d(x, \gamma x)) = \sum_{\{\gamma\}} \sum_{\sigma \in \Gamma_\gamma \backslash \Gamma} \sum_{x \in \Gamma \backslash \tilde{X}} f(d(x, \sigma \gamma \sigma^{-1} x))$$

$$= \sum_{\{\gamma\}} \sum_{x \in \Gamma_\gamma \backslash \tilde{X}} f(d(y, \gamma y)).$$

The Trace Formula

Note that Γ is *strictly hyperbolic*; that is, $\rho \in \Gamma$, $\rho \neq$ the identity element e, implies that ρ is without fixed points. So ρ induces a shift by $v(\rho)$ on a geodesic in X, where

$$v(\rho) = \min_{v \in \tilde{X}} d(v, \rho v) > 1. \tag{7}$$

See Figà–Talamanca and Nebbia [1991].

Definition. We call $\rho \in \Gamma$, $\rho \neq e$, *primitive* if ρ generates the centralizer Γ_ρ. Note that Γ_ρ must always be cyclic since Γ is a free group. Set

$$\mathcal{P}_\Gamma = \{\text{primitive conjugacy classes of } \Gamma \text{ not equal to } \{e\}\}.$$

So our trace formula becomes

$$\sum_{\substack{s_i \text{ from} \\ \text{Spec } A \text{ on } X}} \langle f, h_{s_i} \rangle = f(o)|X| + \sum_{\{\rho\} \in \mathcal{P}_\Gamma} \sum_{r=1}^{\infty} I_{\rho^r}(f), \tag{8}$$

where $I_{\rho^r}(f)$ is the *orbital sum* defined by

$$I_{\rho^r}(f) = \sum_{\Gamma_\gamma \backslash \tilde{X}} f(d(y, \rho^r y)), \tag{9}$$

Next we define the *horocycle transform* by

$$Hf(d) = f(|d|) + (q-1) \sum_{j=1}^{\infty} q^{j-1} f(|d| + 2j), \quad \text{for } d \in \mathbb{Z}. \tag{10}$$

See Terras and Wallace [1998] for a picture of horocycles in a 3-regular tree. Another reference with pictures is Trimble [1993].

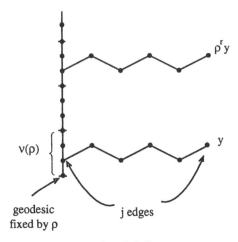

Picture for $\nu(\rho)=3$, $r=2$, $j=5$.

Figure II.16. Finding the fundamental domain $\Gamma_\rho \backslash \tilde{X}$, when $\nu(\rho) = 3$, $r = 2$, $j = 5$.

Lemma 1.

1) **Orbital Sum = Horocycle Transform.** Using the notation (7), (9), and (10), we have

$$I_{\rho^r}(f) = \nu(\rho)Hf(r\nu(\rho)).$$

2) **Inversion for the Horocycle Transform.**

$$f(|d|) = Hf(|d|) - (q-1)\sum_{j=1}^{\infty}(Hf)(|d|+2j).$$

Proof.

1) The fundamental domain for $\Gamma_\rho \backslash \tilde{X}$ is found by looking at the geodesic fixed by ρ modulo ρ. Draw the picture for $\nu(\rho) = 3$, say. See Figure II.16. You have only three inequivalent points on the geodesic. Then the points off the geodesic at a distance $j = 5$ have to be considered. The map ρ^2 moves them up.

   Here $\nu(\rho)$ corresponds to the length of a primitive closed geodesic in X. You can count, for example,

$$\#\{y \mid d(y, \rho^r y) = r\nu(\rho) + 2j\} = \nu(\rho)(q-1)q^{j-1}.$$

2) As in Stark and Terras [1996], you need to separate the even and odd indices. The linear transformations involved are upper triangular with 1s on the diagonal. ∎

Putting everything together, we get the following result:

Theorem 1. **Selberg Trace Formula on the $(q + 1)$-Regular Tree.**

$$\sum_{i=1}^{|X|} \langle f, h_{\lambda_i} \rangle = f(o)|X| + \sum_{\{\rho\} \in \mathscr{P}_\Gamma} v(\rho) \sum_{r \geq 1} Hf(rv(\rho)),$$

where Hf is the horocycle transform defined by (10).

The Ihara Zeta Function

Here we want to use the trace formula to prove that the Ihara zeta function of a $(q+1)$-regular graph is the reciprocal of a polynomial. A different discussion of this fact can be found in Stark and Terras [1996] where the result is generalized to nonregular graphs suggesting that the trace formula should also generalize.

Definition. The *Ihara zeta function* associated to X is a function of a sufficiently small complex variable u given by

$$Z_x(u) = \prod_{\{\gamma\} \in \mathscr{P}_\Gamma} \left(1 - u^{v(\gamma)}\right)^{-1}. \tag{11}$$

We want to use the Selberg trace formula to prove the following result.

Theorem 2. **Ihara's Formula for His Zeta Function.** If X is a connected $(q + 1)$-regular graph with adjacency matrix A and $r = $ rank of fundamental group, then the Ihara zeta function defined by (11) is the reciprocal of a polynomial. More precisely, we have

$$Z_X^{-1}(u) = (1 - u^2)^{r-1} \det(I - Au + qu^2 I).$$

We have $r - 1 = n(q - 1)/2$, where $n = $ the number of vertices in X.

Proof. (Venkov and Nikitin [1994].) We plug

$$(Hf)(d) = g(d) = \begin{cases} u^{|d|-1}, & \text{for } d \neq 0, \\ 0, & \text{for } d = 0 \end{cases}$$

into the Selberg trace formula.

The right-hand side, nonidentity terms are

$$\sum_{\{\rho\}\in\mathcal{P}_\Gamma} \nu(\rho) \sum_{r\geq 1} u^{r\nu(\rho)-1} = \frac{d}{du} \log Z_X(u).$$

The right-hand side identity term is

$$f(o) = g(o) - (q-1)\sum_{j=1}^{\infty} g(2j) = -(q-1)\sum_{j=1}^{\infty} u^{2j-1}$$

$$= -(q-1)\frac{u}{1-u^2}$$

$$= \frac{q-1}{2}\frac{d}{du}\log(1-u^2) = \frac{d}{du}\log\left((1-u)^{(q-1)/2}\right).$$

So combining these results we find that the right-hand side of the trace formula is

$$\frac{d}{du}\log(Z_X(u)(1-u^2)^{r-1}), \tag{12}$$

since $r - 1 = |X|(q-1)/2$.

Now we must work on the left-hand side of the trace formula. We first compute the spherical transform of f:

$$\langle f, h_\lambda \rangle = \sum_{d\in\mathbb{Z}} (Hf)(d)q^{|d|/2}z^d,$$

where Hf is the horocycle transform of (10). To see this, just plug in formula (1) for the spherical function and keep computing.

So we find that the left-hand side of the trace formula is a sum over λ (which are the eigenvalues of A on X) of terms of the form

$$-\frac{d}{du}\log(1-\lambda u + qu^2). \tag{13}$$

Putting all this together gives Ihara's theorem. ∎

Corollary 1. $Z_X(q^{-s})$ satisfies the Riemann hypothesis (meaning that if $\mathrm{Re}\,s \in (0,1)$ and $Z_X(q^{-s})^{-1} = 0$, then $\mathrm{Re}\,s = 1/2$) iff X is a Ramanujan graph.

Proof. Let spec(A) be the set eigenvalues of the adjacency matrix A of X. These eigenvalues are real since A is symmetric. Then

$$Z_X^{-1}(u) = (1-u^2)^{r-1} \prod_{\lambda\in\mathrm{spec}(A)} (1-\lambda u + qu^2).$$

And $1 - \lambda u + qu^2 = (1 - \alpha u)(1 - \beta u)$, where $\alpha\beta = q$ and $\alpha + \beta = \lambda$ and

$$\alpha, \beta = \frac{\lambda \pm \sqrt{\lambda^2 - 4q}}{2}.$$

Thus $|\lambda| \le 2\sqrt{q}$ iff α, β are complex conjugates with absolute value $\sqrt{q}$. If we write $q^s = \alpha, \beta$, with $s = \sigma + it$, we see that $\sigma = 1/2$. When $\lambda = \pm(q+1)$, we have $q^s = \pm 1$ or $\pm q$, that is, $\mathrm{Re}(s) = 0$ or 1. Note that, except for $\lambda = \pm(q+1)$, we have $|\lambda| < q + 1$ and thus

$$1 < |\alpha|, \quad |\beta| < q.$$

The corresponding $q^s = \alpha$ or β then has $0 < \mathrm{Re}(s) < 1$. $\blacksquare$

Next we see that, as with many other zeta functions, there is a functional equation relating $Z_X(u)$ to $Z_X(1/qu)$. You just need to multiply $Z_X(u)(1 - u^2)^{r-1}$ by a polynomial $p(u)$ such that $p(1/qu^2) = \pm(qu^2)^{-n} p(u)$.

Corollary 2. **Functional Equations.** Under the hypotheses of Theorem 1, we have the following functional equations of the Ihara zeta function among others:

1. $\Lambda_X(u) = (1 - u^2)^{n/2+r-1}(1 - q^2u^2)^{n/2} Z_X(u) = (-1)^n \Lambda_X\left(\dfrac{1}{qu}\right).$

2. $\xi_X(u) = (1 + u)^{r-1}(1 - u)^{r-1+n}(1 - qu)^n Z_X(u) = \xi_X\left(\dfrac{1}{qu}\right).$

3. $\Xi_X(u) = (1 - u^2)^{r-1}(1 + qu^2)^n Z_X(u) = \Xi_X\left(\dfrac{1}{qu}\right).$

Proof. We will prove 1 and leave the rest as exercises. To see 1, write

$$\Lambda_X(u) = (1 - u^2)^{n/2}(1 - q^2u^2)^{n/2} \det(I - Au + qu^2 I)^{-1}$$

$$= \left(\frac{q^2}{q^2u^2} - 1\right)^{n/2} \left(\frac{1}{q^2u^2} - 1\right)^{n/2} \det\left(I - A\frac{1}{qu} + I\frac{q}{(qu)^2}\right)^{-1}$$

$$= (-1)^n \Lambda_X\left(\frac{1}{qu}\right).$$

$\blacksquare$

Next we recall a graph theoretical interpretation for a value of the inverse zeta function. A *spanning tree* T in a connected graph X is a tree which is a subgraph of X having as vertices all the vertices of X.

Definition. The *complexity* $\kappa(X)$ of a finite graph X is the number of spanning trees in X.

Biggs [1974] shows that for a $(q+1)$-regular finite graph X with adjacency matrix A, if J is a matrix of 1s, we have

$$\kappa(X) = n^{-2} \det(J + (q+1)I - A). \tag{14}$$

It is easy to go from this to the following corollary. See Hashimoto [1989] who views the corollary as giving an analogue of the class number formula for zeta functions of number fields which comes from taking residues of the Dedekind zeta function (see Terras [1985, Vol. I]).

*Corollary 3. **Complexity is a Value of the Inverse Zeta Function.*** (Hashimoto [1989].) For a finite graph X satisfying the hypotheses of Theorem 2, the complexity can be found as follows:

$$\kappa(X) = \frac{-1}{n(q-1)2^{r-1}} \frac{1}{(1-u)^r Z_X(u)}\bigg|_{u=1}.$$

Exercise. Prove Corollary 3.

See Bass [1992], Hashimoto [1989, 1990, 1992], Stark [in press], and Stark and Terras [1996] for generalizations of the Ihara zeta function to nonregular graphs. See Terras and Wallace [1998] for an application of this chapter to the study of what might be called the induced geodesic flow on $\Gamma \mid \tilde{X} = X$, a finite graph.

References

A. Adolphson, On the distribution of angles of Kloosterman sums, *J. Reine Angew. Math.*, **395** (1989), 214–220.

A. V. Aho, J. E. Hopcroft, and J. D. Ullmann, *The Design and Analysis of Computer Algorithms*, Addison-Wesley, Reading, MA, 1974.

J.-I. Aihara, Why aromatic compounds are stable, *Scientific American*, March, 1992, 62–68.

M. Ajtai, H. Iwaniec, J. Komlós, J. Pintz, and E. Szemerédi, Construction of a thin set with small Fourier coeffficients, *Bull. London Math. Soc.*, **22** (1990), 583–590.

N. Allen, *On the Spectra of Certain Graphs Arising from Finite Fields*, Ph.D. Thesis, Univ. Calif., Santa Cruz, 1996.

J. Angel, *Finite Upper Half Planes over Finite Rings and Their Associated Graphs*, Ph.D. Thesis, Univ. Calif., San Diego., 1993.

J. Angel, Finite upper half planes over finite fields, *Finite Fields Appl.*, **2** (1996), 62–86.

J. Angel, N. Celniker, S. Poulos, A. Terras, C. Trimble, and E. Velasquez, Special functions on finite upper half planes, *Contemporary Math.*, **138**, Amer. Math. Soc., Providence, 1992, 1–26.

J. Angel, S. Poulos, A. Terras, C. Trimble, and E. Velasquez, Spherical functions and transforms on finite upper half planes: eigenvalues of the combinatorial Laplacian, uncertainty, traces, *Contemporary Math.*, **173**, Amer. Math. Soc., Providence, 1994, 15–70.

J. Angel, B. Shook, A. Terras, and C. Trimble, Graph spectra for finite upper half planes over rings, *Linear Algebra and Its Applications*, **226–228** (1995), 423–457.

J. Arthur, The trace formula and Hecke operators, in *Number Theory, Trace Formulas, and Discrete Groups*, K. E. Aubert et al. (eds.), Academic Press, Boston, 1989, 11–27.

K. E. Aubert et al. (eds.), *Number Theory, Trace Formulas, and Discrete Groups*, Academic Press, Boston, 1989.

L. Auslander and R. Tolimieri, Ring structure and the Fourier transform, *Math. Intelligencer*, **7** (1985), 49–52.

P. W. Atkins, *Molecules*, Freeman, New York, 1987.

J. Baggott, *Perfect Symmetry: The Accidental Discovery of Buckminsterfullerene*, Oxford Univ. Press, Oxford, 1996.

G. A. Baker, Drum shapes and isospectral graphs, *J. Math. Physics*, **7** (1966), 2238–2242.

W. W. Rouse Ball and H. S. M. Coxeter, *Mathematical Recreations and Essays*, Dover, New York, 1987.

T. Banchoff, *Beyond the Third Dimension*, Freeman, New York, 1990.

E. Bannai, Orthogonal polynomials in coding theory and algebraic combinatorics, in *Orthogonal Polynomials: Theory and Practice*, P. Nevai and M. E. H. Ismail (eds.), Kluwer, Dordrecht, 1990, 25–53.

E. Bannai, Character tables of commutative association schemes, in *Finite Geometries, Buildings, and Related Topics*, Oxford Science Publ., Oxford, 1990, 105–128.

E. Bannai and T. Ito, *Algebraic Combinatorics I, Association Schemes*, Benjamin/ Cummings, Menlo Park, CA, 1984.

V. Bargmann, Irreducible unitary representations of the Lorentz group, *Annals of Math.*, **48** (1947), 568–640.

H. Bass, The Ihara-Selberg zeta function of a tree lattice, *Int. J. Math.*, **3** (1992), 717–797.

E. Batschelet, Recent statistical methods for orientation (Animal Orientation Sympos. 1970 on Wallops Island), *Amer. Inst. Biol. Sciences*, Washington, D.C., 1971.

K. B. Beauchamp, *Walsh Functions and Their Applications*, Academic, New York, 1975.

E. T. Bell, *Mathematics: Queen and Servant of Science*, Tempus, Redmond, WA, 1989.

E. T. Bell, *The Last Problem*, Math. Assoc. of America, Washington, D.C., 1990.

C. T. Benson and J. B. Jacobs, On hearing the shape of combinatorial drums, *J. Combinatorial Theory (B)*, **13** (1972), 170–178.

B. Berndt and R. Evans, The determination of Gauss sums, *Bull. Amer. Math. Soc.*, **5** (1981), 107–129.

B. Berndt, R. Evans, and K. Williams, *Gauss and Jacobi Sums*, Wiley, New York, 1998.

G. Besseon, On the multiplicity of the eigenvalues of the Laplacian, *Springer Lecture Notes*, 1339, Springer-Verlag, New York, 1988, 32–53.

F. Bien, Construction of telephone networks by group representations, *Notices Amer. Math. Soc.*, **36** (1989), 187–196.

N. Biggs, *Algebraic Graph Theory*, Cambridge Univ. Press, London, 1974.

N. Biggs, E. K. Lloyd, and R. J. Wilson, *Graph Theory: 1736–1936*, Clarendon, Oxford, 1986.

R. E. Blahut, *Theory and Practice of Control Codes*, Addison-Wesley, Reading, MA, 1983.

R.E. Blahut, *Algebraic Methods of Signal Processing and Communications Coding*, Springer-Verlag, New York, 1992.

R. E. Blahut, W. Miller, Jr., and C. H. Wilcox, *Radar and Sonar, I, IMA Vols. in Math. Appl.*, **32**, Springer-Verlag, New York, 1991.

P. Bloomfield, *Fourier Analysis of Time Series: An Introduction*, Wiley, New York, 1976.

E. Bolker, The finite Radon transform, *Contemporary Math.*, **63**, Amer. Math. Soc., Providence, 1987, 27–50.

B. Bollobás, *Graph Theory: An Introductory Course*, Springer-Verlag, New York, 1979.

B. Bollobás, *Random Graphs*, Academic, London, 1985.

Z. I. Borevitch and I. R. Shafarevitch, *Number Theory*, Academic, New York, 1966.

M. Born, *My Life: Recollections of a Nobel Laureate*, Scribner, New York, 1978.

R. Brauer, Representations of finite groups, in *Lectures on Modern Math. I*, T. L. Saaty (ed.), Wiley, New York, 1963, 133–175.

E. O. Brigham, *The Fast Fourier Transform*, Prentice–Hall, Englewood Cliffs, NJ, 1974.

R. D. Brillinger, *Time Series: Data Analysis and Theory*, Holden-Day, San Francisco, 1981.

R. Brooks, Combinatorial problems in spectral geometry, *Lecture Notes in Math.*, **1201**, Springer-Verlag, New York, 1986.

R. Brooks, The spectral geometry of k-regular graphs, *J. d'Analyse*, **57** (1991), 120–151.

R. Brooks, Some relations between graph theory and Riemann surfaces, *Proc. Ashkelon Conf., Israel Math. Conf. Proc.*, **11**, Bar-Ilan Univ., Ramat Gan, 1997, 61–73.

R. Brooks, R. Gornet, and W. Gustafson, Mutually isospectral Riemann surfaces, to appear, in *Adv. in Math.*, 1999.

A. E. Brouwer, A. M. Cohen, and A. Neumaier, *Distance-Regular Graphs*, Springer-Verlag, New York, 1989.

M. Burrow, *Representation Theory of Finite Groups*, Dover, New York, 1993.

P. Buser, Cayley graphs and planar isospectral domains, *Springer Lecture Notes*, **1339**, Springer-Verlag, New York, 1988, 64–77.

P. Buser, *Geometry and Spectra of Compact Riemann Surfaces*, Birkhäuser, Boston, 1992.

F. W. Byron and R. W. Fuller, *Mathematics of Classical and Quantum Physics*, Dover, New York, 1992.

R. Calderbank, A. R. Hammons, Jr., D. V. Kumar, N. J. A. Sloane, and P. Solé, A linear construction for certain Kerdock and Preparata codes, *Bull. Amer. Math. Soc.*, **29** (1993), 218–222.

L. Carlitz, Weighted quadratic partitions over a finite field, *Canadian J. Math.*, **5** (1953), 517–323.

L. Carlitz, Some cyclotomic matrices, *Acta Arith.*, **5** (1959), 293–308.

P. Cartier, Harmonic analysis on trees, *Proc. Symp. Pure Math.*, **26**, Amer. Math. Soc., Providence, 1973, 419–423.

J. W. S. Cassels and A. Fröhlich, *Algebraic Number Theory*, Academic, London, 1967.

N. Celniker, Eigenvalue bounds and girths of graphs of finite, upper half-planes, *Pacific J. Math.*, **166** (1994), 1–21.

N. Celniker, S. Poulos, A. Terras, C. Trimble, and E. Velasquez, Is there life on finite upper half planes?, *Contemporary Math.*, **143** Amer. Math. Soc., Providence, 1993, 65–88.

I. Chavel, *Eigenvalues in Riemannian Geometry*, Academic, New York, 1984.

L. Chihara and D. Stanton, Zeros of generalized Krawtchouk polynomials, *J. Approximation Theory*, **60** (1990), 43–57.

S. D. Chowla, *The Riemann Hypothesis and Hilbert's Tenth Problem*, Gordon and Breach, New York, 1965

F. R. K. Chung, Diameters and eigenvalues, *J. Amer. Math. Soc.*, **2** (1989), 187–196.

F. R. K. Chung, Constructing random-like graphs, *Proc. Symp. Appl. Math.*, **44**, Amer. Math. Soc., Providence, 1991, 21–55.

F. R. K. Chung, *Spectral Graph Theory*, CBMS Regional Conf. Series in Math., No. 92, Amer. Math. Soc., Providence, 1996.

F. R. K. Chung, P. Diaconis, and R. Graham, Random walks arising in random number generation, *Ann. Probability*, **15** (1987), 1148–1165.

F. R. K. Chung and S. Sternberg, Laplacian and vibrational spectra of homogeneous graphs, *J. Graph Theory*, **16** (1992), 609–627.

F. R. K. Chung and S. Sternberg, Mathematics and the buckyball, *American Scientist*, **81** (Jan.–Feb., 1993), 56–71.

F. R. K. Chung and S.-T. Yau, Eigenvalues of graphs and Sobolev inequalities, *Combinatorics, Probability and Computing*, **4** (1995), 11–26.

F. R. K. Chung, A combinatorial trace formula, *Tsing Hua Lectures on Geometry and Analysis* (Hsinchu, 1990–1991), Internat. Press, Cambridge, MA, 1997, 107–116.

B. Cipra, An introduction to the Ising model, *Amer. Math. Monthly*, **94** (1987), 937–959.

B. Cipra, *What's Happening in the Mathematical Sciences*, Vols. I and II, Amer. Math. Soc., Providence, 1993, 1994.

R. Coifman and M. V. Wickerhauser, Wavelets and adapted waveform analysis: mathematical toolkit for signal processing and numerical analysis, *Proc. Symp. Appl. Math.*, **47**, Amer. Math. Soc., Providence, 1993, 119–153.

A. J. Coleman, Induced and subduced representations, in *Group Theory and Its Applications*, Vol. I, M. Loebl (ed.), Academic, New York, 1968.

T. F. Coleman and C. Van Loan, *Handbook for Matrix Computations*, S.I.A.M., Philadelphia, PA, 1988.

J. H. Conway, Monsters and moonshine, *Math. Intelligencer*, **2**, (4) (1980), 165–171.

J. H. Conway, R. T. Curtis, S. P. Norton, R. A. Parker, and R. A. Wilson, *Atlas of Groups*, Oxford Univ. Press, 1985.

J. H. Conway and N. J. A. Sloane, *Sphere Packings, Lattices and Groups*, Springer-Verlag, New York, 1993.

J. W. Cooley and J. W. Tukey, An algorithm for the machine calculation of complex Fourier series, *Math. Computation*, **19** (1965), 297–301.

C. A. Coulson, B. O'Leary, and R. B. Mallion, *Hückel Theory for Organic Chemists*, Academic, New York, 1978.

R. Courant and D. Hilbert, *Methods of Mathematical Physics*, Vol. I, Wiley-Interscience, New York, 1953.

H. S. M. Coxeter, *J. Math. Phys.*, **12** (1933), 334–335.

H. S. M. Coxeter, *Regular Complex Polytopes*, Cambridge Univ. Press, Cambridge, 1991.

R. F. Curl and R. E. Smalley, Fullerenes, *Scientific American*, **265** (Oct., 1991), 54–63.

C. W. Curtis, Representations of finite groups of Lie type, *Bull. Amer. Math. Soc.*, **1** (1979), 721–757.

C. W. Curtis, Representation theory of finite groups: from Frobenius to Brauer, *Math. Intelligencer*, **14**, (4) (1992), 48–57.

C. W. Curtis and I. Reiner, *Representation Theory of Finite Groups and Associative Algebras*, Wiley, New York, 1966.

C. W. Curtis, *Methods of Representation Theory*, Vols. I and II, Wiley, New York, 1981, 1987.

D. Cvetković, M. Doob, and H. Sachs, *Spectra of Graphs*, Academic, New York, 1980.

G. Daniels, A city girl discovers the forest, in *Another Wilderness: New Outdoor Writing by Women*, S. F. Rogers (ed.), Seal Press, Seattle, 1994, 68–89.

G. C. Danielson and C. Lanczos, Some improvements in practical Fourier analysis and their application to x-ray scattering from liquids, *J. Franklin Institute*, **233** (1942), 365–380, 435–452.

I. Daubechies, *Ten Lectures on Wavelets*, CBMS-NSF Regional Conference Series in Applied Math., **61**, Soc. Industrial and Appl. Math, Philadelphia, PA, 1992.

H. Davenport, *Multiplicative Number Theory*, 2nd ed., Springer-Verlag, New York, 1980.

H. Davenport, *The Higher Arithmetic*, Dover, New York, 1983.

G. Davidoff and P. Sarnak, *An Elementary Approach to Ramanujan Graphs*, preprint.

P. J. Davis, *Circulant Matrices*, Wiley, New York, 1979.

M. DeDeo, *Graphs over the Ring of Integers modulo 2^r*, Ph.D. Thesis, Univ. Calif., San Diego, 1998.

P. Deligne, *Cohomologie étale (SGA 4 1/2), Springer Lecture Notes*, **569**, Springer-Verlag, Berlin, 1977.

A. K. Dewdney, *The Turing Omnibus: 61 Excursions in Computer Science*, Computer Science Press, Rockville, MD, 1989.

P. Diaconis, *Group Representations in Probability and Statistics*, Institute of Math. Statistics, Hayward, CA, 1988.

P. Diaconis and R. L. Graham, The Radon transform on $\mathbb{Z}_2^k$, *Pacific J. Math.*, **118** (1985), 323–345.

P. Diaconis and D. Rockmore, Efficient computation of the Fourier transform on finite groups, *J. Amer. Math. Soc.*, **3** (1990), 297–332.

P. Diaconis and L. Saloff-Coste, Moderate growth and random walk on finite groups, *Geom. Funct. Anal.*, **4** (1994), no. 1, 1–36.

P. Diaconis and L. Saloff-Coste, Random walks on finite groups: a survey of analytic techniques, *Probability Measures on Groups and Related Structures*, **XI** (Oberwolfach, 1994), World Sci. Publishing, River Edge, NJ, 1995, 44–75.

P. Diaconis and M. Shashahani, Products of random matrices as they arise in the study of random walks on groups, *Contemporary Math.*, **50**, Amer. Math. Soc., Providence, 1986, 185–195.

P. Diaconis and D. Stroock, Geometric bounds for the eigenvalues of Markov chains, *Ann. Appl. Probability*, **15** (1987), 1148–1165.

F. Digne and J. Michel, *Representations of Finite Groups of Lie Type*, Cambridge Univ. Press, Cambridge, 1991.

D. L. Donoho and P. B. Stark, Uncertainty principles and signal recovery, *S.I.A.M. J. Appl. Math.*, **49** (1989), 906–931.

L. Dornhoff, *Group Representation Theory, Parts A and B*, Dekker, New York, 1971, 1972.

L. Dornhoff and F. Hohn, *Applied Linear Algebra*, MacMillan, New York, 1978.

P. Doyle and J. L. Snell, *Random Walks and Electric Networks*, Math. Assoc. of America, Washington, D.C., 1984.

D. S. Dummit and R. M. Foote, *Abstract Algebra*, Prentice–Hall, Englewood Cliffs, NJ, 1991.

A. Dvoretzky and J. Wolfowitz, Sums of random integers reduced modulo m, *Duke Math. J.*, **18** (1951), 501–507.

H. Dym and H. P. McKean, *Fourier Series and Integrals*, Academic, New York, 1972.

F. J. Dyson, Applications of group theory in quantum physics, *S.I.A.M. Review*, **8** (1966), 1–10.

J. Elkington, J. Hailes, and J. Makower, *The Green Consumer*, Penguin, New York, 1990.

D. F. Elliott and K. R. Rao, *Fast Transforms: Algorithms, Analyses, Applications*, Academic, New York, 1982.

J. Elstrodt, Die Selbergsche Spurformel für Kompakte Riemannsche Flächen, *Jahresbericht d. Deutch. Math.-Verein*, **83** (1981), 45–77.

R. Evans, Hermite character sums, *Pacific J. Math.*, **122** (1986), 357–390.

R. Evans, Character sums over finite fields, *Lecture Notes in Pure and Appl. Math.*, **141**, Dekker, New York, 1993, 57–73.

R. Evans, Character sums as orthogonal eigenfunctions of adjacency operators for Cayley graphs, *Cont. Math.*, **168** (1994), 33–50.

R. Evans, Spherical functions for finite upper half planes with characteristic 2, *Finite Fields Appl.*, **1** (1995), 376–394.

A. Fässler and E. Stiefel, *Group Theoretical Methods and Their Applications*, Birkhäuser, Boston, 1992.

W. Feller, *Introduction to Probability Theory and Its Applications*, Vol. I, Wiley, New York, 1968.

A. Figá-Talamanca and M. A. Picardello, *Harmonic Analysis on Free Groups*, Dekker, New York, 1983.

A. Figá-Talamanca and C. Nebbia, *Harmonic Analysis and Representation Theory for Groups Acting on Homogeneous Trees*, Cambridge Univ. Press, Cambridge, 1991.

J. Fill, The Radon transform on $\mathbb{Z}_n$, *S.I.A.M. J. Discrete Math.*, **2** (1989), 262–283.

N. J. Fine, On the Walsh functions, *Trans. Amer. Math. Soc.*, **65** (1949), 372–414.

N. J. Fine, The generalized Walsh functions, *Trans. Amer. Math. Soc.*, **69** (1950), 66–77.

M. Fisher, On hearing the shape of a drum, *J. Combinatorial Theory*, **1** (1966), 105–125.

K. Flornes, A. Grossman, M. Holschneider, and B. Torresani, Wavelets on discrete fields, *Applied and Computational Harmonic Analysis*, **1**, No. 2 (1994), 137–146.

G. E. Forsythe and W. R. Wasow, *Finite-Difference Methods for PDEs*, Wiley, New York, 1960.

J. Friedman (ed.), *Expanding Graphs, DIMACS Series in Discrete Math. and Theor. Comp. Sci.*, Vol. 10, Amer. Math. Soc., Providence, 1993.

W. Fulton and J. Harris, *Representation Theory: A First Course*, Springer-Verlag, New York, 1991.

J. A. Gallian, *Contemporary Abstract Algebra*, D.C. Heath, Lexington, MA, 1990.

R. Gangolli and V. S. Varadarajan, *Harmonic Analysis of Spherical Functions on Real Reductive Groups*, Springer-Verlag, Berlin, 1988.

R. Garabedian, *Partial Differential Equations*, Wiley, New York, 1964.

I. M. Gelfand, *Collected Papers*, Vols. I–III, Springer-Verlag, New York, 1988.

I. M. Gelfand, M. I. Graev, and I. I. Piatetski-Shapiro, *Representation Theory and Automorphic Forms*, Academic, Boston, 1990.

I. M. Gelfand, M. I. Graev, and N. Vilenkin, *Generalized Functions*, Vol. 5, Academic, New York, 1966.

P. Gerl, A local central limit theorem on some groups, *Lecture Notes in Statistics*, Vol. 8, Springer-Verlag, New York, 1981, 73–82.

M. Gerstenhaber, The 152nd proof of the law of quadratic reciprocity, *Amer. Math. Monthly*, **70** (1963), 397–398.

W. J. Gilbert, *Modern Algebra with Applications*, Wiley, New York, 1976.

J. Ginibre, General formulation of Griffiths inequalities, *Comm. Math. Phys.*, **16** (1970), 310–328.

C. D. Godsil, *Algebraic Combinatorics*, Chapman and Hall, New York 1993.

P. Goetgheluck, Fresnel zones on the screen, *Experimental Math.*, **2** (4) (1993), 301–309.

L. J. Goldstein, Density questions in algebraic number theory, *Amer. Math. Monthly*, **78** (1971), 342–351.

I. J. Good, The interaction algorithm and practical Fourier analysis, *J. Royal Stat. Soc., Ser. B*, **20** (1958), 361–372, addendum, **22** (1960), 372–375.

C. S. Gordon, When you can't hear the shape of a manifold, *Math. Intelligencer*, **11** (1989), 39–47.

C. S. Gordon and D. L. Webb, You can't hear the shape of a drum, *American Scientist*, **84** (Jan.–Feb., 1996), 46–55.

C. S. Gordon, D. L. Webb, and S. Wolpert, Isospectral plane domains and surfaces via Riemannian orbifolds, *Inventiones Math.*, **110** (1992), 1–22.

D. Gordon, Perfect multiple error-correcting arithmetic codes, *Math. Computation*, **49** (1987), 621–633.

D. Gorenstein, *Finite Groups*, Harper and Row, New York, 1968.

D. Gorenstein, *Finite Simple Groups: An Introduction to Their Classification*, Plenum, New York, 1982.

D. Gorenstein, The enormous theorem, *Scientific American*, **253** (6), (1985), 104–115.

R. Graham, M. Grötschel, and L. Lovász (eds.), *Handbook of Combinatorics*, Vols. I and II, M.I.T. Press. and Elsevier, Cambridge, MA, 1995.

C. W. J. Granger and P. Newbold, *Forecasting Economic Time Series*, Academic, San Diego, 1986.

J. A. Green, The characters of the finite general linear groups, *Trans. Amer. Math. Soc.*, **80** (1955), 402–447.

A. Greenhalgh, *Random Walks on Groups and Subgroup Invariance Properties*, Ph.D. Thesis, Stanford Univ., 1987.

D. Greenspan, *Discrete Models*, Addison-Wesley, Reading, MA, 1973.

D. Greenspan, *Discrete Numerical Methods in Physics and Engineering*, Academic, New York, 1974.

I. Grossman and W. Magnus, *Groups and Their Graphs*, Math. Assoc. of America, Washington, D.C., 1992.

L. C. Grove, *Groups and Characters*, Wiley, New York, 1997.

F. A. Grünbaum, Trying to beat Heisenberg, *Lecture Notes in Pure and Appl. Math.*, **122**, Dekker, New York, 1990, 657–665.

F. A. Grünbaum, M. Bernfeld, and R. E. Blahut, *Radar and Sonar, II, IMA Vols. in Math. Appl.*, **39**, Springer-Verlag, New York, 1992.

V. Guillemin, Perspectives in integral geometry, *Contemporary Math.*, **63**, Amer. Math. Soc. Providence, 1987, 137–150.

D. Gurarie, *Symmetries and Laplacians*, North-Holland, Amsterdam, 1992.

M. Gutzwiller, *Chaos in Classical and Quantum Mechanics*, Springer-Verlag, New York, 1990.

R. Guy, *Reviews in Number Theory: 1973–83*, Amer. Math. Soc., Providence, 1984.

R. Haberman, *Elementary Applied Partial Differential Equations with Fourier Series and Boundary Value Problems*, Prentice–Hall, Englewood Cliffs, NJ, 1983.

J. Hadamard, Résolution d'une question rélative aux déterminants, *Bull. Soc. Math.* (2), **17** (1893), 240–248.

M. Hall, Jr., *The Theory of Groups*, MacMillan, New York, 1959.

M. Hamermesh, *Group Theory and its Application to Physical Problems*, Addison-Wesley, Reading, MA, 1962.

F. Harary (ed.), *Graph Theory and Theoretical Physics*, Academic, London, 1967.

G. H. Hardy, *Ramanujan, Twelve Lectures on Subjects Suggested by His Life and Work*, Cambridge Univ. Press, Cambridge, 1940.

Harish-Chandra, Eisenstein series over finite fields, in *Collected Papers*, Vol. 4, Springer-Verlag, New York, 1984, 8–21.

H. F. Harmuth, *Transmission of Information by Orthogonal Functions*, Springer-Verlag, New York, 1972.

M. A. Harrison, *Introduction to Switching and Automata Theory*, McGraw-Hill, New York, 1965.

K. Hashimoto, Zeta functions of finite graphs and representations of p-adic groups, in *Automorphic Forms and Geometry of Arithmetic Varieties, Advanced Studies in Pure Math.*, K. Hashimoto and Y. Namikawa (eds.), **15**, Academic, New York, 1989, 211–280.

K. Hashimoto, On zeta and L-functions of finite graphs, *Int. J. Math.*, **1** (1990), 381–396.

K. Hashimoto, Artin-type L-functions and the density theorem for prime cycles on finite graphs, *Int. J. Math.*, **3** (1992), 809–826.

M. T. Heideman, D. H. Johnson, and C. S. Burrus, Gauss and the history of the fast Fourier transform, *IEEE ASSP Magazine*, **1** (4) (1984), 14–21.

H. Heilbronn, Zeta-functions and L-functions, in *Algebraic Number Theory*, J. W. S. Cassels and A. Fröhlich eds., Academic, London, 1967, 204–230.

D. Hejhal, The Selberg trace formula and the Riemann zeta function, *Duke Math. J.*, **43** (1976a), 441–482.

D. Hejhal, *The Selberg Trace Formula for PSL (2 ℝ), I, II, Lecture Notes in Math.*, Vols. 548 and 1001, Springer-Verlag, New York, 1976b, 1983.

D. Hejhal, On Eigenvalues of the Laplacian for Hecke triangle groups, in *Advanced Studies in Pure Math., Vol. 21, Zeta Functions in Geometry*, Math. Soc. of Japan, 1992, 359–408.

D. Hejhal and B. Rackner, On the topography of Maass wave forms for PSL (2. ℤ), *Experimental Math.*, **1** (1992), 275–305.

S. Helgason, Duality and Radon transform for symmetric spaces, *Amer. J. Math.*, **85** (1963), 667–692.

S. Helgason, Lie groups and symmetric spaces, in *Battelle Rencontres*, C. M. DeWitt and J. A. Wheeler (eds.), Benjamin, New York, 1968, 1–71.

S. Helgason, *The Radon Transform*, Birkhaüser, Boston, 1980a.

S. Helgason, The X-ray transform on a symmetric space, *Lecture Notes in Math.*, **838**, Springer-Verlag, New York, 1980b.

S. Helgason, *Groups and Geometric Analysis*, Academic, New York, 1984.

S. Helgason, Some results on Radon transforms, Huygen's principle, and X-ray transforms, *Contemp. Math.*, **63** (1987), 151–177.

S. Helgason, The totally geodesic Radon transform on constant curvature spaces, *Contemporary Math.*, **113**, Amer. Math. Soc. Providence, 1990, 141–149.

A. Helversen-Pasotto, Character sum identities in analogy with special function identities, *Adv. Math.*, **128** (1997), 186–189.

I. N. Herstein, *Topics in Algebra*, Ginn, Waltham, MA, 1964.

M. Hildebrand, Random processes of the form $x_{n+1} \equiv a_n x_n + b_n (\mathrm{mod}\, p)$ *Ann. Prob*, **21** (1993), 710–720.

M. Hildebrand, Random walks supported on random points of $\mathbb{Z}/n\mathbb{Z}$, *Probability Theory Related Fields*, **100** (1994), 191–203.

A. Hodges, *Alan Turing: The Enigma*, Simon and Schuster, New York, 1983.

M. Holschneider, Wavelet analysis over Abelian groups, *Appl. Comp. Harmonic Analysis*, **2** (1995), 52–60.

M. Holschneider, *Wavelets: An Analysis Tool*, Oxford Univ. Press, Oxford, 1995.

C. Hooley, On Artin's Conjecture, *J. Reine Angew. Math.*, **225** (1967), 209–220.

R. A. Horn and C. R. Johnson, *Topics in Matrix Analysis*, Cambridge Univ. Press, Cambridge, 1991.

L. K. Hua, *Introduction to Number Theory*, Springer-Verlag, Berlin, 1982.

T. W. Hungerford, *Algebra*, Springer-Verlag, New York, 1974.

N. E. Hurt, *Geometric Quantization in Action*, Reidel, Dordrecht, 1983.

M. N. Huxley, Introduction to Kloostermania, in *Elementary and Analytic Theory of Numbers*, H. Iwaniec (ed.), Polish Scientific Publ., Warsaw, 1985, 217–306.

Y. Ihara, On discrete subgroups of the 2×2 projective linear group over p-adic fields, *J. Math. Soc. Japan*, **18** (1966), 219–235.

K. Ireland and M. Rosen, *A Classical Introduction to Modern Number Theory*, Springer-Verlag, New York, 1982.

I. M. Isaacs, *Character Theory of Finite Groups*, Academic, San Diego, 1976.

H. Iwaniec, Selberg's lower bound of the first eigenvalue for congruence groups, in *Number Theory, Trace Formulas, and Discrete Groups*, K. Aubert, E. Bombieri, D. Goldfeld (eds.), Academic, Boston, 1989, 371–375.

S. Iyanaga and Y. Kawada, *Encyclopedic Dictionary of Math.*, Vols. I and II, M.I.T. Press, Cambridge, MA, 1980.

N. Jacobson, *Basic Algebra*, Vols. I and II, W. H. Freeman, New York, 1980, 1985.

G. James and M. Liebeck, *Representations and Characters of Groups*, Cambridge Univ. Press, Cambridge, 1993.

M. Kac, Random walks and the theory of Brownian motion, *Amer. Math. Monthly*, **54** (1947), 369–391.

M. Kac, Can you hear the shape of a drum?, *Amer. Math. Monthly*, **73** (1966), 1–23.

N. Katz, *Sommes Exponentielles, Astérisque*, **79** (1980), 1–209.

N. Katz, *Gauss Sums, Kloosterman Sums, and Monodromy Groups*, Princeton Univ. Press, 1988.

N. Katz, Estimates for Soto–Andrade sums, *J. Reine Angew. Math.*, **438** (1993), 143–161.

N. Katz, A note on exponential sums, *Finite Fields Appl.*, **1** (1995), 395–398.

J. P. Keener, *Principles of Applied Mathematics: Transformation and Approximation*, Addison-Wesley, Redwood City, CA, 1988.

J. G. Kemeny and J. L. Snell, *Finite Markov Chains*, Van Nostrand, Princeton, 1960.

S. F. A. Kettle, *Symmetry and Structure*, Wiley, Chichester, 1985.

R. W. Keyes, The future of the transistor, *Scientific American*, June, 1993, 70–78.

M. Klawe, Limitations on explicit constructions of expanding graphs, *S.I.A.M. J. Comput.*, **13** (1984), 156–166.

F. Klein and R. Fricke, *Vorlesungen über die Theorie der Elliptische Modulfunktionen*, Vol. I, Johnson Reprint Corp., New York, 1966.

H. D. Kloosterman, The behavior of general theta functions under the modular group and the characters of binary modular congruence groups, I, II, *Ann. Math.*, **47** (1946), I: 317–375; II: 376–447.

R. S. Knox and A. Gold, *Symmetry in the Solid State*, Benjamin, New York, 1964.

D. E. Knuth, *Art of Computer Programming, Vol II, Semi-Numerical Algorithms*, Addison-Wesley, Reading, MA, 1981.

T. W. Körner, *Fourier Analysis*, Cambridge Univ. Press, Cambridge, 1988.

D. E. Koshland, Jr., Molecule of the year, *Science*, **254** (Dec. 20, 1991), 1705.

B. Kostant, The graph of the truncated icosahedron and the last letter of Galois, *Notices A. M. S.*, **42** (Sept. 1995), 959–968.

A. Krieg, *Hecke Algebras, Memoirs of Amer. Math. Soc.*, **435**, Amer. Math. Soc., Providence, 1990.

H. W. Kroto and D. R. M. Walton, *The Fullerenes: New Horizons for the Chemistry, Physics and Astrophysics of Carbon*, Cambridge Univ. Press, Cambridge, 1993.

J. Kuang, A natural orthogonal basis of eigenfunctions of the Hecke algebra acting on Cayley graphs, *Proc. Amer. Math. Soc.*, **123** (1995), 3615–3622.

J. Kuang, On eigenvalues related to finite Poincaré planes, *Finite Fields Appl.*, **3** (1997), 151–158.

N. Kurokawa and T. Sunada (eds.), *Zeta Functions in Geometry, Advanced Studies in Pure Math.*, Vol. 21, Kinokuniya, Japan, 1992.

W. M. Kwok, Character tables of association schemes of affine type, *Eur. J. Combinatorics*, **13** (1992), 167–182.

G. Lachaud, Distribution of the weights of the dual of the Melas code, *Discrete Math.*, **79** (1989/90), 103–106.

G. Lachaud, and J. Wolfman, Sommes de Kloosterman, courbes elliptiques et codes cycliques en characteristique 2, *Comptes Rendus Acad. Sci. Paris*, **305** (1987), 881–883.

J. Laffery and D. Rockmore, Fast Fourier analysis for SL_2 over a finite field and related numerical experiments, *Experimental Math.*, **1** (1992), 115–139.

T. Y. Lam, Representations of finite groups: A hundred years, I, II, *Notices of the A. M. S.*, **45** (3,4) (1998), 361–312, 465–474.

E. Landau, *Vorlesungen über Zahlentheorie*, Leipzig, 1927.

S. Lang, *Undergraduate Analysis*, Springer-Verlag, New York, 1983.

S. Lang, *Algebra*, Addison-Wesley, Redwood City, CA, 1984.

S. Lang, $SL_2(\mathbb{R})$, Springer-Verlag, New York, 1985.

S. Lang, *Linear Algebra*, Springer-Verlag, New York, 1987.

N. Lebedev, *Special Functions and Their Applications*, Dover, New York, 1972.

D. H. Lehmer, Mathematical methods in large-scale computing units, in *Proceedings of a Second Symposium on Large-Scale Digital Calculating Machinery*, Harvard Univ. Press, Cambridge, MA, 1951, 141–146.

H. W. Lenstra, Perfect arithmetic codes, *Sém. Delange-Pisot-Poitou* (1977–1978), exp. 15, 1–14.

W. LeVeque, *Reviews in Number Theory*, Amer. Math. Soc., Providence, 1974.

W. Li, Character sums and Ramanujan graphs, *J. Number Theory*, **41** (1992), 199–217.

W. Li, Number-theoretic constructions of Ramanujan graphs, *Astérisque*, **228**, Soc. Math. de France, Paris, 1995.

W. Li, A survey of Ramanujan graphs, in R. Pellikaan, M. Perret, and S. G. Vladut (eds.), *Arithmetic, Geometry, and Coding Theory, Proc. Conf. at CIRM, Luminy, France,* de Gruyter, Berlin, 1996a.

W. Li, *Number Theory with Applications*, World Scientific, Singapore, 1996b.

W. Li, Eigenvalues of Ramanujan graphs, *IMA Volumes in Math. and its Applications*, **109** Springer-Verlag, New York, in press.

K. Likharev and T. Claeson, Single electronics, *Scientific American*, June, 1992, 80–85.

J. S. Lomont, *Applications of Finite Groups*, Dover, New York, 1993.

O. Loos, *Symmetric Spaces*, Vols. I and II, Benjamin, New York, 1969.

L. Lovász, Spectra of graphs with transitive groups, *Periodica Mathematica Hungarica*, **6** (1975), 191–196.

A. Lubotzky, *Discrete Groups, Expanding Graphs and Invariant Measures*, Birkhäuser, Basel, 1994.

A. Lubotzky, Cayley graphs: eigenvalues, expanders and random walks, in *Surveys in Combinatorics, London Math. Soc. Lect. Notes*, **218**, Cambridge Univ. Press, 1995, 155–189.

A. Lubotzky, R. Phillips, and P. Sarnak, Ramanujan graphs, *Combinatorica*, **8** (1988), 261–277.

J.-M. Luck, P. Moussa, and M. Waldschmidt (eds.), *Number Theory and Physics, Springer Proc. in Physics*, Vol. 47, Springer-Verlag, New York, 1990.

I. MacDonald, *Spherical Functions on a Group of p-Adic Type*, Ramanujan Institute for Advanced Study, Univ. of Madras, Madras, India, 1971.

G. Mackey, *The Theory of Group Representations*, University of Chicago Press, Chicago, 1976.

G. Mackey, *Unitary Group Representations in Physics, Probability and Number Theory*, Benjamin/Cummings, Reading, MA, 1978a.

G. Mackey, *Harmonic Analysis as the Exploitation of Symmetry – A Historical Survey*, Rice University Studies, **64**, Houston, 1978b.

F. J. MacWilliams and N. J. A. Sloane, *The Theory of Error-Correcting Codes*, North Holland, Amsterdam, 1988.

H. B. Mann (ed.), *Error Correcting Codes, Proc. Symp. U.S. Army Math. Res. Center*, Univ. Wisconsin, Wiley, New York, 1968.

K. V. Mardia, *Statistics of Directional Data*, Academic, New York, 1972.

G. A. Margulis, Arithmetic groups and graphs without short cycles, *6th International Symposium on Information Theory, Tashkent (1984), Abstracts 1*, 123–5 (In Russian).

S. L. Marple, Jr., *Digital Spectral Analysis with Applications*, Prentice–Hall, Englewood Cliffs, NJ, 1987.

O. Martin, A. M. Odlyzko, and S. Wolfram, Algebraic properties of cellular automata, *Comm. Math. Phys.*, **93** (1984), 219–258.

M. Martinez, *Homogeneous Spaces of Subgroups of $GL(n, \mathbb{F}_q)$ and Related Hypergraphs*, Ph.D. Thesis, Univ. Calif., San Diego, 1998.

W. Massey, *Algebraic Topology: An Introduction*, Springer-Verlag, New York, 1967.

Math Works, *Matlab Version 4 User's Guide*, Prentice–Hall, Englewood Cliffs, NJ, 1995.

K. Maurin, *General Eigenfunction Expansions and Unitary Representations of Topological Groups*, Polish Scientific Publ., Warsaw, 1968.

B. D. McKay, The expected eigenvalue distribution of a large regular graph, *Linear Algebra and Its Appl.*, **40** (1981), 203–216.

H. P. McKean, Kramers–Wannier duality for the 2-dimensional Ising model as an instance of the Poisson sum formula, *J. Math. Phys.*, **5** (1964), 775–776.

A. Medrano, *Super-Euclidean Graphs and Super-Heisenberg Graphs: Their Spectral and Graph-Theoretic Properties*, Ph.D. Thesis, Univ. Calif., San Diego, 1998.

A. Medrano, P. Myers, H. Stark, and A. Terras, Finite analogues of Euclidean space, *J. Comp. Appl. Math.*, **68** (1996), 221–238.

A. Medrano, P. Myers, H. Stark, and A. Terras, Finite Euclidean graphs over rings, *Proc. Amer. Math. Soc.*, **126** (1998), 701–710.

M. L. Mehta, *Random Matrices and the Statistical Theory of Energy Levels*, Academic, New York, 1967.

A. J. Menezes (ed.), *Applications of Finite Fields*, Kluwer, Boston, 1993.

C. Moreno, *Algebraic Curves over Finite Fields*, Cambridge Univ. Press, Cambridge, 1991.

A. Mukhophyay (ed.), *Recent Developments in Switching Theory*, Academic, New York, 1971.

P. Myers, *Euclidean and Heisenberg Graphs: Spectral Properties and Applications*, Ph.D. Thesis, Univ. Calif., San Diego, 1995.

P. Myers, Finite Heisenberg graphs, preprint.

T. Nagell, *Introduction to Number Theory*, Chelsea, New York, 1964.

M. A. Naimark and A. I. Stern, *Theory of Group Representations*, Springer-Verlag, New York, 1982.

Y. Nambu, Field theory of Galois fields, in *Quantum Field Theory and Quantum Statistics*, Vol. I, I. A. Batalin et al (eds.), Hilger, Bristol, 1987, 625–636.

W. Narkiewicz, *Elementary and Analytic Theory of Algebraic Numbers*, Polish Sci. Publ., Warsaw, 1974.

P. M. Neumann, Finite permutation groups, edge-coloured graphs and matrices, in *Topics in Group Theory and Computation*, M. P. J. Curran (ed.), Academic, London, 1977, 82–118.

D. E. Newland, *An Introduction to Random Vibrations, Spectral and Wavelet Analysis*, Longman, Essex, England, 1993.

E. Noether, *Collected Papers*, Springer-Verlag, New York, 1983.

H. J. Nussbaumer, *Fast Fourier Transform and Convolution Algorithms*, Springer-Verlag, New York, 1982.

R. W. K. Odoni, On Gauss sums (mod p^n), $n \geq 2$, *Bull London Math. Soc.*, **5** (1973), 325–327.

T. Ono, *An Introduction to Algebraic Number Theory*, Plenum, New York, 1990.

E. Pennisi, Buckyballs still charm, *Science News*, **140** (Aug. 24, 1991), 120–123.

R. Penrose, *The Emperor's New Mind: Concerning Computers, Minds, and the Laws of Physics*, Oxford Univ. Press, New York, 1989.

A. Perelomov, *Generalized Coherent States and Their Applications*, Springer-Verlag, Berlin, 1986.

R. Perlis, On the equation $\zeta_K(s) = \zeta_{K'}(s)$, *J. Number Theory*, **9** (1977), 342–360.

I. I. Piatetski-Shapiro, *Complex Representations of $GL(2, K)$ for Finite Fields K*, *Contemporary Math.*, **16**, Amer. Math. Soc., Providence, 1983.

M. Piercy, *Circles on the Water*, Knopf, New York, 1990.

V. Pless, *Introduction to the Theory of Error-Correcting Codes*, Wiley, New York, 1989.

A. Poli and L. Huguet, *Error Correcting Codes: Theory and Applications*, Prentice–Hall, Hertfordshire, UK, 1992.

G. Pólya, *Collected Papers*, III, IV, M.I.T. Press, Cambridge, MA, 1984.

J. C. T. Pool, *Mathematical Aspects of Statistical Mechanics, S.I.A.M.-A.M.S. Proc.*, Vol. 5, Amer. Math. Soc., Providence, 1972.

S. Poulos, *Graph-Theoretic and Spectral Properties of Finite Upper Half-Planes*, Ph.D. Thesis, Univ. Calif., San Diego, 1991.

D. L. Powers, *Boundary Value Problems*, 3rd ed., Saunders (Harcourt Brace), Ft. Worth, TX, 1987.

T. Powers, *Heisenberg's War*, Knopf, New York, 1993.

W. H. Press, B. P. Flannery, S. A. Teukolsky, and W. T. Vetterling, *Numerical Recipes: The Art of Computing*, Cambridge Univ. Press, Cambridge, 1986.

O. Pretzel, *Error-Correcting Codes*, Oxford Univ. Press, Oxford, 1992.

G. Quenell, Spectral diameter estimates for k-regular graphs, *Adv. Math.*, **106** (1994), 122–148.

H. Rademacher, *Lectures on Elementary Number Theory*, Blaisdell, New York, 1964.

C. M. Rader, Discrete Fourier transform when the number of data samples is prime, *Proc. I.E.E.E.*, **5, 6** (1968), 1107, 1108.

R. A. Rankin, *Modular Forms and Functions*, Cambridge Univ. Press, Cambridge, 1977.

I. Richards, Number theory, in *Mathematics Today*, L. A. Steen (ed.), Vintage, 1980, 37–64.

K. Rosen, *Elementary Number Theory and Its Applications*, Addison-Wesley, Reading, MA, 1993.

R. Rucker, *The Fourth Dimension: A Guided Tour of the Higher Universes*, Houghton-Mifflin, Boston, 1984.

D. Ruelle, Zeta functions and statistical mechanics, *Astérisque*, **40** (1976), 167–176.

D. Ruelle, *Dynamical Zeta Functions for Piecewise Monotone Maps of the Interval*, *CRM Monograph Series*, Vol. 4, Amer. Math. Soc., Providence, 1994.

C. Runge, *Zeit. Math. Physik*, **48, 53** (1903, 1905), 433, 117.

C. Runge and H. König, *Die Grundlehren der Mathematischen Wissenschaften, Vorlesungen über Numerischen Rechnen*, Vol. II, Berlin, J. Springer, 1964.

H. Salié, Über die Kloostermanschen Summen S (u,v;q), *Math. Z.*, **34** (1932), 91–109.

P. Samuel, *Algebraic Theory of Numbers*, Houghton-Mifflin, Boston, 1970.

J. T. Sandefur, *Discrete Dynamical Systems: Theory and Applications*, Oxford Univ. Press, Oxford, 1990.

P. Sarnak, *Some Applications of Modular Forms*, Cambridge Univ. Press, Cambridge, 1990.

P. Sarnak, Arithmetic quantum chaos, *Israel Math. Soc. Conf. Proc., Schur Lectures*, Bar-Ilan Univ., Ramat-Gan, Israel, 1995.

P. Sarnak, The spacing distribution between zeros of zeta functions, Lecture at IMA, Minneapolis, MN, Summer, 1996.

P. Sarnak, Spectra and eigenfunctions of Laplacians, in *Partial Differential Equations and Their Applications, CRM Proc. Lecture Notes*, **12**, Amer. Math. Soc., Providence, 1997, 261–276.

W. Schempp, *Harmonic Analysis on the Heisenberg Nilpotent Lie Group*, Longman, Essex, England, 1986a.

W. Schempp, Group theoretical methods in approximation theory, elementary number theory and computational signal geometry, in *Approximation Theory, V*, C. K. Chui, L. L. Schumaker, and J. D. Ward (eds.), Academic, Orlando, FL, 1986b, 129–171.

W. Schempp and B. Dreseler, *Einführung in die Harmonische Analyse*, Teubner, Stuttgart, 1980.

W. Schmidt, *Equations over Finite Fields: An Elementary Approach, Lecture Notes in Math.*, **536**, Springer-Verlag, New York, 1976.

M. Schönert et al. *GAP-Groups, Algorithms, and Programming*, Lehrstuhl D für Math., Rheinisch-Westf. Tech. Hoch, Aachen, 1995.

I. J. Schoneberg, The finite Fourier series and elementary geometry, *Amer. Math. Monthly*, **57** (1950), 390–404.

O. Schreier and E. Sperner, *Modern Algebra and Matrix Theory*, Chelsea, New York, 1959.

M. R. Schroeder, *Number Theory in Science and Communication*, Springer-Verlag, Berlin, 1986.

W. R. Scott, *Group Theory*, Dover, New York, 1987.

J. J. Seidel, Graphs and their spectra, *Combinatorics and Graph Theory*, PWN-Polish Scientific Publishers, Warsaw, 1989, 147–162.

A. Selberg, *Collected Papers*, Vol. I, Springer-Verlag, New York, 1989.

J.-P. Serre, *A Course in Arithmetic*, Springer-Verlag, New York, 1973.

J.-P. Serre, *Linear Representations of Finite Groups*, Springer-Verlag, New York, 1977.

J.-P. Serre, *Trees*, Springer-Verlag, New York, 1980.

D. Shanks, Five number-theoretic algorithms, *Proc. 2nd Manitoba Conf. on Numerical Math.*, 1972, 51–70.

D. Shanks, Calculation and applications of Epstein zeta functions, *Math. Comput.*, **29** (1975), 271–287.

D. Shanks, *Solved and Unsolved Problems in Number Theory*, Chelsea, New York, 1985.

W. T. Shaw and J. Tigg, *Applied Mathematica: Getting Started, Getting It Done*, Addison-Wesley, Reading, MA, 1994.

A. Silberger, An elementary construction of the representations of SL(2,GF(q)), *Osaka J. Math.*, **6** (1969), 329–338.

J. A. Silverman, *A Friendly Introduction to Number Theory*, Prentice–Hall, Upper Saddle River, NJ, 1997.

S. Simpson, Where bears walk, in *Another Wilderness: New Outdoor Writing by Women*, S. F. Rogers (ed.), Seal Press, Seattle, 1994, 10–22.

N. J. A. Sloane, An introduction to association schemes and coding theory, in *Theory and Application of Special Functions*, R. A. Askey (ed.), Academic, New York, 1975, 225–260.

C. Small, *Arithmetic of Finite Fields*, Dekker, New York, 1991.

K. T. Smith, The uncertainty principle on groups, *S.I.A.M. J. Appl. Math.*, **50** (1990), 876–882.

J. L. Snell, *Introduction to Probability Theory with Computing*, Prentice–Hall, Englewood Cliffs, NJ, 1975.

J. Soto-Andrade, Geometrical Gel'fand models, tensor quotients, and Weil representations, *Proc. Symp. Pure Math.*, **47**, Amer. Math. Soc., Providence, 1987, 305–316.

B. Srinivasan, *Representations of Finite Chevalley Groups: A Survey, Lecture Notes in Math.*, Vol. 764, Springer-Verlag, Berlin, 1979.

D. Stanton, *Finite Groups, Induced Representations and Orthogonal Polynomials*, Univ. Calif., San Diego Lecture Notes, 1981.

D. Stanton, Orthogonal polynomials and Chevalley groups, in *Special Functions: Group Theoretic Aspects and Applications*, R. Askey et al. (eds.), Reidel, Dordrecht, 1984, 87–128.

D. Stanton, An introduction to group representations and orthogonal polynomials, in *Orthogonal Polynomials: Theory and Practice*, P. Nevai and M. E. H. Ismail (eds.), Kluwer, Dordrecht, 1990, 419–433.

H. Stark, *An Introduction to Number Theory*, M.I.T. Press, Cambridge, MA, 1978.

H. Stark, Modular forms and related objects, *Canadian Math. Soc. Conf. Proc.*, **7** (1987), 421–455.

H. Stark, Galois theory, algebraic number theory and zeta functions, in *From Number Theory to Physics*, M. Waldschmidt et al. (eds.), Springer-Verlag, Berlin, 1992.

H. Stark, Multipath zeta functions of graphs, *IMA Volumes in Math. and Its Applications*, **109**, Springer-Verlag, New York, in press.

H. Stark and A. Terras, Zeta functions of finite graphs and coverings, *Adv. Math.*, **121** (1996), 124–165.

M. E. Starzak, *Mathematical Methods in Chemistry and Physics*, Plenum, New York, 1989.

R. Steinberg, The representations of GL(3, q), GL(4, q), PGL(3, q), and PGL(4, q), *Canadian. J. Math.*, **3** (1951), 225–235.

S. Sternberg, *Group Theory and Physics*, Cambridge Univ. Press, Cambridge, 1994.

W. Stevens, *The Palm at the End of the Mind: Selected Poems and a Play*, Vintage Books, New York, 1971.

I. Stewart, *Game, Set, and Math: Enigmas and Conundrums*, Blackwell, Oxford, 1989.

G. Strang, *Linear Algebra and Its Applications*, Academic, New York, 1976.

G. Strang, *Introduction to Applied Math.*, Wellesley–Cambridge Press, Wellesley, MA, 1986.

G. Strang and T. Nguyen, *Wavelets and Filter Banks*, Wellesley–Cambridge Press, Wellesley, MA, 1996.

R. S. Strichartz, *A Guide to Distribution Theory and Fourier Transforms*, CRC Press, Boca Raton, FL, 1994.

J. C. Strikwerda, *Finite Difference Schemes and Partial Differential Equations*, Wadsworth and Brooks/Cole, Belmont, CA, 1989.

K. Stumpff, *Tafeln und Aufgaben für Harmonischen Analyse und Perrodogrammrechnung*, J. Springer, Berlin, 1939.

T. Sunada, L-functions in geometry and some applications, *Lecture Notes in Math.*, **1201**, Springer-Verlag, New York, 1986, 266–284.

T. Sunada, Fundamental groups and Laplacians, *Lecture Notes in Math.*, **1339**, Springer-Verlag, New York, 1988, 248–277.

N. S. Szabó and R. J. Tanaka, *Residue Arithmetic and Applications to Computer Technology*, McGraw-Hill, New York, 1967.

T. Tamagawa, On Selberg's trace formula, *J. Fac. Sci. U. Tokyo, I*, **8** (1960), 363–386.

S. Tanaka, On irreducible unitary representations of some special linear groups of the second order, I, II, *Osaka J. Math.*, **3** (1966), I: 217–227, II: 229–242.

G. Taubes, The disputed birth of buckyballs, *Science*, **253** (Sept. 17, 1991), 1476–1479.

A. Terras, Noneuclidean harmonic analysis, *SIAM Review*, **24** (1982), 159–193.

A. Terras, *Harmonic Analysis on Symmetric Spaces and Applications*, Vols. I and II, Springer-Verlag, New York, 1985, 1988.

A. Terras, Eigenvalue problems related to finite analogues of upper half planes, in *40 More Years of Ramifications: Spectral Asymptotics and Its Applications*, S. A. Fulling and F. J. Narcowich (eds.), in the series *Discourses in Math.*, **1** (1991), Math. Dept., Texas A&M, College Station, TX.

A. Terras, Survey of spectra of Laplacians on finite symmetric spaces, *Experimental Math.*, **5** (1) (1996), 15–32.

A. Terras, A survey of discrete trace formulas, *IMA Volumes in Math. and Its Applications*, Vol. 109, Springer-Verlag, New York, in press.

A. Terras and D. Wallace, Ergodic properties of quotients of flows on trees, preprint, 1998.

L. H. Thomas, Using a computer to solve problems in physics, in *Applications of Digital Computers*, W. Freiberger and W. Prager (eds.), Guin, Boston, 1963.

A. D. Thomas and G. V. Wood, *Group Tables*, Shiva Publ. Ltd., Orpington, Kent, UK, 1980.

R. Tolimieri, M. An, and C. Lu, *Algorithms for Discrete Fourier Transforms and Convolutions*, Springer-Verlag, New York, 1989.

C. Trimble, *Some Special Functions on p-adic and Finite Analogues of the Poincaré Upper Half Plane*, Ph.D. Thesis, Univ. Calif., San Diego, 1993.

A. Turing, Finite approximations to Lie groups, *Ann. Math.*, **39** (1938), 105–111.

A. Valette, Graphes de Ramanujan et applications, *Sém. Bourbaki, exp.* 829, March, 1997.

B. L. van der Waerden, *Algebra*, Vols. I and II, Springer-Verlag, New York, 1991.

J. H. van Lint, *Introduction to Coding Theory*, Springer-Verlag, New York, 1982.

E. Velasquez, *The Radon Transform on Finite Groups*, Ph.D. Thesis, Univ. Calif., San Diego, 1991.

E. Velasquez, Radon transform on finite symmetric spaces, *Pacific J. Math.*, **177** (1997), 369–376.

E. Velasquez, The uncertainty principle on Cayley graphs, *Pacific J. Math.*, **184** (1998), 367–379.

E. Velasquez, The heat equation on finite upper half planes, preprint.

A. B. Venkov and A. M. Nikitin, The Selberg trace formula, Ramanujan graphs and some problems of mathematical physics, *Petersburg Math. J.*, **5** (3) (1994), 419–484.

M.-F. Vignéras, L'équation fonctionelle de la fonction zêta de Selberg de la groupe modulaire PSL(2,$\mathbb{Z}$), *Astérisque*, **61** (1979), 235–249.

M.-F. Vignéras, Variétés riemanniennes isospectrales et non isométriques, *Ann. Math.* (2), **112** (1980), 21–32.

M.-F. Vignéras, An elementary introduction to the local trace formula of J. Arthur. The case of finite groups, *Jahresbericht Deutsch. Math.-Verein. Jubiläumstagung 1990* (1992), 281–296.

N. J. Vilenkin, *Special Functions and the Theory of Group Representations, Transl. Math. Monographs*, **22**, Amer. Math. Soc., Providence, 1968.

N. W. Vilenkin, On a class of complete orthogonal systems, *Izv. Akad. Nauk. S.S.S.R. Ser. Math.*, **11** (1947), 363–400 (in Russian).

J. F. Wakerly, *Digital Design Principles and Practices*, Prentice-Hall, Englewood Cliffs, NJ, 1994.

J. S. Walker, *The Fast Fourier Transform*, CRC Press, Boca Raton, FL, 1991.

D. I. Wallace, The Selberg trace formula for SL(3, $\mathbb{Z}$)\ SL(3, $\mathbb{R}$)/SO(3, $\mathbb{R}$), *Trans. Amer. Math. Soc.*, **345** (1994), 1–36.

J. L. Walsh, A closed set of orthogonal functions, *Amer. J. Math.*, **45** (1923), 5–24.

A. Wawrzyńczyk, *Group Representations and Special Functions*, Reidel, Boston, 1984.

A. Weil, *Basic Number Theory*, Springer-Verlag, Berlin, 1973.

A. Weil, *Collected Papers*, Vols. I–III, Springer-Verlag, New York, 1979.

P. D. Welch, The use of fast Fourier transform for the estimation of power spectra: a method based on time-averaging over short, modified, periodograms, *IEEE Trans. Audio Electroacoust., AU-15* (June, 1967), 70–73.

H. Weyl, *Symmetry*, Princeton Univ. Press, Princeton, NJ, 1952.

H. Weyl, *Gesammelte Abhandlungen*, Vols. I–III, Springer-Verlag, New York, 1968.

J. A. Wolf, Uncertainty principles for Gelfand pairs and Cayley complexes, in *75 Years of Radon Transforms*, S. Gindikin and P. Michor (eds.), International Press, Boston, 1994, 271–292.

S. Wolfram, *The Mathematica Book*, Wolfram Media & Cambridge Univ. Press, Champaign, IL and Cambridge, 1996.

F. Yates, The design and analysis of factorial experiments, Commonwealth Agriculture Bureaux, Burks, England, Farnam Royal, 1937.

M. Zack, *Convergence to Uniform on the Finite Heisenberg Group and Applications to Random Number Generators*, Ph.D. Thesis, Univ. Calif., San Diego, 1989.

Index

Printed in the United States
By Bookmasters